Recreational Fisheries

Fish and Aquatic Resources Series

Series Editor: Tony J. Pitcher

Director, Fisheries Centre, University of British Columbia, Canada

The *Blackwell Publishing Fish and Aquatic Resources Series* is an initiative aimed at providing key books in this fast-moving field, published to a high international standard.

The Series includes books that review major themes and issues in the science of fishes and the interdisciplinary study of their exploitation in human fisheries. Volumes in the Series combine a broad geographical scope with in-depth focus on concepts, research frontiers and analytical frameworks. These books will be of interest to research workers in the biology, zoology, ichthyology, ecology and physiology of fish and the economics, anthropology, sociology and all aspects of fisheries. They will also appeal to non-specialists such as those with a commercial or industrial stake in fisheries.

It is the aim of the editorial team that books in the *Blackwell Publishing Fish and Aquatic Resources Series* should adhere to the highest academic standards through being fully peer reviewed and edited by specialists in the field. The Series books are produced by Blackwell Publishing in a prestigious and distinctive format. The Series Editor, Professor Tony J. Pitcher is an experienced international author, and founding editor of the leading journal in the field of fish and fisheries.

The Series Editor and Publisher at Blackwell Publishing, Nigel Balmforth, will be pleased to discuss suggestions, advise on scope, and provide evaluations of proposals for books intended for the Series. Please see contact details listed below.

Titles currently included in the Series (Full details at www.FishKnowledge.com)

1. *Effects of Fishing on Marine Ecosystems and Communities* (S. Hall) 1999
2. *Salmonid Fishes* (Edited by Y. Altukhov *et al.*) 2000
3. *Percid Fishes* (J. Craig) 2000
4. *Fisheries Oceanography* (Edited by P. Harrison & T. Parsons) 2000
5. *Sustainable Fishery Systems* (A. Charles) 2000
6. *Krill* (Edited by I. Everson) 2000
7. *Tropical Estuarine Fishes* (S. Blaber) 2000
8. *Recreational Fisheries* (Edited by T.J. Pitcher & C.E. Hollingworth) 2002

For further information concerning books in the series, please contact:

Nigel Balmforth, Professional Division, Blackwell Publishing, Osney Mead, Oxford OX2 0EL, UK

Tel: +44 (0) 1865 206206; Fax +44 (0) 1865 721205

e-mail: nigel.balmforth@blacksci.co.uk

Recreational Fisheries:

Ecological, Economic and Social Evaluation

Edited by

Tony J. Pitcher

Fisheries Centre, UBC, Canada

and

Charles E. Hollingworth

Bangor, Wales, UK

b

Blackwell Science

Editorial Offices:
Osney Mead, Oxford OX2 0EL
25 John Street, London WC1N 2BS
23 Ainslie Place, Edinburgh EH3 6AJ
350 Main Street, Malden
MA 02148 5018, USA
54 University Street, Carlton
Victoria 3053, Australia
10, rue Casimir Delavigne
75006 Paris, France

Other Editorial Offices:

Blackwell Wissenschafts-Verlag GmbH
Kurfürstendamm 57
10707 Berlin, Germany

Blackwell Science KK
MG Kodenmacho Building
7–10 Kodenmacho Nihombashi
Chuo-ku, Tokyo 104, Japan

Iowa State University Press
A Blackwell Science Company
2121 S. State Avenue
Ames, Iowa 50014–8300, USA

First published 2002

Set in 10/13 pt Times
by Sparks Computer Solutions Ltd, Oxford
http://www.sparks.co.uk
Printed and bound in Great Britain by
MPG Books, Bodmin, Cornwall

DISTRIBUTORS
Marston Book Services Ltd
PO Box 269
Abingdon
Oxon OX14 4YN
(*Orders:* Tel: 01235 465500
Fax: 01865 465555)

USA and Canada
Iowa State University Press
A Blackwell Science Company
2121 S. State Avenue
Ames, Iowa 50014-8300
(*Orders:* Tel: 800-862-6657
Fax: 515-292-3348
Web www.isupress.com
email: orders@isupress.com

Australia
Blackwell Science Pty Ltd
54 University Street
Carlton, Victoria 3053
(*Orders:* Tel: 03 9347 0300
Fax: 03 9347 5001)

A catalogue record for this title is available from the British Library

ISBN 0-632-06391-2

Library of Congress
Cataloging-in-Publication Data
is available

For further information on Blackwell Science, visit our website:
www.blackwell-science.com

Contents

List of Contributors

Oystein Aas
Norwegian Institute for Nature Research (NINA)
Division for Man–Environment Studies
Fakkelgarden
N–2624 Lillehammer
Norway
E-mail: Oystein.Aas@ninalil.ninaniku.no

Barbara Calvert
National Fisheries Policy Framework
Fisheries and Oceans Canada
200 Kent Street
Ottawa, Ontario
Canada K1A 0E6
E-mail: calvertb@dfo-mpo.gc.ca

David Campbell
David Campbell and Associates, Fishery Economists
PO Box 228
Kippax, ACT 2615
Australia

Anne P.M. Coleman
Department of Primary Industry and Fisheries
GPO Box 990
Darwin, NT 0810
Australia
E-mail: Anne.Coleman@dpif.nt.gov.au

Sean Cox
Center for Limnology
University of Wisconsin
Madison, Wisconsin 53706
USA
E-mail: spcox@facstaff.wisc.edu

Robert B. Ditton
Department of Wildlife and Fisheries Sciences
Texas A&M University
College Station, Texas 77843–2258
USA
E-mail: rditton@unix.tamu.edu

John W. Duffield
Department of Economics
University of Montana
Missoula, Montana 59812
USA
E-mail: bioecon@montana.com

Henk Folmer
Mansholt Graduate School, Wageningen University
Hollandseweg 1
6706 KN Wageningen
The Netherlands
E-mail: Henk.Folmer@Alg.ech.wau.nl

Brad Gentner
NOAA, NMFS, NMFS F/ST1
1315 East West Highway
Silver Spring, Maryland 20910
USA
E-mail: brad.gentner@noaa.gov

Stephen Gledhill
Environment Agency, Head Office
Rio House, Waterside Drive
Aztec West
Almondsbury, Bristol, BS32 4UD
United Kingdom
E-mail: stephen.gledhill@environment-agency.gov.uk

Marc H. Griffiths
Linefish Section
Marine and Coastal Management
Private Bag X2 Roggebai
8012 Cape Town
South Africa
E-mail: mgriffit@mcm.wcape.gov.za

Gary W. Henry
New South Wales Fisheries
PO Box 21
Cronulla, NSW 2230
Australia

Phil Hickley
National Coarse Fisheries Centre
Environment Agency
Arthur Drive, Hoo Farm Industrial Estate
Worcester Road
Kidderminster, DY11 7RA
United Kingdom
E-mail: phil.hickley@environment-agency.gov.uk

Charles E. Hollingworth
School of Biological Sciences
University of Wales Bangor
Bangor, Gwynedd LL57 2UW
United Kingdom
E-mail: c.e.hollingworth@bangor.ac.uk

Robert E. Kearney
Applied Ecology Research Group
University of Canberra
ACT 2601
Australia
E-mail: kearney@scides.canberra.edu.au

G. Cornelis van Kooten
Department of Applied Economics and Statistics
University of Nevada – Reno
Mail Stop 204
Reno, Nevada 89557–0105
USA
E-mail: kooten@cabnr.unr.edu

Steve J. Lamberth
Linefish Section
Marine and Coastal Management
Private Bag X2 Roggebai
8012 Cape Town
South Africa
E-mail: lamberth@mcm.wcape.gov.za

Alan Lowther
NOAA, NMFS, NMFS F/ST1
1315 East West Highway
Silver Spring, Maryland 20910
USA
E-mail: alan.lowther@noaa.gov

Jeremy M. Lyle
Tasmanian Aquaculture and Fisheries Institute
University of Tasmania
Marine Research Laboratories
Nubeena Crescent
Taroona, TAS 7053
Australia
E-mail: jeremy.lyle@dpiwe.tas.gov.au

Jim Lyons
Environment Agency, Midland Region
Trentside Offices
Scarrington Road, West Bridgford
Nottingham, NG2 5FA
United Kingdom
E-mail: jim.lyons@environment-agency.gov.uk

Margaret F. Merritt
Commissioner's Office
Alaska Department of Fish & Game
1300 College Rd
Fairbanks, Alaska 99701
USA
E-mail: pmerritt@ak.net

J. Walter Milon
Department of Economics
University of Central Florida
Orlando, Florida 32816
USA
E-mail: milon@fred.ifas.ufl.edu

Christopher J. Neher
Bioeconomics
315 S Fourth East
Missoula, Montana 59801
USA
E-mail: bioecon@montana.com

Tony J. Pitcher
Fisheries Centre
2204 Main Mall
University of British Columbia
Vancouver, BC
Canada V6T 1Z4
E-mail: t.pitcher@fisheries.ubc.ca

David Policansky
National Research Council
2001 Wisconsin Ave. NW
Washington, DC 20418
USA
E-mail: DPolican@nas.edu

Murray A. Rudd
Mansholt Graduate School, Wageningen University
Hollandseweg 1
6706 KN Wageningen
The Netherlands
E-mail: marudd@interchange.ubc.ca

Werner Steffens
German Angler Association
Weissenseer Weg 110
D–10369 Berlin-Lichtenberg
Germany, EU
E-mail: Ulf.Steffens@t-online.de

Ussif Rashid Sumaila
Chr. Michelsen Institute, Bergen, Norway and Fisheries Centre
University of British Columbia
2204 Main Mall
Vancouver, BC
Canada V6T 1Z4
E-mail: r.sumaila@fisheries.ubc.ca

Carol E. Thailing
Human Dimensions of Fisheries Lab
Department of Wildlife and Fisheries Sciences
Texas A&M University
210 Nagle Hall
2258 TAMU
College Station, TX 77843–2258
Phone (979) 845–4283
Fax (979) 845–3786
E-mail: cet2730@unix.tamu.edu

Eric Thunberg
NMFS, NEFSC, Social Sciences Branch
166 Water Street
Woods Hole, Massachusetts 02543–1026
USA
E-mail: Eric.Thunberg@noaa.gov

Anna-Liisa Toivonen
Finnish Game and Fisheries Research Institute
POB 6
Helsinki, FIN–00721
Finland, EU
E-mail: anna-liisa.toivonen@rktl.fi

Carl Walters
Fisheries Centre,
University of British Columbia
2204 Main Mall
Vancouver, BC
Canada V6T 1Z4
E-mail: c.walters@fisheries.ubc.ca

Laurie West
Kewagama Research
42 Waterside Court
Noosa Waters, QLD 4566
Australia

Michael Winkel
German Angler Association
Weissenseer Weg 110
D–10369 Berlin-Lichtenberg
Germany, EU
E-mail: info@anglerverband.com

Series Foreword

Fish researchers (a.k.a. fish freaks) like to explain, to the bemused bystander, how fish have evolved an astonishing array of adaptations, so much so that it can be difficult for them to comprehend why anyone would study anything else. Yet, at the same time, fish are among the last wild creatures on our planet that are hunted by humans for sport or food. As a consequence, few today would fail to recognise that the reconciliation of exploitation with the conservation of biodiversity provides a major challenge to our current knowledge and expertise. Even evaluating the trade-offs that are needed is a difficult task. Moreover, solving this pivotal issue calls for a multidisciplinary conflation of fish physiology, biology and ecology with social sciences such as economics and anthropology, in order to probe new frontiers of applied science. The *Blackwell Publishing Fish and Aquatic Resources Series* is an initiative aimed at providing key, peer-reviewed texts in this fast-moving field.

While bony fish stem from a great radiation that followed the invention of the swim bladder in the Cretaceous period 100 million years ago, some fish groups, such as the sharks, lungfish and sturgeons, are more ancient beasts. Survivors from earlier eras may be more widespread than we think: the deep-sea coelacanths, formerly known only from the Indian Ocean, have recently turned up in Indonesia. Also, these fishes may be more effectively adapted to specialised niches than their ancient body plans would suggest. For example, rays and angel sharks have perfected the art of the ambush predator, while most cartilaginous fishes can detect electric discharges in the nerves of their prey. Bony fish themselves have evolved into an amazing array of habitats and niches. As well as the open sea, there are fish in lakes, ponds, rivers and rock pools; in deserts, forests, mountains, the great deeps of the sea, and the extreme cold of the Antarctic; in warm waters of high alkalinity or of low oxygen; and in habitats like estuaries or mudflats, where their physiology is challenged by continuous change. Air-breathing climbing perch (regularly found up trees), walking catfish and mangrove mudskippers are currently repeating the land invasion of their Carboniferous ancestors. We can marvel at high-speed swimming adaptations in the fins, tails, gills and muscles of marlins, sailfish and warm-blooded tunas; gliding and flapping flight in several groups of fish; swinging, protrusible jaws providing suction-assisted feeding that have evolved in parallel in groupers, carps and cods; parental care in mouth-brooding cichlids; the birth of live young in many sharks, tooth carps, rockfish and blennies; immense migrations in salmon, shads and tunas; and even the so-called four-eyed fish, with eyes divided into upper air and lower water-adapted sections.

In addition to providing food, recreation (and inspiration for us fish freaks), it has, recently been realised that fish are essential components of aquatic ecosystems that provide vital services to human communities. Sadly, virtually all sectors of the stunning biodiversity of fishes are at risk from human activities. In freshwater, for example, the largest mass extinction event since the end of the dinosaurs occurred when the Nile perch was introduced into Lake Victoria, eliminating over 100 species of endemic haplochromine fish. But, at the same time, precious food and income from the Nile perch fishery was created in a miserably poor region. In the oceans, we have barely begun to understand the profound changes that have accompanied a vast expansion of human fishing over the past 100 years.

Recreational Fisheries: Ecological, Economic and Social Evaluation, edited by myself and Charles Hollingworth, is the eighth book in the *Blackwell Publishing Fish and Aquatic Resources Series*. It comprises 17 chapters written by 34 authors drawn from 9 countries worldwide. The book derives from a selection of papers presented at an international conference on evaluating the benefits of sport fishing held at the Fisheries Centre, University of British Columbia, in June 1999 (Pitcher 1999). Some additional chapters have been solicited by the editors. Since I am one of the book editors, this series editor's introduction is unusually brief.

My hope is that this book will begin the vital process of evaluating the potential of sport fisheries in the light of the well-documented disastrous state of world commercial fisheries. It is clear that sport fisheries have the potential for spreading a profound, and relatively new, ethos for conservation ('non-consumptive fisheries' to use the rather ugly jargon), expressed as catch-and-release, strict catch limits, and support for habitat restoration and fish population conservation. Hence, in the future, sport fisheries may come to play a vital role in the preservation of our aquatic ecosystems by helping to balance exploitation with conservation.

Professor Tony J. Pitcher
Editor, Blackwell Publishing Fish and Aquatic Resources Series
Director, Fisheries Centre, University of British Columbia, Vancouver, Canada
October 2001

Reference

Pitcher, T. J. (1999) *Evaluating the benefits of recreational fisheries*. Fisheries Centre Research Reports 7 (2).

Preface

I sent a message to the fish:
I told them 'This is what I wish.'
The little fishes of the sea,
They sent an answer back to me.
The little fishes' answer was
'We cannot do it, Sir, because – '
'I'm afraid I don't quite understand' said Alice.
'It gets easier further on' Humpty Dumpty replied.
I sent to them again to say
'It will be better to obey.'
The fishes answered with a grin,
'Why, what a temper you are in!'

Lewis Carroll (Charles Lutwidge Dodgson) (1871)
Through the Looking-Glass and What Alice Found There
Macdonald, London (Chapter 6)

Perhaps relationships between humans and fishes are sometimes a touch strained (Humpty Dumpty's fishes are trying to avoid being cooked), but the world will always have its malcontents. Recreational fishing, commonly regarded as benign, is important for three principal reasons: economic (it is a billion-dollar business), social (fishing is deeply embedded in the cultures of many nations), and ecological (fishing affects the environment and food webs, and may promote environmental awareness).

However, the costs and benefits of sport fishing have proved surprisingly hard to evaluate. Accordingly, the first international conference on Evaluating the Benefits of Recreational Fishing was held at the Fisheries Centre, Vancouver, Canada, from 1 to 3 June 1999 (Papers and discussions are published as *Fisheries Centre Research Reports,* Vol. 7(2), 1999). At the conference it rapidly became evident that, despite much work and strong opinions, the field tended to lack consistency of approach and methodological rigour. However, and more importantly, it was also obvious that these weaknesses could be overcome. So there was a clear need for a book that would identify the key issues in evaluating recreational fisheries, draw together state-of-the-art methodology, and in particular adopt a world-wide scope.

In compiling this book, we have not only assembled selected keynote and contributed papers from the Vancouver meeting, but have also commissioned chapters by recognised authorities from around the world. In so doing, we asked Robert Kearney from Australia to write a historical introduction to establish the state-of-the-art in sport fishery research, and Øystein Aas from Norway to write a wrap-up chapter, looking to the future. (Dr Aas read all chapters in draft.) The chapters have been refereed to international journal standards, revised and edited. We have sought – as far as possible – for generalities or at least for findings and techniques that might be widely applied in designing the next generation of research projects in sport fisheries. That is easy to say and surprisingly hard to do, but we believe we have made a start. We shall have more to say about this in Chapter 1.

Acknowledgements

Joelle Row (Cape Town) had the original idea of holding the Vancouver conference and played an enthusiastic role in planning and fund-raising. We thank Gunna Weingartner for organising the conference and Janice Doyle for administrative assistance later. We are grateful to the conference sponsors: Atlantic Salmon Federation; Fisheries and Oceans Canada; Ministry of Fisheries, Government of British Columbia; National Oceanic and Atmospheric Administration, US Government; South African Deep Sea Angling Association. Especially we thank the authors for supporting this project, meeting deadlines, and putting up with our editorial 'suggestions'. We are more grateful than we can say to the eminent referees for their thoughtful and constructive reports that did so much to broaden some chapters and focus others. We applaud our publisher Nigel J. Balmforth and his staff at Blackwell Science (Oxford) for publishing the book. We also thank the many others – not least our families – whose contributions were so important to the genesis and completion of this volume.

Tony J. Pitcher, Vancouver, BC, Canada
Chuck Hollingworth, Bangor, North Wales, UK

Chapter 1
Fishing for Fun: Where's the Catch?

Tony J. Pitcher and Charles E. Hollingworth

Abstract

This book is concerned with evaluating recreational fishing, distinguished from commercial fishing, which is primarily for profit, and subsistence fishing, primarily for food, by being performed mainly for fun. We review what lessons may be learned from the history of sport fishing. It turns out that evaluating sport fishing's balance sheets of costs and benefits is a surprisingly complex task that has rarely been tackled consistently. Hence, methodological essays and case studies comprise the major portions of this book. Chapters in the book address a range of seemingly simple questions such as: What is the catch? How many anglers are there? How many will there be in the future? What is sport fishing worth? Why do anglers do it?

A range of policy issues are discussed such as conflict with the commercial and artisanal fishery sectors; ways of dealing with the 'common property' problem of anglers being attracted to well-managed sport fishing localities and crowding them so that 'success breeds failure'. In some cases, sport fisheries can even affect conservation goals. The book presents two different perspectives on catch-and-release, seen on the one hand as a tool of ethically responsible conservation and on the other as pandering to base human instincts. Finally, we argue that conflating the management of fish and people in sport fisheries requires interdisciplinary research and consilience between the social and natural sciences.

Key words: catch-and-release, fishery management, interdisciplinary research, sport fisheries.

Definitions of sport fishing

Recreational fishing can be defined as catching fish for fun. Added to the fun, there may be auxiliary benefits such as profit, food, and exercise. Nevertheless, the fun part is mandatory. Commercial or subsistence fisheries, on the other hand, are work, and even though work can be fun too (sometimes), it is not the primary motivation.

So, this is not to say that hooking a 1-metre-long croaker for dinner from the shore in Cameroon, or using GPS in the Norwegian Sea to steer a 100-metre-long midwater trawler onto a 50-tonne herring shoal swimming at 100 metres depth, is not fun. The important point is that the primary motivation of the hungry African is a dramatic input of food to his family's stomachs. The Norwegian skipper's motivation is a dramatic input of cash to his local economy. On the other hand, a big-game angler landing a 1-tonne marlin, which he/she may well tag-and-release after 5 hours of exhausting but exhilarating sport, is mainly concerned with fun and, in fact, the big-game angler probably experiences a dramatic outflow of cash from his local economy.

We shall not split hairs and attempt to distinguish sport fishing from recreational fishing: for our purposes, they are synonymous. However, lack of an agreed definition can make it difficult for sport fishers to assert legal rights in allocation disputes (Kearney 2002). The most useful definitions for fisheries scholars impute a recreational motivation to fishing (Policansky 2002). However, divining the true intentions of us humans makes for slippery edges to such categories (and rich lawyers).

Fun is what recreational fishers will pay for (Pitcher 1999). Although what is perceived as fun may seem a little perverse in some cases (*vide* our sweaty marlin fisher above), the task of evaluating the benefits of sport fishing – the theme of this book – is confounded by the problem of enumerating these diverse auxiliary benefits. Simply deciding on the units to be employed to measure the benefits is not trivial, as there is far more to this than economics and money. Who benefits and who pays? The accounting process raises problems of definitions and crosses disciplinary boundaries.

What can history tell us about sport fishing?

Fishing for fun turns out to have a history as long as human civilisation itself. Moreover, this history illustrates most of the problems with which we grapple in this book. Sport fishers are commonly termed anglers, a word that derives from an ancient Aryan root 'ank', meaning a bend or hook (and found in the Egyptian 'ankh' – one of the forebears of the Christian cross as a religious symbol). In Britain, 'angler' has been used for one who catches fish with a hook, rod and line (*OED*) from at least Anglo-Saxon times (*OED* cites King Alfred in AD 880). In the Middle Ages 'anglers' were also thieves armed with rods, lines and metal hooks, who leaned out of windows and hooked valuables from the pockets of passers-by. But apart from rods, sport fisheries can include spear fishing, which, although it now has the connotation of 'high tech' scuba diving, is in fact the oldest known human fishing technique (90 000 BP according to Yellen *et al.* 1995). Fish hooks and gorges were made from at least 50 000 BP, initially from bone and stone, later from metals (e.g. copper in Mesopotamia 5000 BP: Sahrange and Lundbeck 1992). Fishing nets likely date from about 40 000 BP (Pringle 1997). These techniques were probably invented to catch fish for food, but it was probably fun too.

The first known image of pole, hook and line fishing dates back to the tomb of two wealthy Egyptian manicurists from around 4000 BP (Fig. 1.1), although such fishing may have been primarily for food. However, it certainly looks fun, and when it was painted this Middle Kingdom mural reflected a fashion of recalling the gentle domestic ethic of the Old Kingdom in Egyptian history.

A thousand years later, we find a more urbane Egypt, where fishing for fun has become the preserve of a rich and powerful elite. Fish were kept in large stone tanks, and nobles seated in comfortable chairs caught them on rod and line, possibly with insects as bait (Fig. 1.2).

Fig. 1.1 Egyptian painted mural of rod, hook and pole fishing from the tomb of one of the Pharaoh's manicurists, Knumhotep, from Beni Hasan, Middle Kingdom, 4000 BP. Fun, food or profit?

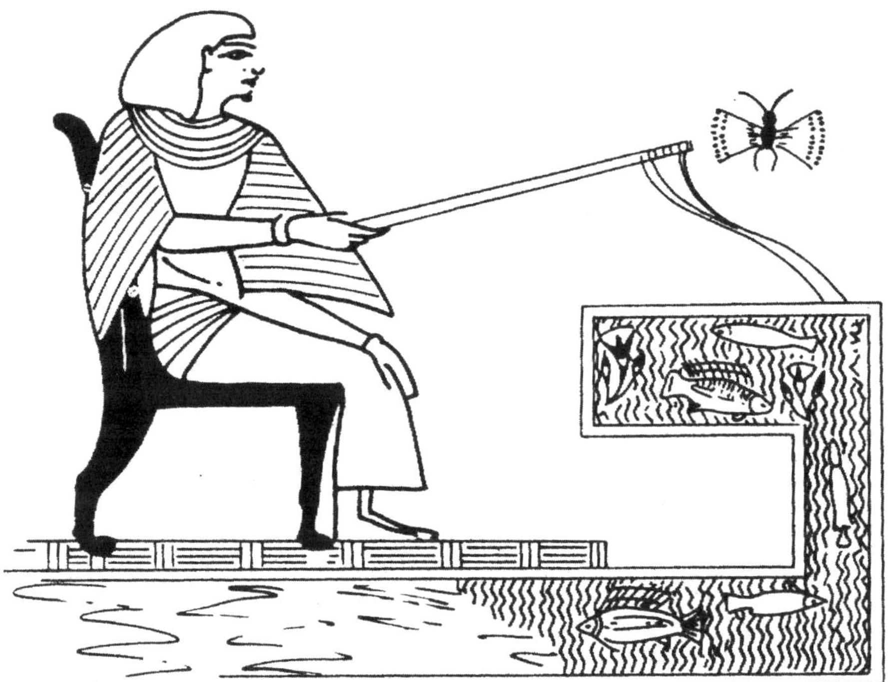

Fig. 1.2 Egyptian noble sport fishing tilapia, from tomb of Nebwenef, a High Priest under Ramses the Great, Thebes, 3290 BP. As well as looking like a lot of fun, and reflecting the immense power of the rich in maintaining such resources, this image also has religious significance, since, long before Christians employed a fish logo, tilapia were an Egyptian symbol of rebirth.

Constructing and maintaining such arrangements would have provided quite a boost for the local economy. Fun, food, and power drove this recreational fishing.

The 'angler's art' is defined by Plato (*The Sophist*, 2400 BP) as 'the kind which strikes with a hook and draws the fish from below upwards' but – unlike hunting, which was given divine sanction by Artemis (the original Greek version of the Roman goddess Diana, who is better known to us today) and a pack of lesser Greek hunting deities – the ancient Greeks tended to regard fishing as a lowly occupation fit only for slaves and children, and sport fishing as temptation to be avoided. This means the slaves and children likely had all the fun. For the rest it must have been just food and profit, although once you have them, keeping all those youngsters and slaves busy doing something useful all day could be reckoned a benefit (Fig. 1.3).

The Romans, who conquered the Greeks, but then adopted Greek culture as fashionable, had a similar attitude. Oppianus, writing a long Greek poem (*Halieutica*) in about AD 220, reports what appears to be fishing for fun, but mainly describes a remarkable diversity of commercial fishing techniques. Indeed, fishing was an important area of Roman commerce and cuisine (Fig. 1.4), but evidently not regarded as fun, at least not among the literate elite who have left us records of their culture. Many coastal cities, such as the ill-fated Pompeii, specialised in fishing and exported aromatic fish sauce, but there are only a few records of anyone fishing for fun. Claudius Aelianus (AD 220), writing in archaic Greek, describes fly-fishing in Macedonia. The outrageous emperor Nero (lived AD 37–68) fished mullet with golden nets and fed slaves to pet moray eels, but maybe for a people who divested the entire

Fig. 1.3 Greek boy (and an octopus) having fun fishing with a rod and line. Ambrosios, Greece, 2480 BP.

Fig. 1.4 Roman pole, hook and line fishers from North Africa, mosaic, Sousse, 1950 BP. What a diversity of fish! Food and profit for sure, but fun only for these boys, while their parents were probably engaged in gorier recreational pursuits.

Mediterranean basin of its large carnivores to set them upon criminals and Christians, mere sport fishing did not provide sufficient entertainment.

One of the few Roman accounts of fishing for (some kind of) fun, concerns Mark Antony and his passion for Cleopatra, the queen of Ptolemaic Egypt (2050 BP). The Roman historian Plutarch writes in the second century AD:

> One time he went fishing and had the misfortune not to catch anything while Cleopatra was present. So he ordered the fisherman secretly to dive underneath and attach fish that had already been caught to his hooks, but Cleopatra was not fooled after she saw him pull up two or three. She pretended to be amazed and told her friends and invited them come as observers on the next day. After a large audience had gathered on the fishing boats and Antony had lowered his line, Cleopatra told one of her slaves to get in ahead of the others and attach a salted fish from the Black Sea to his hook. When Antony thought he had caught something he pulled it up, and when (as might be expected) loud laughter followed, she said 'General, leave the fishing rod to us, the rulers of the Pharos and Canopus; your game is cities and kingdoms and countries.'

After the Romans popular culture became oral, and only the monastic tradition created written works, so, although there is frequent mention of fishing for food, records of sport fishing through the dark ages are almost nonexistent (Hoffmann 1999). Anglers who carried on their art in those troubled times left no traces. The civilised Arabs, who made huge advances in mathematics, technology and the arts, seem not to mention sport fishing, and it is not until well after the European Renaissance that we find books dealing with sport fishing. The first major work appears in England as 'Treatyse of Fysshynge With an Angle', included in the second edition of *The Boke of St. Albans* printed by Wynkyn de Worde in 1496. Sport fishing rods were described as 18–22 feet long with lines made of plaited horsehair. Artificial flies described here are surprisingly modern and many are still in use. The author and origins of this work remain a mystery. An attribution to a Dutch nun called Juliana Berners seems to have been a sixteenth century invention. Almost certainly the material came from medieval Europe and can be traced back though a booklet on 'How to Catch Fish' published by Jacob Köbel in Heidelberg in 1493, to manuscripts from Strasbourg in the 1440s (Hoffmann 1999).

Sometimes the 'fun' aspect of recreational fishing can be hard for the uninitiated to recognise. In the Middle Ages, this point was amusingly discussed in a *Dialogue between a Hunter and a Fisher*, published by Fernando Basurto in Spain in 1539.

> *Fisher:* ... fishing has great advantages over hunting for the reasons I have given you; for the pre-eminence of the soul, not to mention the health of the body.
> *Hunter:* Fisher, you know a lot, you tell me good things in praise of your fishing, and because I accept them as such it could be that you will convert me.
> *Fisher:* Sir, persistence kills the stag. Do not fail to believe me ... the calm you will bring by practising my activity and leaving yours, which is warlike ... and unhealthy for the body.
> *Hunter:* Everything you have said is well taken, but I beg you to tell me how your health is.
> *Fisher:* Why do you ask?
> *Hunter:* Do you not know why? Because I see you very relaxed sitting in this dampness, and I think you must have a pain in the gut, or other pains which enter into human bodies when it gets cold.
> *Fisher:* Sir, you do not know what you say, for accustomed pain preserves the passion of the patient. When one starts to use the arms they tire and hurt, but once they have been used the fatigue lessens. ... the start of a voyage causes sea sickness but once the first fury has passed it leaves the patient with hunger. ... after I began to enjoy this human glory, the laborious feelings were banished, and there was no pain from sitting, especially when the fish are biting, for then there is no thirst nor hunger nor heat nor cold. ... what other activity can take its place?

The hunter goes onto complain that fishing is not good for what are nowadays called relationships, to which the fisher replies that it does not matter because fishing, 'by immediate pleasure, makes one forget all that is absent.'

Here, echoes of the martyr ethic, the individualist, and the chauvinist are found in our wet and bedraggled sport fisher (Fig. 1.5) defending his sport against 'military' hunting.

One recurrent theme in recreational fishing, proving the 'fishing is for fun' axiom, is that it costs far more to catch fish than they can possibly be worth as marketed commodities. Again we can find a medieval example, published in that treatise on how to catch fish by Jacob Köbel (1493), where expensive culinary ingredients are used to prepare bait to catch trout or grayling:

Fig. 1.5 Recreational fishing in the Middle Ages of Europe could be expensive and cold. This pole, hook and line fisher having fun near Köln in Germany is well wrapped. There is probably food in the pail. From a German edition of *How to Catch Fish* printed by Johann von Aich in Köln, AD 1550.

> Take a black chicken and yolks of three eggs and a pea sized amount of saffron. Then take the chicken and make a hole in it and press all the listed material into it and sew the hole up again. Then place the chicken in a pile of horses' manure for three or four days and as many weeks as it takes the chicken to become rotten. Then you will find little yellow worms in it. Put these on the hook each time and keep the others in a little closed box. Thus you will experience marvels.

The same author promises that 'great marvels' will come to fishers who smear hands, shins, front and back with a mash of camphor, wheat flour, heron grease, crushed heron bone and olive oil. Remember, this is fun!

In the Middle Ages, freshwater fish (e.g. pike and carp) were introduced to Britain for both food and fun. Indeed, until very recently, fish introductions have been a major part of sport fishing around the world that has brought Northern Hemisphere salmonids to such unlikely habitats as the highlands of East Africa, Australia and South America. The benefits and costs of such introductions are a challenge to evaluate.

In more recent history, big-game fishing, made feasible by powered boats, appears to have been pioneered in 1898 by C.F. Holder, who caught bluefin tuna off Santa Catalina Island,

California. For those interested in the history of sport fishing, in this book Policansky (2002) and Kearney (2002) include more examples, comprehensive material may be found in Sahrange and Lundbeck (1992), and some recent archaeological findings are summarised in Pitcher (2001).

What sport fishing needs

The four principal needs of recreational fisheries are, first and second, to obtain more accurate, detailed evaluation of direct and indirect ecological and economic impacts; thirdly, to systematise and carefully evaluate the social impacts; and fourthly, to implement adaptive management plans using information gathered by sport fishers, thereby bridging the gap to fishery scientists.

It was evident from the Vancouver conference that the principal requirements of recreational fisheries today are, first, for recreational fisheries to receive equitable treatment with commercial and other fishery sectors (native, artisanal). Also, there should be proper recognition of the conservation benefits of catch-and-release and of compliance with bag limits. Furthermore, there are clear scientific benefits to be had from sport fishers' careful monitoring of abundance and change, and from well-managed data records and tag returns. For a human activity evidently worth some billions of dollars world-wide, surprisingly few of these issues are currently addressed by research into sport fisheries. Hence, it is our hope that the material in this book may stimulate more research in this area.

Balance sheets for sport fishing

We are primarily concerned with three types of benefits and costs that arise from recreational fisheries: economic balance sheets, where benefits are desired but generally poorly measured; ecological balance sheets, which have only recently been recognised; and social balance sheets, which have rarely been considered at all.

In this book, Bob Kearney (2002) draws up detailed balance sheets for each of these fields. It is alarming, as evident from the state-of-the-art chapters in this book, that in each field rather few of the key issues have been rigorously addressed through research. From the ecological balance sheet 8/15 (53%) issues have been tackled; from the economic balance sheet, only 3/10 (30%) are addressed; and from the social balance sheet, 5/18 (27%) issues receive some mention. Overall, only 37% of the key issues identified by Kearney appear to have been the subjects of substantive research. Moreover, even quite simple questions about sport fisheries have only recently begun to be addressed.

Answering simple questions about sport fishing

How many people fish for fun? How much and what do they catch? How much is sport fishing worth to the economy? Why do anglers do it? A series of fundamental questions like this must be answered if we are to perform ecological, economic and social evaluations of the impacts of recreational fisheries.

Answering these simple questions is not as easy as it might seem. Much of this book is therefore devoted to describing details of methodology and assessing the scope of the results. Survey methods themselves receive a lot of attention. Sport fishing surveys can be run by phone (Australia: Lyle *et al.* 2002), mail (Scandinavia: Toivonen 2002), interception at the fishing location (USA: Gentner and Lowther 2002) or at airports (Australia: Lyle *et al.* 2002) and keeping a diary of various kinds (Australia: Lyle *et al.* 2002). Surveys may be combined with a catch and effort census, and may also ask willingness-to-pay questions (Alaska: Duffield *et al.* 2002). Methods of analysis vary from simple compilations of percentages to very complex statistical models (some of which don't look like much fun – a sigh from the editors!): the whole range of possibilities is expressed in this book. Further recent examples may be found in the Vancouver conference proceedings (e.g. walleye in Tennessee, USA: O'Bara 1999). Many of the lessons to be learned with regard to survey methodology are succinctly summarised in the final chapter by Aas (2002).

What is the catch?

To perform an ecological evaluation of the impact of sport fisheries, we need to know how much recreational fisheries catch. In some countries, such as the USA (Gentner and Lowther 2002) and Germany (Steffens and Winkel 2002), these data are readily available, but elsewhere complete information has not yet been gathered.

Although sport fish catches are generally much smaller than those of commercial fisheries (e.g. South African swordfish: Griffiths and Lamberth 2002), and therefore may not be a primary concern for conservation, this may not be so for some species and locations. In the USA, catches of six of the top ten sport fishing species exceed the commercial catch – by an order of magnitude in the case of dolphin fish (*Coryphaena*), yellowtail (*Seriola*), spotted sea trout (*Cynoscion*) and red drum (*Sciaenops*) (Gentner and Lowther 2002). In such cases, in the interests of conservation and rebuilding, sport fisheries have begun to accept the kind of closures and restrictions long borne by commercial fisheries. In British Columbia, for example, the commercial fishery for heavily depleted coho and chinook salmon has recently been closed or greatly restricted, but sport fishery catches were of sufficient concern to cause complete closure during the 1999 season.

How many anglers?

For economic and social evaluations, we need to know how many anglers there are. The numbers of sport fishers varies among countries, and can itself be tricky to estimate (Kearney 2002).

Chapters in this book show that in many developed nations, the figure is around 2–4% of the population (e.g. Germany: 2.1%; USA: 2.8%; UK: 3.5% (freshwater only); Europe: 4.7%), while in Scandinavian countries the figures appear much higher: Denmark: 12.5%; Iceland: 31.5%; Sweden: 35%; Finland: 40%; Norway: 50% (Toivonen 2002). Although fewer young anglers appear to be joining this popular sport, the US figure is projected to increase to about 10% by 2025 (Thunberg and Milon 2002). In some countries, exact figures are hard to obtain and often depend on the survey methods employed: for example, in Australia figures between 4% and 26% are quoted (Kearney 2002, Lyle *et al.* 2002). Complete figures for developing nations are harder to come by: for example in the South African line fishery, only half of one percent of the population appears to be involved in the principal sport fishery (Griffiths and Lamberth 2002).

What is sport fishing worth?

Surprisingly, even with the use of complex numerical estimation techniques and massive surveys, it is often not possible to answer what appear at first sight to be simple questions such as 'what is the total value of sport fisheries?' (see Gentner and Lowther 2002). Value itself turns out to be a complex question (Rudd *et al.* 2002). You have to distinguish between value obtained by adding up the various market transactions involved in sport fisheries (payment of licence fees, boat and guide hire, sales of rods, travel costs, magazines, etc.) from the value as perceived by the angler (often determined by willingness to pay, or allied survey techniques: Rudd *et al.* 2002, Duffield *et al.* 2002).

There are a few figures for total annual market values (money turned over in the economy). In this book we have from Germany US$1bn (Steffens and Winkel 2002); US$3.4bn for freshwaters in England and Wales (Lyons *et al.* 2002); Sweden US$281m; Denmark US$60m; US$250m for one South African sport fishery (Griffiths and Lamberth 2002).

Large-scale surveys cost a lot (Lyle *et al.* 2002, Duffield *et al.* 2002, Gentner and Lowther 2002) and 'apart from plain money and work, a multinational survey takes more time than you think' (Toivonen 2002). The chapters reporting the American and Australian regional and national surveys provide the examples of the latest techniques for estimating this information, while Nordic countries, the UK and Germany provide some 'big-picture' results.

Additional WTP (willingness to pay) for fishers is about 42% of expenditure for five Nordic countries, but WTP for nonfishers is about equal to fishers' expenditure (Toivonen 2002). On the whole, it seems from the material presented in this book that estimates from contingent valuation methods (CVM) produce lower values for sport fisheries than expenditure. This may be because CVM are not robust in the face of changes in earning power – for example the amount you might pay this week for a fishing option may not be the same next week after you have lost your job. Contingent valuation method is especially suspect in the face of international currency exchange rates. For example, a US big-game fisher going on holiday in Mexico or Kenya (Fig. 1.6) may spend more money fishing for one day (say $1000) than many local people, including the boat crew or resort staff, might earn in 5 years, making it impossible to include both types of person in a survey aimed at evaluating policy options for the sport fishery.

A more robust methodology may eventually be found in an alternative approach that uses the 'method of paired comparisons' (Thurstone 1927, David 1988, Dunn-Rankin 1983), which has recently been applied to environmental valuation, where it has been termed a 'damage schedule' (Knetsch 1994). Participants in a 'damage schedule' survey are asked to choose the more desirable of a pair of alternatives, but no direct monetary value questions are asked. By taking all possible pairs, the rank order and degree by which the participants prefer each option in the list of alternatives can be worked out statistically. An example applied to coastal zone management options in Thailand is given by Chuenpagee *et al.* (2001). In this book,

Fig. 1.6 Sport fishing for marlin, swordfish and sailfish on the Kenyan coast is not only a lot of fun but it brings in much valuable foreign exchange to the Kenyan economy and provides jobs for local people. In Kenya, much of the sport fishery catch is sold for food on local markets.

Rudd *et al.* (2002) mention an example where money values may be included in the choice questions, but the technique likely deserves further exposure, especially perhaps when evaluating sport fisheries in developing countries.

Although few data are yet available about the importance of sport fisheries in developing nations, this must not be overlooked. For example, in Zimbabwe the annual Lake Kariba tiger fish contest brings in a great deal of wealth to the relatively poor northern region of the country (Machena 1989). Likewise, in Kenya, both freshwater angling for Nile perch in Lake Kariba and marine big-game angling bring in valuable foreign exchange (Abuodha 1999; Hemphill 1995). In this book, Griffiths and Lamberth (2002) cover the South African marine rod-and-line fishery and Sumaila (2002) discusses the Namibian silver kob fishery. Rudd *et al.* (2002) supply examples from Costa Rica, Puerto Rico and the Turks and Caicos Islands. In Belize, parts of Mexico and elsewhere in the Caribbean, former commercial artisanal fishers can make more money guiding nonconsumptive sport fishing tourists, bringing ecological as well as economic benefits (Rachel Graham and Will Heyman, University of York, UK, personal communication). It is evident that more work is urgently needed on sport fishing impacts in developing countries.

Why do anglers do it?

What is the motivation for humans to go sport fishing? The simple and obvious answer 'to catch fish for fun' turns out to be more complex on deeper analysis. Our historical examples above show that humans can get their fun in different ways. Nowadays different anglers can cite opposing motivations for fishing, for example companionship and privacy. The ancient human instinct of hunting for food contrasts with pleasure in learning about and not harming the natural world. Relaxation and nostalgia, conjuring up images of a sleepy summer riverbank straight out of Kenneth Grahame's *The Wind in the Willows*, contrasts with the thrill of competition, or the exhilaration of pitting one's skill and muscles against a huge game fish, more worthy of the action-packed novel *The Eye of the Tiger* by Wilbur Smith (1975). Other motivations are the use of expensive high-tech equipment, exploration and exercise (Kearney 2002; Lyons *et al.* 2002). All these reasons, we guess, are fun, not food or work.

However, food is undeniably an important issue in sport fishing. South African inshore sport fishers take their catch home to eat (Griffiths and Lamberth 2002), and 92% of Germans eat the fish they have caught (Steffens and Winkel 2002). Yet, at least in some places, non-consumptive anglers may contribute more to the economic value of sport fisheries (Calvert 2002). In South Africa, off-shore recreational fishers regard themselves as the guardians of conservation for broadbill swordfish because of extensive use of catch-and-release.

Catch-and-release

Australian aborigines competed in spear fishing contests at tribal gatherings (Kearney 2002), but there are many cultures that regard fishing for fun as distinctly odd, if not sacrilegious. For example, First Nation groups on the coast of British Columbia have a hard time coming to terms with sport fisheries for salmon, especially catch-and-release, since their ethics specifically exclude harming creatures other than for food, and you are certainly not supposed to have fun doing such things (e.g. Jones and Williams-Davidson 2000). This view finds an echo in the more puritan wing of the European preservationist movement, which regards sport fishing as a form of fish torture (see Policansky 2002). At the Vancouver conference, recreational fishers were not only urged to take the potential political power of such lobby groups seriously, but also to consider the physiological study of the effects of catch-and-release (Cooke *et al.* 1999). Catch-and-release, seen on the one hand as expressing a powerful conservation ethic ('all fish grow if you let them go') and on the other as infliction of unnecessary pain and suffering, is discussed in two chapters in this book, one by David Policansky (2002) emphasising the North American experience and another, from a European perspective, by Øystein Aas *et al.* (2002).

Some policy issues

Conflict between sport fisheries and the commercial fishery sector is notorious. In South Africa, commercial fishers (and unreported fishing) appear virtually to have destroyed the swordfish population, long the preserve of sport fishers who largely practise catch-and-release (Griffiths and Lamberth 2002). However, the ecological interactions between these two sectors are not always simple: exclusion of commercial gill netters from the Mary River in Australia's Northern Territories did not lead to an expected increase in sport fish catch rates (Griffin and Walters 1999). Sumaila (2002) shows how coexistence between sport and commercial fishers may be found in Namibian fisheries using the 'Nash equilibrium' of game theory. Evaluations that try to account for conflict and quantify its reduction under alternative policies are in their infancy. One technique might be a modification of the 'Rapfish' multidisciplinary fishery evaluation method (Pitcher and Preikshot 2001).

In South Africa, a complex web of political and social inter-relationships crystallises around sport fishing – a pathological legacy of years of apartheid. Demands for equity and restitution after years of deprivation are not only understandable but also essential, yet they are placing great pressure on marine resources that have, by world standards, been well-managed for sustainability. The chapter by Griffiths and Lamberth (2002) provides several poignant examples of the dilemmas faced by the new democratic government in a fishery that has overlapping commercial, subsistence and recreational sectors.

The US anglers' attitude of being able to fish where and when they damn well please (Cox and Walters 2002) is almost supported by constitutional right (see Policansky 2002), but German anglers have to pass an exam before they are allowed to have fun fishing (Steffens and Winkel 2002).

About a third of anglers do not catch anything at all, but so many are attracted to good fishing sites that successful management may breed failure (Cox and Walters 2002). One solution to overcrowding that might preserve high catch rates is a lottery (Policansky 2002).

Some categories of recreational fishers may need special consideration in management and legislation. The requirements of specialised anglers, such as fly fishers, big-game anglers, carp pike and walleye anglers are often discussed, but other categories of people – for example, children, the elderly, disabled people, women and minorities – are not often reflected in sport fishing management plans (Kearney 2002, Lyons *et al.* 2002).

How can we manage sport fisheries?

What about management of sport fisheries? Usually management decisions have to be taken in the absence of the information that is really needed. Partly this may be a consequence of the right questions not having been asked, but, just as in commercial fisheries, researchers are notorious in not being able to bring themselves to provide simple big-picture answers even after receiving successive waves of heavy funding. So managers must often take decisions in ignorance but, we may hope, in good faith. Evidently, it does not work very well (a clear example from South Africa is recounted in Griffiths and Lamberth 2002). The alternative 'mushroom' technique of management – 'keep them in the dark and pour manure on their heads' – probably has just as good (or bad) a track record.

Some are suspicious of the results of surveys because 'anglers lie and cheat' and illegal catches can be a serious problem. Using material from sport fisheries in Alberta, Canada, Sullivan (1999) writes:

> There may be a strong negative exponential correlation between catch rates and illegal harvest. Because of low catch rates, enforcement officers would seldom encounter anglers with illegal fish, resulting in an apparent paradox of high illegal harvests with a small percentage of anglers breaking the law. ... Exaggeration by anglers was not constant, but strongly increased with declining catch rates.

At the Vancouver conference, this work was mentioned as a study 'to be read again and again'.

The generally unrecognised reason for failures in sport fishery management, like commercial fisheries, is that there is no Holy Grail of an 'optimal solution'. (Cries of 'heresy!')

We think that searches for the magic technique to achieve this are illusory, and yet waste a huge amount of resources and time in sport fishery research. The best we can hope for is an explicitly stated series of trade-offs among competing interests and issues, with uncertainties fully expressed and an adaptive management plan to push back the tide of our ignorance.

At the end of this book, Øystein Aas (2002) emphasises how research methods need improving. Moreover, there needs to be more awareness of the interdisciplinary nature of fisheries research, especially the human dimension. Indeed, the bad news appears to be the dawning realisation that sustainable management of living natural resources may well be inherently incompatible with human hunting behaviour (e.g. Pitcher and Pauly 1998). In addition, if human behaviour is to be effectively managed, we will have to find ways of reconciling two fundamentally divergent world views or paradigms. The social sciences try to find out how (and why) people feel about things, whereas natural science tries to describe what things actually are. The good news is that consilience (Wilson 1998) between these approaches appears to be the genesis of a new form of management science that will reconcilc human motivation – fishing for fun – with the sustainable exploitation of living natural resources – having fish in the future.

References

Aas, Ø. (2002) The next chapter: multicultural and cross-disciplinary progress in evaluating recreational fisheries. In: *Recreational Fisheries: Ecological, Economic and Social Evaluation* (eds Pitcher, T.J. & Hollingworth, C.E.), Chapter 17, this volume. Blackwell Science, Oxford, UK.

Aas, Ø., Thailing, C. & Ditton, R.B. (2002) Controversy over catch-and-release recreational fishing in Europe. In: *Recreational Fisheries: Ecological, Economic and Social Evaluation* (eds Pitcher, T.J. & Hollingworth, C.E.), Chapter 7, this volume. Blackwell Science, Oxford, UK.

Abuodha, P. (1999) Status and trends in Kenyan recreational marine fisheries. In: *Evaluating the Benefits of Recreational Fisheries* (ed. Pitcher, T.J.), Fisheries Centre Research Reports, Volume 7 Number 2, pp. 46–50. University of British Columbia, Vancouver.

Aelianus, Claudius (c. 220) *On the Characteristics of Animals*. [Schofield, A.F. (Translator). 1958–89 3 volumes, Loeb Classical Library, Harvard University Press, Cambridge, MA.]

Basurto, F. (1539) *Dialogo que agora se hazia* ... [Dialogue between a hunter and a fisher ...] Zaragoza: G. Coci. [English translation by Adrian Schubert, Thomas V. Cohen, & Richard Hoffman.]

Brewer, D.J. & Freidman, R.F. (1989) *Fish and fishing in ancient Egypt*. American University in Cairo Press, Cairo.

Calvert, B. (2002) The importance of angler motivations in sport fishery management. In: *Recreational Fisheries: Ecological, Economic and Social Evaluation* (eds Pitcher, T.J. & Hollingworth, C.E.), Chapter 9, this volume. Blackwell Science, Oxford, UK.

Chuenpagee, R., Knetsch, J.L. & Brown, T.C. (2001) Environmental damage schedules: community judgments of importance and assessments of losses. *Land Economics*, **77**(1), 1–11.

Cooke, S.J., Schreer, J.F. & McKinley, R.S (1999) Evaluating catch-and-release angling practices from the fish's point of view. In: *Evaluating the Benefits of Recreational Fisheries* (ed. Pitcher, T.J.), Fisheries Centre Research Reports, Volume 7, Number 2. pp. 78–87. University of British Columbia, Vancouver.

Cox, S. & Walters, C. (2002) Maintaining quality in recreational fisheries: how success breeds failure in management of open-access sport fisheries. In: *Recreational Fisheries: Ecological, Economic and Social Evaluation* (eds Pitcher, T.J. & Hollingworth, C.E.), Chapter 8, this volume. Blackwell Science, Oxford, UK.

David, H.A. (1988) *The Method of Paired Comparisons*. London: Charles Griffin & Co.

Donati, A. & Pasin, P. (1997) *Pesca e Pescatori nell'antichita*. Arte, Milan.

Duffield, J. W., Merritt, M.F. & Neher, C. J. (2002) Valuation and policy in Alaskan sport fisheries. In: *Recreational Fisheries: Ecological, Economic and Social Evaluation* (eds Pitcher, T.J. & Hollingworth, C.E.), Chapter 13, this volume. Blackwell Science, Oxford, UK.

Dunn-Rankin, R. (1983) *Scaling Methods*. Lawrence Erlbaum: New Jersey.

Gentner, B. & Lowther, A. (2002) Evaluating marine sport fisheries in the USA. In: *Recreational Fisheries: Ecological, Economic and Social Evaluation* (eds Pitcher, T.J. & Hollingworth, C.E.), Chapter 14, this volume. Blackwell Science, Oxford, UK.

Grahame, K. (1908) *The Wind in the Willows*. Methuen.

Griffin, R.K. & Walters, C.J. (1999) Recreational fishery response to rebuilding and reallocation of the barramundi (*Lates calcarifer*) stocks in Australia's Northern Territory. In: *Evaluating the Benefits of Recreational Fisheries* (ed. Pitcher, T.J.), Fisheries Centre Research Reports, Volume 7, Number 2. pp. 101–104. University of British Columbia, Vancouver.

Griffiths, M.H. & Lamberth, S. J. (2002) Evaluating a marine sport fishery in South Africa. In: *Recreational Fisheries: Ecological, Economic and Social Evaluation* (eds Pitcher, T.J. & Hollingworth, C.E.), Chapter 16, this volume. Blackwell Science, Oxford, UK.

Hemphill, S. (1995) The ecology and exploitation of yellowfin tuna, *Thunnus albacares* (Bonnaterre 1788) in the Pemba Channel, Kenya. DPhil Thesis, University of Wales, Bangor, UK.

Hoffmann, R.C. (1999) *Fishers' craft and lettered art: tracts on fishing from the end of the middle ages*. University of Toronto Press, Toronto.

Jones, R. & Williams-Davidson, T.-L. (2000) Applying Haida ethics in today's fishery. In: *Just Fish: the ethics of Canadian fisheries* (eds Coward, H., Ommer, R. & Pitcher, T.J.) pp. 101–115. Institute of Social and Economic Research Press, St John's, Newfoundland.

Kearney, R.E. (2002) Recreational fishing: value is in the eye of the beholder. In: *Recreational Fisheries: Ecological, Economic and Social Evaluation* (eds Pitcher, T.J. & Hollingworth, C.E.), Chapter 2, this volume. Blackwell Science, Oxford, UK.

Knetsch, J.L. (1994) Environmental valuation: some problems of wrong questions and misleading answers. *Environmental Values* **3**, 351–366.

Köbel, J. (1493) *Die Kunst wie man Fisch und Vögel fahren soll*. [How to catch fish and birds.] Heidelberg. [English translation Richard C. Hoffman.]

Lyle, J.M., Coleman, A.P.M., West, L., Campbell, D. & Henry, G.W. (2002) New large-scale survey methods for evaluating sport fisheries. In: *Recreational Fisheries: Ecological, Economic and Social Evaluation* (eds Pitcher, T.J. & Hollingworth, C.E.), Chapter 15, this volume. Blackwell Science, Oxford, UK.

Lyons, J., Hickley, P. & Gledhill, S. (2002) An evaluation of recreational fishing in England and Wales. In: *Recreational Fisheries: Ecological, Economic and Social Evaluation* (eds Pitcher, T.J. & Hollingworth, C.E.), Chapter 12, this volume. Blackwell Science, Oxford, UK.

Machena, C. (ed) (1989) *Annual Report of the Lake Kariba Fisheries Research Institute*, Kariba, Zimbabwe.

O'Bara, C.J. (1999) Economic benefits and value of a localized and seasonal walleye fishery. In: *Evaluating the Benefits of Recreational Fisheries* (ed. Pitcher, T.J.), Fisheries Centre Research Reports, Volume 7, Number 2. pp. 124–128. University of British Columbia, Vancouver.

OED (1999) *Oxford English Dictionary* (2nd edn, CD-Rom). Oxford University Press, Oxford, UK.

Oppianus of Anazarbus (c. 220) *Halieutica*. [Mair A.W. (Translator) (1928) Oppian, Colluthus, Tryphiodorus. Loeb Classical Library, No. 219. Harvard University Press, Cambridge, MA.]

Pitcher, T.J. (1999) Fishing for fun. In: *Evaluating the Benefits of Recreational Fisheries* (ed. Pitcher,

T.J.), Fisheries Centre Research Reports, Volume 7, Number 2. pp. 5–8. University of British Columbia, Vancouver.

Pitcher, T.J. (2001) Fisheries managed to rebuild ecosystems: reconstructing the past to salvage the future. *Ecological Applications*, **11**(2), 601–617.

Pitcher, T.J. & Pauly, D. (1998) Rebuilding ecosystems, not sustainability, as the proper goal of fishery management. In: *Reinventing Fisheries Management* (eds Pitcher, T.J., Hart, P.J.B. & Pauly, D.), pp. 311–329. Chapman & Hall, London.

Pitcher, T.J. & Preikshot, D.B. (2001) Rapfish: a rapid appraisal technique to evaluate the sustainability status of fisheries. *Fisheries Research*, **49**(3), 255–270.

Plato (360 BC) *The Sophist*. [Fowler, H.N. (Translator) (1988) Plato: Theaetetus Sophist. Harvard University Press.]

Plutarch (2nd cent. AD) A portrait of Cleopatra. Sections 25.5–28.1 in *The Life of Mark Antony*, 29.

Policansky, D. (2002) Catch-and-release recreational fishing: a historical perspective. In: *Recreational Fisheries: Ecological, Economic and Social Evaluation* (eds Pitcher, T.J. & Hollingworth, C.E.), Chapter 6, this volume. Blackwell Science, Oxford, UK.

Pringle, H. (1997) Ice age communities may be earliest known net hunters. *Science*, **277**, 1203.

Rudd, M.A., Folmer H. & van Kooten, G.C. (2002) Economic evaluation of recreational fishery policies. In: *Recreational Fisheries: Ecological, Economic and Social Evaluation* (eds Pitcher, T.J. & Hollingworth, C.E.), Chapter 3, this volume. Blackwell Science, Oxford, UK.

Sahrange, D. & Lundbeck, J. (1992) *A History of Fishing*. Springer, Berlin.

Smith, W. (1975) *The Eye of the Tiger*. William Heinemann. London.

Steffens, W. & Winkel, M. (2002) Evaluating recreational fishing in Germany. In: *Recreational Fisheries: Ecological, Economic and Social Evaluation* (eds Pitcher, T.J. & Hollingworth, C.E.), Chapter 10, this volume. Blackwell Science, Oxford, UK.

Sullivan, M. (1999) How and Why Fishers Lie and Cheat. In: *Evaluating the Benefits of Recreational Fisheries* (ed Pitcher, T.J.), pp.165–166. Fisheries Centre Research Reports, Volume 7, Number 2, University of British Columbia, Vancouver.

Sumaila, U. R. (2002) Recreational and commercial fishers in the Namibian silver kob fishery. In: *Recreational Fisheries: Ecological, Economic and Social Evaluation* (eds Pitcher, T.J. & Hollingworth, C.E.), Chapter 4, this volume. Blackwell Science, Oxford, UK.

Toivonen, A.-L. (2002) A survey of the economic value of Nordic recreational fisheries. In: *Recreational Fisheries: Ecological, Economic and Social Evaluation* (eds Pitcher, T.J. & Hollingworth, C.E.), Chapter 11, this volume. Blackwell Science, Oxford, UK.

Thunberg, E.M. & Milon, J.W. (2002) Projecting recreational fishing participation. In: *Recreational Fisheries: Ecological, Economic and Social Evaluation* (eds Pitcher, T.J. & Hollingworth, C.E.), Chapter 5, this volume. Blackwell Science, Oxford, UK.

Thurstone, L.L. (1927) A law of comparative judgment. *Psychology Review*, **34**, 273–286.

Wilson, E.O. (1998) *Consilience: the unity of knowledge*. Alfred A. Knopf, New York.

Yellen, J.E., Brooks, A.S., Cornelissen, E., Mehlman, M.J. & Stewart, K. (1995) A middle stone-age worked bone industry from Katanda, Upper Semliki Valley, Zaire. *Science*, **268**, 553–556.

Chapter 2
Recreational Fishing: Value is in the Eye of the Beholder

Robert E. Kearney

Abstract

In the absence of agreed evaluation the history of recreational fishing is reviewed and the ecological, economic and social issues that impact value are discussed. It is argued that the broad spectrum of reasons why individuals value recreational fishing and the differences in the perceptions of participants make it difficult, if not impossible, to derive any single measure of value. Individuals frequently attach great value to recreational fishing but the reasons that underpin this evaluation are often idiosyncratic and even vary with time for each individual. The lack of agreed definition of value makes it difficult for anglers collectively to assert rights to the resource and justify allocation and management priority.

Key words: ecological, economic, property rights, recreational fishing, social value.

Introduction

Fish is a major food staple for hundreds of millions of people. The world's commercial catch is well described, about 86 million tonnes reported in 1998 (FAO 2000), valued at approximately \$100bn at the first point of sale.

Recreational fishing is fun. Millions of people (possibly hundreds of millions) fish recreationally. But there is no estimate of the total catch and no agreed method for valuing the catch or the activity that produces it.

Most of the value of commercial capture production is realised as human food; the majority of the remainder coming from animal food and fertiliser. The nature of trade in such products makes valuation straightforward and relatively precise. The bulk of the recreational catch also goes directly for human consumption, but food is often, and increasingly, not the primary objective or output of recreational fishing. Evaluation is complex and imprecise, even in the few situations where catches are known.

The tension between the users of fish for food or for recreation is not restricted to contrasting commercial and recreational uses of resources, but rather is also a primary factor within the broad spectrum of people classified as recreational fishers. The range of such players is great, from those focused on maximising landings (for food or barter) through the many gradations of those who value sport and hunting, to those who believe the resource is to be enjoyed for what it is and not killed at all (Kearney 1999a).

Recreational fishing has increased in developed countries in line with increased leisure time and affluence. In effect, subsistence activities have progressively given way to recreational pursuits in these countries.

Not only have anglers' numbers increased, but the distribution has continued to swing towards those who favour recreation over food gathering. Increased participation coupled with increased affluence and technological advances in gear have also inevitably impacted the underlying resource base and necessitated management. The need for management coupled with the diversity of interests and paucity of information on who are recreational fishers, why do they fish, what do they catch, what is their impact and what is their value, have led to increased urgency in the quest for answers.

Since the mid-1990s, the need to better define recreational fisheries and the management thereof has gained expression through growing scientific literature and international conferences (for example, Hancock 1995, Hickley and Tomkins 1996 and Pitcher 1999a). The most recent of these concentrated on evaluating the benefits of recreational fishing (Pitcher 1999a), and while considerable progress was made the diversity of opinion was more a feature than was agreement on a definitive methodology. The lack of precision in knowledge, the difficulty in distinguishing perception from reality, bias in the available data and past failure to regard recreational fisheries assessments as serious science all contributed to the poor state of recreational fisheries evaluations at the conclusion of the twentieth century.

In the absence of agreed evaluation this chapter reviews the history of recreational fishing and discusses the ecological, economic and social issues that impact value. It is argued that the broad spectrum of reasons why individuals value recreational fishing and the differences in the perceptions of participants make it difficult, if not impossible, to derive any single measure of value. Individuals frequently are passionate about recreational fishing but the reasons that underpin this commitment (evaluation) are often idiosyncratic and even vary with time for each individual. The lack of agreed definition of value in turn makes it difficult for anglers to assert rights to the resource and to justify allocation and management priority. The need to do so is highlighted.

Some historical perspectives

The first evidence of human association with fish in such a manner that fishing is at least implied (fish spears) appears in the fossil record of approximately 90 000 years ago. Records of more recent periods have unearthed more diverse tools, such as fish hooks, net sinkers and ropes (Diamond 1991). This early evidence confirms the role of fishing in the hunter-gatherer existence that dominated most societies.

Lee (1968) reports that approximately one third of the world's ancient hunter-gatherer societies were dependent upon fishing, half on gathering and one sixth on hunting larger mammals. The successful fisher, or hunter, had access to a greater variety, and probably more and better food with resulting nutritional benefits. Laughlin (1968) and Diamond (1991) argue fishing and hunting success influenced a person's standing in society and hence, sexual success. Good fishers were likely to have more offspring who could, in turn, be anticipated to have a greater chance of survival and procreation. Genetic selection for fishing commitment and aptitude seems inevitable.

Pitcher and Hollingworth (2002) attribute the first record of pole and hook and line fishing to an Egyptian mural in a tomb dated approximately 4000 BP. They acknowledge that such fishing may have been for food 'but it certainly looks fun' (see also Pitcher 1999b, p. 5). Another tomb engraving almost a thousand years later leaves no doubt that fishing for fun was being practised in Egypt at that time, even though it was presumably the 'preserve of a rich and powerful elite' (Pitcher 1999b, p. 5).

While an advantage remained with those capable of accumulating fish for food, the sporting aspects of hunting and fishing progressively emerged. Butler (1930) attributes the first written record of acknowledgement of the joys of recreational fishing to approximately 2500 BP; 'the sheer enjoyment of matching and overcoming the strength, swiftness and cunning of the quarry' (Butler 1930, p. 14).

Butler (1930) also reports that the ancient Greeks required fishing to serve a lengthy apprenticeship as a utility before being accepted as a sport. Plato's *Book of Laws* written around 2400 BP considers the value of sport for the education of the country's elite young men. Of fishing, Plato says 'I pray that no yearning, no passionate love, of sea fishing or of angling may ever take possession of you, my young friends' (cited by Butler 1930, p. 127). Plato's concern was that fishing was a temptation which wasted time and dissipated the energy of those who should have been preparing for the service of their country by more intellectual pursuits and more physical forms of sport (Butler 1930). Obviously Plato didn't value recreational fishing highly, even though he appreciated its appeal.

Oppian (1820 BP) describes fishers as placid figures sitting by the sea waiting for fish to oblige; fishing is thus a gentle art (Butler 1930). However in a later comparison of sports he acknowledges the dangers of fishing thus: 'the perils of storm and wreck at sea are far more formidable than any perils of the land' (Butler 1930, p. 16).

Fly fishing, which one could be forgiven for assuming was practised at least partially for sport, was first reported from Macedonia in 1800 BP (Dunn 1991). In England more than a thousand years later, 1496, the famous *Treatyse of Fysshinge Wyth an Angle* detailed different types of flies for different conditions (Dunn 1991).

In North America the period of post-European settlement saw the continuation of sport, including fishing, being included in daily routines (Struna 1996). The dual roles of fishing, as sport and as a source of food, again became confounded. Rights to fish became an issue with local governments determining that hunting and fishing on one's property were inviolable rights, with even Sunday ordinances, which restricted many pastimes, being waived for fishing and fowling (Struna 1996). The 1730s saw the establishment of the first formal angling club, the Schuylkill Fishing Company outside Philadelphia (Struna 1996). Through to the early twentieth century, hunting and fishing continued to attract participants from most levels of society in North America, the notable exception being the lower classes from the city (Spears and Swanson 1978).

In 1876 Rowan wrote that the angling in Canada was the finest in the world. There was only one drawback to the perfect happiness of the angler on Canadian rivers – the flies: 'I suppose they are sent to prevent him from being too happy' (Rowan 1876, p. 379). Happiness was not solely determined by catch.

Even in those early days, access and property rights issues were apparent (Rowan 1876). Access was restricted by the use of open and closed seasons for salmon and trout fishing. Previously all rivers in Canada had been open to all anglers, however, too many people were attracted to fishing. In an attempt to shift the burden of protecting rivers to the private sector, the Government decided to lease the rivers, with politics playing a major role in the allocation of leases. This led to an early use of property rights and discontent among many anglers and the general public (Rowan 1876).

Fish were an important part of the staple food of many Australian Aboriginal tribes, especially those in northern New South Wales and south-east Queensland where fish were crucial during the winter months. There is evidence that during pre-European times tribes would travel large distances to attend gatherings that included activities such as fishing (Frankel 1991). As late as the mid to late 1800s Aborigines fished for sport and competition during at least some of these large gatherings (Gilmore 1963). Stories of tribal competitions, which included tests of skill and endurance, both features of Aboriginal life, are woven into Aboriginal legends. Examples of such competitions included men from two tribes contesting a spear fishing contest (Smith 1930).

Current state of knowledge

The most recent international review of the ecological, economic and social costs and benefits of recreational fishing was that afforded by the June 1999 Vancouver conference (Pitcher 1999a). While information presented at that conference represented a considerable advance on what was previously published, the available data and analyses are clearly insufficient. Mainstream scientific literature contains few assessments of recreational fisheries even though the popular angling press would have us believe that recreational fisheries are far more valuable than their commercial counterparts, on which the scientific literature is voluminous.

In Vancouver I lamented the lack of scientific rigour in Australia's approach to assessing recreational fisheries (Kearney 1999a). My more recent efforts to unearth published assessments for other countries have been similarly unrewarding. What literature exists is almost exclusively restricted to North American, European, South African, New Zealand and Australian experience. Accepting that regarding fishing as a recreational rather than food-gathering activity is largely a developed country perspective (except for the privileged minority in developing countries) it is logical to assume developed country domination of published assessments. But the scarcity of quality assessments even for the above-mentioned nations is indicative of a lack of management (government) use of, and therefore need for, such information. It also reflects associated apathy from those nations' leading fisheries scientists towards rigorous collection and processing of data on recreational fishing. This could be interpreted as a lack of value by the scientific minority.

The literature on evaluating recreational fishing in North America is predominantly related to factors that influence angler satisfaction (Moeller and Engelken 1972; Holland and Ditton 1992; Spencer 1993) and the importance of keeping and consuming fish (Fedler and Ditton 1986; Matlock *et al.* 1988) or practising catch-and-release (Hunt 1991). In combination these works confirm that there is a large number of reasons why people go recreational fishing and many factors which influence the priority given to individual reasons. Consensus on why is elusive.

Holland and Ditton (1992) review eleven previous studies which prioritised a total of 15 reasons (nature, escape, sociability, relaxation, catching fish, equipment, nostalgia, exercise, eating fish, achievement, challenge, learning, exploration, privacy, and catching trophy fish). Their review suggests that appreciation of nature, relaxation and escape are more valued than catching fish, but they do highlight the results of Graefe and Fedler (1986) who concluded that catch was most important, particularly catch in relation to expectations. These studies and their counterparts in other countries have confirmed the variety of reasons why individuals value fishing. Agreed reasons why communities or countries should value such activities remain poorly described.

Australian data unfortunately provide an excellent example of current imprecision and even disagreement in national evaluations of recreational fishing. Surveys conducted for state government and recreational industry groups have given participation figures (those that fish at least once a year) of approximately 30% of the total population (PA Management Consultants 1984; VIFTA 1997), yet national Bureau of Statistics data puts the participation rate at 4.8% (Australian Bureau of Statistics 1998). It appears that this enormous discrepancy can be attributed to bias in the way that the respective surveys were conducted. Surveys based on a questionnaire that begins by announcing that the survey is about recreational fishing give high figures, presumably resulting from the prompt, while those that ask which sports interviewees participate in give low representation to fishing, presumably because many people participate for reasons other than sport, or mention only their most prominent sports.

A major current Australian national survey has preliminary figures of 27% participation (Henry 2000, personal communication), which also appears high. Statistics from the state of Victoria based on the number of anglers who purchased licences to fish recreationally in that state, corrected for compliance rates reported by licence enforcement officers, suggest the correct figure is closer to 14% (McDonald 2000, personal communication).

Clearly the reasons individuals value recreational fishing vary greatly, but many individuals have a strong commitment to whatever aspect(s) motivates them. Passion is common among devotees. Governments, and their agents, are far less certain and committed. What support there is appears more a response to a less-than-convincing perception that a lot of votes might be involved if the collective individual passions should ever be unified for a single cause.

Knowledge, or lack thereof, which influences the current balance sheet, reasons for conflict with other resource interests, and present management practices are reviewed in following sections.

Factors for the balance sheet

In any evaluation there are two sides to the ledger. In evaluating recreational fishing there is also need to note that benefits and costs for individuals and for society are not always aligned. Tables 2.1–2.3 provide lists of positive and negative ecological, economic and social attributes, not in priority order but with those of greater value to individuals towards the top.

Ecological issues are clearly of general community impact even though this impact usually comes from the collective influence of many individuals. On the whole, the items on the balance sheet can only be measured imprecisely and have multiple implications. For example, the negative effects of fish mortality will include more than just decreased availability of target species; secondary effects such as distortion of food chains, webs and ecosystems will inevitably follow. Relative abundance of species will also be influenced by those species that are targeted and those individuals that are retained.

Even though the economic aspects appear relatively straightforward, economists have struggled to describe an effective formula for evaluating recreational fishing. Part of this problem can be attributed to the difficulty of quantifying, in economic terms, the social and environmental factors that are interwoven with the more obviously economic attributes. I have argued that the major benefits of recreational fisheries are social (see 'Current state of knowledge' above), and economic rationalism does not well accommodate social issues (Connor 2000).

There are many social factors that have complex benefits for individuals and for the broader community. For example the education a person gains about the complexities of ecosystems and inter-relationships between habitats, vegetation, invertebrates and fish is of great value to that individual, and through the actions of well educated citizens, to the whole of society. The benefits to society from having healthy and content citizens are also real.

Numerous issues can have both positive and negative attributes. Advantages to one individual, for example the privacy and solitude, can be seen by others, even members of the same family, to be antisocial.

The dangers of recreational fishing are not always obvious. Rock-fishing from open headlands is regarded as one of Australia's most dangerous sports with numerous fatalities reported each year (Kearney 1999a).

Social class divisions are prominent in freshwater fisheries where rights to lakes and streams are controlled by privileged minorities, notably in Europe and the United Kingdom.

Table 2.1 Ecological balance sheet for recreational fishing.

Positive	Negative
Education	Fish mortality
Promotes environmental responsibility	Fish behavioural disturbances
Aids environmental monitoring	Pest species introductions
Engenders support for restoration	Native species translocations
Data collection	Manipulation of gene pool
Aids surveillance of environmental vandalism	Spreads diseases
	Pollution
	Environmental disturbance
	Wilderness violation

Table 2.2 Economic balance sheet for recreational fishing.

Positive	Negative
Quality food	Access costs
Employment	Inefficient method of harvesting food
Tourism	Reduces commercial fisheries
Decentralisation	Management and compliance costs
Promotes exports, for some countries	Increases imports (equipment) for some countries

Table 2.3 Social balance sheet for recreational fishing.

Positive	Negative
Privacy	Dangerous
Relaxation	Requires enforcement
Exercise	Antisocial
Sport	Social class divisions
Camaraderie	Violates animal rights
Competition	
Bonding (parent/child)	
Sustenance	
Quality nutrition	
Employment	
Education	
Environmental monitoring	
Decentralisation	

Competition and conflict

The mention to recreational fishers of competition and conflict with other resource users normally evokes some derogatory response about our commercial colleagues. Yet anglers, as a group, are in competition to the point of conflict with a number of other groupings with an interest in fisheries resources, for example indigenous societies, urban developers and water access owners. Furthermore the diversity of values and interests within the group labelled as recreational fishers is such that competition among categories within the group can lead to greater diminution of some individual benefits than comes from external influence.

Within the recreational sector

The recreational fishing fraternity ranges from those whose pursuit of large catches, and methods of subsequent disposal, brands them unlicensed professionals, through the many variants of hunters, sport fishers, adventurers and social participants, to those who enjoy the open spaces and if they have a line in the water, do so merely to make an excuse for being there (Kearney 1995a). Recreational fishers value their participation in the above range of activities for many different and varying reasons (see, for example, Holland and Ditton 1992 and Kearney 1999a). Not surprisingly the activities of some will be in conflict with the aspirations of others. The person in search of solitude and serenity is not amused when the two-day hike to a wilderness trout stream is met by a helicopter load of affluent, tourist anglers.

Mostly the conflict is more subtle. Enthusiasts at different levels of the spectrum are mostly ignorant of the level of their competition with other levels. The average angler catches few fish. Many studies have shown a small percentage takes large catches (for example Steffe *et al.* 1996) and as many as one third of anglers catch no fish at all (Sport Fishing Institute 1964 cited by Moeller and Engelken 1972). However, in many cases, particularly in marine fisheries, an angler is more likely to blame commercial operators than other anglers. This tendency exists even if total anglers' catches exceed the commercial take. Or even if his colleague fishing alongside him in his own boat represents the single greatest threat to his immediate chances of catching a fish.

As values change with our moods, from day to day, so our aspirations are challenged. How many of us have sought the solace and gratification provided at our favourite, and relatively secret, fishing hole, only to find it occupied by 'friends' or relatives, to whom we had introduced it on an earlier social excursion?

Collectively we espouse equity and resource sharing, and yet knowing that a few take most of the catch still does not overcome our aversion to restrictive bag limits, particularly for marine and estuarine species. Responsible citizens who have never caught a bag limit, let alone exceeded it, can wax lyrical about the evils of restricting anglers when commercial catches go unchecked. The popular angling press fuels this perception (see, for example, Harnwell 2000 and Kaberry 2001).

With the commercial fishing sector

Without diminishing the seldom-appreciated tensions within the recreational fishing community it is necessary to acknowledge external pressures which result in numerous, more obvious, conflicts.

Anglers' disputes with commercial fishers are legend, and they continue. Human nature creates the expectation of conflict between large numbers of people exploiting a common, finite resource (Hardin 1968). World-wide there are problems. Specific cases display individual nuances but the principles vary little. A recent high profile dispute between anglers and commercial fishers in nine bays and inlets of the state of Victoria, Australia provides a typical example: anglers had gained the attention of the press and the state's leader of Government (the Premier) by asserting the following evils (perceptions) of commercial fishing (from VIFTA 1997):

(1) Commercial catches had over-exploited target species such that collapse was imminent.
(2) As a result of (1), recreational catches had declined and the recreational industry, which, it was argued, was worth in excess of a billion Australian dollars annually, was threatened.
(3) Large numbers of under-sized fish were killed by commercial netting.
(4) Large commercial catches of baitfish species reduce the availability of food for key angling species and other predators, such as dolphins and penguins, which have high conservation and tourism value.
(5) Commercial haul netting damages sea grass beds and other important habitats.

Commercial fishers' concerns (perceptions) about anglers included (from WBM 1997):

(1) Some anglers ignore bag limits resulting in excessive take. Some even sell their catch, representing illegal competition for licensed operators.
(2) The large numbers of anglers, and the resulting combined effort and catch, give rise to significant mortality of under-sized fish, retained illegally, or killed accidentally.
(3) Large numbers of recreational boats and ill-disciplined behaviour by some, result in damage to sea grass beds (by propellers and anchors).
(4) Anglers' boats on commercial hauling grounds interfere with commercial fishing.
(5) Recreational bait diggers have a negative effect on fish habitats.

The Fisheries Co-Management Council of Victoria commissioned a series of five studies to investigate the above claims and to gather appropriate socioeconomic and environmental data to aid overall assessment. In its conclusions the Council noted that recreational catch rates were probably declining, with the primary causes likely to be increased total effective recreational fishing effort, and resource effects of environmental degradation. Total catches by the commercial and recreational sectors varied between species but were of similar magnitude for most species that were targeted by both groups (WBM 1997). The total economic values of the two sectors were not dissimilar (Hundloe 1997). Environmental damage by either group was not great but pollution from bait wrappers and fuel emissions from the recreational sector were of concern (WBM 1997). Council highlighted the conclusion that environmental degradation posed a far more serious threat to long-term resource security for both sectors than the direct effects of fishing (Fisheries Co-Management Council 1998).

After considering all resource and socioeconomic assessments, the Fisheries Co-Management Council recommended no change to the allocations between the recreational and commercial sectors but noted that redistribution of resources within the recreational sector (by alterations to bag and size limits and greater compliance with regulations and codes of practice) was more urgently required (Fisheries Co-Management Council 1998).

With customary resource owners (users)

Allocation of resources is the most pressing single issue facing fisheries managers in many countries. Determination of ownership and rights of access are central to the allocation process. With wider recognition of the rights of aboriginal peoples throughout the world have come increased claims by those people for resources that traditionally underpinned their existence. The latter part of the twentieth century saw numerous pivotal decisions in many countries. Two that significantly impact fisheries resource allocations were the New Zealand Treaty of Waitangi Settlement Act in 1992 (Treaty of Waitangi 1992) and the Mabo decision in the Australian courts in the same year (Mabo 1992). New Zealand's handling of customary fishing rights and their impact on recreational fishers provides an excellent case study.

New Zealand manages most of its fisheries resources through assessing total allowable catches (TACs) and subsequent quota allocations. For each key species in each management area the Fisheries Minister sets a TAC based on scientific advice on the maximum sustainable level of exploitation. The TAC is then allocated, by the Minister, to the customary commercial and recreational sectors. The customary share is allocated to Maori in two parts: firstly a commercial component in the form of quota for key species, and secondly, a noncommercial component more aligned with traditional use. The commercial share (noncustomary) is allocated in well-defined individual transferable quotas (ITQs). The recreational share is allocated as a loosely defined allowance, which lacks the security of a well-described and legislatively protected share of the resource. It has, however, historically been set at levels that accommodate the expectations of the sector, based on previous years' catches (McMurran 2000). As such it has served recreational fishers well. However as competition for stable, or declining, resources increases, challenges and conflicts appear certain to increase (Kearney 2001).

In addition to direct competition between traditional resource users and anglers there are less obvious, but not necessarily less important, potential disagreements; the attitudes of aboriginal communities towards any alien access to some sacred sites could well restrict access to key angling locations. Social attitudes to practices such as catch-and-release, which some, such as aboriginal peoples in the Pacific Northwest, regard as disrespectful (Jones & Williams-Davidson 2000), may well curtail activities.

In freshwater fisheries there has been long-standing conflict between landholders and anglers seeking access. National laws governing such matters vary enormously and local traditions in parts of countries create further complexities. Any process which restricts access and/or affords it only to a privileged (affluent) few is certain to remain a source of conflict.

With conservation imperatives

Continued declines in the world's fisheries resources coupled with increased acceptance of the value of biodiversity have prompted many governments to look to new conservation measures. Marine protected areas (MPAs) are being heralded as a solution to both. Many countries are initiating, or at least investigating, representative series of MPAs as environmental conservation measures (see, for example, ANZECC 1999).

While the rational use of MPAs should be seen as a benefit to the long-term security and quality of recreational fisheries there are many tensions. Government consultative processes tend to be less than perfect and even if recreational angling representatives are consulted it is often impossible for such individuals to truly represent the opinions of the spectrum of recreational interests. It is also argued that governments are pushing ahead with restrictions on the use of marine areas because of their need to be seen to be acting to conserve biodiversity when their efforts to restrict terrestrial activities have been singularly unsuccessful because of the relative strength of property rights for farmers and graziers (see 'value and resource security', below).

The degree to which MPAs impact recreational fisheries will be determined by a large number of factors, such as the size, location, purpose and restrictions peculiar to each reserve. But there is little doubt some will impact recreational fishers and at least some anglers will be unhappy.

Government regulations relating to wilderness areas also impact recreational enthusiasts to a variable extent.

With animal rights activists

The growth in commitment to political and social responsibilities in the last few decades has caused re-evaluation of many human activities that impact other citizens and other living organisms. In many cases management practices have had to be dramatically revised; fisheries management of by-catch of marine mammals, sea birds and turtles provides examples (AFMA 2001, Harris and Ward 1999).

Management changes have not been restricted to by-catch species; the harvest of fur seals in the Northern Hemisphere and duck hunting in Australia are but two examples. Animal ethics committees now play a key role in most research institutions and businesses where live animals are involved.

Recreational fishing is coming under increased scrutiny. Not only is the way in which fish are being handled (particularly as live-bait) and killed being questioned but practices such as catch-and-release now have their opponents. It is indeed significant that catch-and-release has been banned in some countries (Cowx, in press) [see also Steffens and Winkel 2002: eds]. The underlying philosophy is that if you do not need the fish to eat you should not be tormenting them [see further comments by Aas *et al.* 2002 and Policansky 2002: eds]. Recent suggestions that many fish that are captured and released do change their behaviour, and some may forgo spawning, are sure to intensify the debate.

Does management practice reflect value?

As was concluded at the Vancouver conference the three major categories of value of recreational fishing, ecological, economic and social are respectively, only recently recognised, poorly measured or rarely valued at all (Kearney 1999b, Pitcher 1999a). Not surprisingly then, current management practice is seldom deliberately aligned with a comprehensive assessment of real values.

If, as I have previously argued (Kearney 1999a), the real value is social then management should reflect social benefits. It should encourage those beneficial activities on the positive side of the ledger (Table 2.3) and counter those on the negative.

The great variance in the ways in which individuals derive benefit and in the disparity that often occurs between benefits to individuals and to society as a whole, makes it difficult to assess current practice against stated objectives. Most management agencies with responsibilities for recreational fisheries espouse goals that establish and maintain quality fisheries. Quality is not often defined but it is usually implied to relate to maximising the total recreational catch and/or offering a realistic expectation of encountering a trophy fish. In Australian saltwater fisheries bag limits are usually so high that they restrict only the very best anglers and do little to improve the lot of the majority whose realistic goal is to catch one fish (a 1997 contingent valuation of Victorian anglers' willingness to pay for one extra fish was \$4–\$8 whereas the second fish was valued at only \$0.40–\$0.80 (Kinhill 1997)). Surely bag limits should be designed to get optimum value for as many people as possible, not to accommodate sizeable catches by the vocal minority?

Economic modelling has recently shown that expenditure is an inappropriate indicator of value (Hundloe 1997), particularly in countries like Australia where approximately 80% of expenditure is on imports (Kearney 1999a). O'Bara (1999) demonstrates that the economic value of tourists fishing a freshwater fishery in Tennessee USA is greater than that from locals participating in the same fishery. Combining such data sets could lead an economic rationalist to conclude that recreational fishing in Australia, at least, should be reserved for tourists! (Kearney 1999b).

Recreational fisheries are seldom managed to optimise the ecological balance sheet (Table 2.1). However, many countries are taking increasing measures to reduce the negative issues, particularly species introductions and translocations, pollution and environmental disturbance. Most governments do not utilise the power of the recreational fishing community for championing and monitoring the conservation of aquatic habitats (Kearney 1995b).

Value and resource security

I have recently argued that disputes between fisheries resource users will progressively be concentrated on the rights of users to long-term access and allocation (property rights) (Kearney 2001). Much of what follows in this section is modified from Kearney (2001), *Marine Policy*, 25: 49–59 with permission from Elsevier Science.

Complementary pressures of increased demand for fish (human population growth, increased appreciation of sea-food and globalisation of markets), increased power and effectiveness of fishing (more and better vessels and ever-improving information and technology) and relatively less effective compliance with conservation imperatives (reduced community conscience, high costs and ineffectiveness of enforcement and increased incentives to exploit, including growing pressures for continuous cash flows and profits) led to growing calls for management that increases security for resource users.

By the 1980s the growing influence of economic rationalism strengthened the supposition that efficient natural resource management in modern society required the allocation of property rights and responsibilities. Under the property rights paradigm resources are secured and economic efficiency optimised when the whole of the resource is allocated in a manner that allows market forces to dominate. Realisation of long-term value of sustainable use, and generation of maximum returns from short-term harvest, are assumed to result from the more responsible use, husbandry and protection that comes with ownership. While not a panacea, property rights management, often in the form of ITQs, has been sufficiently successful to assume its use will increase (for comprehensive coverage of recent discussion on the use of property rights in fisheries management see Shotton 2000a and b).

The impacts of human population growth and increased demand for fish have not been limited to commercial fisheries. Continued growth in leisure time and disposable income in the developed world has greatly increased angler participation and the combined effective fishing effort of the community. Continued conflict between the competing users of finite fisheries resources is inevitable.

Thus at the beginning of the twenty-first century we are facing:

(1) Increased competition for diminishing resources
(2) A lack of agreed principles for the allocation of resources (access and use) between competing groups (commercial, customary, subsistence, recreational and nonconsumptive) (see Scott 2000 and Shotton 2000c)
(3) Growing definition of the rights of commercial and customary users and allocations based on these rights (for example McMurran 2000)
(4) Remaining uncertainty over the value of recreational fisheries (Pitcher 1999a)
(5) Little definition of the rights of recreational users and erosion of these rights (discussed below)

It appears inevitable that governments will try to pass more and more of the responsibilities and costs for resource conservation and management to users, under the premise that user pays (see, for example, France and Excel 2000). Such devolution seems certain to involve increased property rights for commercial fishers, even if only to coerce them to pay. Consideration of the possible benefits and drawbacks of the increased use of property rights on the recreational sector quickly highlights the lack of definition of who comprises this sector, what do they value (collectively) and what rights do they have.

Anglers assume they have, as a birthright, access to fish resources [see also Pitcher and Hollingworth 2002; Cox and Walters 2002 regarding common law rights – eds]. However, collectively their rights to, or responsibility for, a share of fisheries resources have not been defined. There is a nonspecified common law right of all individuals to fish in the ocean (Palmer 2000), as there is a right to do anything that is not prohibited. However this right to fish is being increasingly reduced by statutes, such as bag and size limits, which modify what is legal. The introduction of recreational licences removes the right of all to fish for free, and therefore, arguably changes the nature of the fundamental right to fish.

New Zealand is one of very few countries that is attempting more clearly to define the right of recreational fishers. The Ministry of Fisheries acknowledges that the rights of recreational fishers remain poorly defined even though by world standards its recreational fisheries are superb and knowledge of catch and effort and the allocation process (McMurran 2000) all appear exemplary. The Ministry is now working with the New Zealand Recreational Fishing Council Inc to improve the definition of property rights. Key objectives are:

- 'Better defining the recreational right by introducing a proportional-share arrangement' (McMurran 2000 p. 185), which is a percentage of the TAC that is not varied from year to year at the pleasure of the Minister.
- 'Enhancing recreational fishers' rights to directly manage their share' (McMurran 2000 p. 185), that is a greater involvement of representatives of the recreational sector in the management of recreational fishing practices. Regionally based groups have been proposed for this purpose (McMurran 2000).

It is significant that many New Zealand recreational fishers believe their right to the resource has preference over that of commercial fishers. They cite a 1989 National Policy for Marine Recreational Fisheries which states 'Preference will be given to non-commercial fishing in areas readily accessible to and popular with the public, where a species is not sufficiently abundant to support both non-commercial and commercial fishing' (NZ Ministry of Fisheries 1998 p. 4). However, a number of court cases since 1989 have demonstrated that the Minister is not required to give preference to either recreational or commercial interests (NZ Recreational Fishing Council 1999). This reaffirms Palmer's reminder that the definition, and hence the value of a right, is a legal issue and no matter what fisheries managers' intentions may be it is ultimately up to the courts to decide the nature and value of the right (Palmer 2000).

It appears recreational fishers have little choice but to become more involved in the race for property rights. Our current lack of commitment appears to be the result of uncertainty over who we are (collectively), what we value, why society should value us and why we should be given rights.

Conclusions

There can be little doubt that the first human attempts at fishing would have been for food. It is also given that success would have brought pleasure, even enjoyment. Once success became less precarious one can assume at least some participants found it fun. The first archaeological record of promoting primarily the fun side of fishing attributes this distinction to the Egyptians almost three thousand years ago (Pitcher 1999b).

Since the beginning of fishing, values have evolved; from essential for survival, to a primary source of food but also used for barter, then of subsistence value, a food supplement but also fun, to fun and also food and finally to just fun (catch-and-release). In the modern developed country all of these values, with the possible exception of the first, still have their devotees. The problem for management, and for society, is that supporters are being progressively polarised into different categories. With diminishing fish resources and growing human population pressures the categories are progressively finding themselves in increasing competition and even conflict.

Tensions between commercial and recreational users of resources have probably existed since fishing was first regarded as recreation. In recent years tensions with other parties, notably customary resource users and owners, preservationists and conservationists and animal rights activists have begun to force the recreational fishing community to defend its assumed right to largely unrestricted access to fish resources. In preparing this defence we have become aware of the diversity of ecological, economic and social reasons why fishing is valued, or disliked, by individuals and the broader community. We have also realised how far we are from universal agreement, or even approximate consensus, on what aspects should be most valued and what methodology should be used to derive a composite assessment of worth.

While we procrastinate, others are gaining more securely defined rights, which are increasingly assuming the characteristics of either private property, legislated restriction zones or politically correct restraints on practices.

The immense diversity in the recreational fishing community has not engendered comprehensive or good representation. Even basic knowledge about who we are, what we do and what we want, is poor. Until we answer these questions how do we determine appropriate representation? Yet if we remain divided on what we value and why we deserve to be involved in the management process the balance sheet will not be one of which we can be proud, or on which we can rely.

Acknowledgements

I thank the Editors, Tony and Chuck for the opportunity to contribute this chapter and for the constructive comments from an anonymous reviewer. Kylie Peterson provided excellent assistance with literature searches and Deborah Carraro with the finalisation of the manuscript.

References

Aas, Ø, Thailing, C. & Ditton, R.B. (2002) Controversy over catch-and-release recreational fishing in Europe. In: *Recreational Fisheries: Ecological, Economic and Social Evaluation* (eds Pitcher, T.J.

& Hollingworth, C.E.), Chapter 7, this volume. Blackwell Science, Oxford, UK.

AFMA (2001) By-catch Action Plans (numerous), Australian Fisheries Management Authority, Canberra, Australia.

ANZECC (1999) Australian and New Zealand Environment Conservation Council Task Force on Marine Protected Areas 1999, *Strategic Plan of Action for the National Representative System of Marine Protected Areas: A Guide for Action by Australian Governments*. Environment Australia, Canberra.

Australian Bureau of Statistics (1998) *Participation in sport and physical activities, Australia, 1997–98*. ABS Catalogue Number 4177.0. ABS, Canberra.

Butler, A.J. (1930) *Sport in classic times*. William Kaufmann Inc, CA.

Connor, R. (2000) Are ITQs Property Rights? Definition, Discipline and Disclosure. In: *Use of property rights in fisheries management. Proceedings of the FishRights99 Conference*. Workshop presentations (ed Shotton, R), pp. 29–38. FAO Fisheries Technical Paper 404/2, FAO, Rome.

Cowx, I.G. (in press) Recreational fisheries. In: The Fisheries Handbook (eds Hart, P. & Reynolds, J.). Blackwell Science, Oxford.

Cox, S. & Walters, C. (2002) Maintaining quality in recreational fisheries: how success breeds failure in management of open-access sport fisheries. In: *Recreational Fisheries: Ecological, Economic and Social Evaluation* (eds Pitcher, T.J. & Hollingworth, C.E.), Chapter 8, this volume. Blackwell Science, Oxford, UK.

Diamond, J.M. (1991) *The Rise and Fall of the Third Chimpanzee*. Radius, London.

Dunn, B. (1991) *Angling in Australia: its history and writings*. David Ell Press, Balmain.

FAO (2000) *FAO Fisheries Department summary tables of fishery statistics*. Updated 21 March 2000. Available: http://www.fao.org/fi/statist/summtab/pr_a1a.asp.

Fedler, A.J. & Ditton, R.B. (1986) A framework for understanding the consumptive orientation of recreational fishers. *Environmental Management,* **10**, 221–227.

Fisheries Co-Management Council (1998) *Bay and inlet fisheries study – executive summary of findings*. Fisheries Co-Management Council, Melbourne.

France, M. & Excel, M. (2000) No Rights, No Responsibility. In: *Use of property rights in fisheries management. Proceedings of the FishRights99 Conference*. Mini-course lectures and Core Conference presentations (ed. Shotton, R.), pp 235–240. FAO Fisheries Technical paper 404/1, FAO, Rome.

Frankel, D. (1991) *Remains to be seen: Archaeological insights into Australian prehistory*. Longman Cheshire, Melbourne.

Gilmore, M. (1963) *Old Days – Old Ways: A book of recollections*. Angus & Robertson, Sydney.

Graefe, A.R. & Fedler, A.J. (1986) Situational and subjective determinants of satisfaction in marine recreational angling. *Leisure Sciences*, **8**, 275–295.

Hancock, D.A. (ed) (1995) Recreational fishing: What's the catch? *Australian Society for Fish Biology Workshop Proceedings*. 30–31 August, 1994, Canberra. Australian Society for Fish Biology, Canberra.

Hardin, G. (1968) The tragedy of the commons. *Science*, **162**, 1243–1248.

Harnwell, J. (2000) Save our stripies. *Fishing World*, March 2000:9.

Harris, A. & Ward, P. (1999) *Non-target Species in Australia's Commonwealth Fisheries. A Critical Review*. Bureau of Rural Sciences, Canberra.

Henry, G. (2000) Personal communication. New South Wales Fisheries, Cronulla, Australia.

Hickley, P. & Tomkins, H. (1996) *The Social, Economic and Management Aspects of Recreational Fisheries*. Fishing News Books. Blackwell Science, Oxford.

Holland, S.M. & Ditton, R.B. (1992) Fishing trip satisfaction: A typology of anglers. *North American Journal of Fisheries Management,* **12**, 28–33.

Hundloe, T. (1997) *Report to the Victorian Fisheries Co-management Council on the allocation of fish between commercial and recreational fishers*. Fisheries Co-Management Council, Melbourne.
Hunt, R.L. (1991) Evaluation of a catch-and-release fishery for brown trout regulated by an unprotected slot length. *Technical Bulletin No. 173*. Department of Natural Resources, Madison, WI.
Jones, R. & Williams-Davidson, T-L. (2000) Applying Haida ethics in today's fishery. In: *Just Fish: the ethics of Canadian fisheries* (eds Coward, H., Ommer, R. & Pitcher, T.J.), pp. 101–115. Institute of Social and Economic Research Press, St John's, Newfoundland.
Kaberry, P. (2001) Bluewater madness. *Fishing World*, May 2001:23.
Kearney, R.E. (1995a) Recreational fishing: what's the catch. In: *Recreational Fishing: What's the Catch* (ed Hancock, D.A.), pp. 10–23. Australian Society for Fish Biology Workshop Proceedings, Canberra, 30–3 August 1994, Australian Society for Fish Biology, Canberra.
Kearney, R.E. (1995b) Biodiversity and fisheries management: the implications of extracting maximum yields from interactive ecosystems. In: *Conserving Biodiversity: Threats and Solutions* (eds Bradstock, R.A. *et al.*), pp. 300–306. Surrey Beatty & Sons, Chipping Norton.
Kearney, R.E. (1999a) Evaluating recreational fishing: managing perceptions and/or reality. In: *Evaluating the Benefits of Recreational Fisheries* (ed Pitcher, T.J.), pp. 9–14. Fisheries Centre Research Reports, Volume 7, Number 2. University of British Columbia, Vancouver.
Kearney, R.E. (1999b) Perspectives on evaluating fisheries from the keynote speakers – Bob Kearney. In: *Evaluating the Benefits of Recreational Fishing* (ed. Pitcher, T.J.), pp. 44–45. Fisheries Centre Research Reports, Volume 7, Number 2, University of British Columbia, Vancouver.
Kearney, R.E. (2001) Fisheries property rights and recreational/commercial conflict: implications of policy developments in Australia and New Zealand. *Marine Policy,* **25**, 49–59.
Kinhill (1997) *Socio-economic impacts of future management options for scalefish in Victoria's Bays and Inlets*. Fisheries Co-Management Council, Melbourne, 69.
Laughlin, W.S. (1968) Hunting: An integrating biobehaviour system and its evolutionary importance. In: *Man the Hunter* (eds Lee, R.B. & DeVore, I.), pp. 304–320. Aldine Publishing Company, Chicago.
Lee, R.B. (1968) What hunters do for a living, or, how to make out on scarce resources. In: *Man the Hunter* (eds Lee, R.B. & DeVore, I.), pp. 30–48. Aldine Publishing Company, Chicago.
Mabo v. *Queensland*, 1992, *Australian Law Reports*, **107**, Australia
Matlock, G.C., Saul, G.E. & Bryan, C.E. (1988) Importance of fish consumption to sport fishers. *Fisheries,* **13**, 25–26.
McDonald, M. (2000) Personal communication. Victorian Department of Natural Resources and Environment, Melbourne, Australia.
McMurran, J. (2000) Property rights and recreational fishing: never the twain shall meet? In: *Use of property rights in fisheries management. Proceedings of the FishRights99 Conference*. Workshop Presentations (ed. Shotton, R), pp. 184–187. FAO Fisheries Technical Paper 404/2, FAO, Rome.
Moeller, G.H. & Engelken, J.H. (1972) What fishers look for in a fishing experience. *Journal of Wildlife Management,* **36**, 1253–1257.
NZ Ministry of Fisheries (1998) *Improving the management of recreational fisheries*. NZ Ministry of Fisheries, July 1998, Wellington.
NZ Recreational Fishing Council (1999) Recreational Fishing Council Conference: Workshop Papers – draft. Joint New Zealand Recreational Fishing Council and Ministry of Fisheries Working Group.
O'Bara, C.J. (1999) Economic benefits and value of a localized and seasonal walleye fishery. In: *Evaluating the Benefits of Recreational Fishing* (ed Pitcher, T.J.), pp. 124–129. Fisheries Centre Research Reports, Volume 7, Number 2. University of British Columbia, Vancouver.
PA Management Consultants (1984) *National survey of participation in recreational fishing*. Report number 1, The Australian Recreational Fishing Confederation, Melbourne.

Palmer, W. (2000) Legal planning for management of fisheries using property rights. In: *Use of property rights in fisheries management. Proceedings of the FishRights99 Conference.* Mini-course lectures and Core Conference presentations (ed Shotton, R.). FAO Fisheries Technical Paper 404/1. FAO, Rome.

Pitcher, T.J. (ed) (1999a) *Evaluating the Benefits of Recreational Fisheries*. Fisheries Centre Research Reports Volume 7, Number 2. University of British Columbia, Vancouver.

Pitcher, T.J. (1999b) Director's Foreword. In: *Evaluating the Benefits of Recreational Fisheries* (ed T J. Pitcher), pp. 5–8. Fisheries Centre Research Reports, Volume 7, Number 2. University of British Columbia, Vancouver.

Pitcher, T.J. & Hollingworth, C.E. (2002) Fishing for fun: where's the catch? In: *Recreational Fisheries: Ecological, Economic and Social Evaluation* (eds Pitcher, T.J. & Hollingworth, C.E.), Chapter 1, this volume. Blackwell Science, Oxford, UK.

Policansky, D. (2002) Catch-and-release recreational fishing: a historical perspective. In: *Recreational Fisheries: Ecological, Economic and Social Evaluation* (eds Pitcher, T.J. & Hollingworth, C.E.), Chapter 6, this volume. Blackwell Science, Oxford, UK.

Rowan, J.J. (1876) *The emigrant and sportsman in Canada*. Edward Standford, London.

Scott, A. (2000) Introducing property rights in fisheries management. In: *Use of property rights in fisheries management. Proceedings of the FishRights99 Conference.* Mini-course lectures and Core Conference presentations (ed Shotton, R.), pp. 1–13. FAO Fisheries Technical paper 404/1, FAO, Rome.

Shotton, R. (ed) (2000a) *Use of property rights in fisheries management. Proceedings of the FishRights99 Conference.* Mini-course lectures and Core Conference presentations. FAO Fisheries Technical paper 404/1, FAO, Rome.

Shotton, R. (ed.) (2000b) *Use of property rights in fisheries management. Proceedings of the FishRights99 Conference.* Workshop presentations. FAO Fisheries Technical Paper 404/2, FAO, Rome.

Shotton, R. (2000c) Current property rights systems in fisheries management. In: *Use of property rights in fisheries management. Proceedings of the FishRights99 Conference.* Mini-course lectures and Core Conference presentations (ed Shotton, R.), pp. 45–50. FAO Fisheries Technical paper 404/1, FAO, Rome.

Smith, W.R. (1930) *Myths and legends of the Australian aboriginals*. Harrap, London.

Spears, B.M. & Swanson, R.A. (1978) *History of sport and physical activity in the United States*. Wm. C. Brown Co., Iowa.

Spencer, P.D. (1993) Factors influencing satisfaction of anglers on Lake Miltona, Minnesota. *North American Journal of Fisheries Management,* **13**, 201–209.

Steffe, A.S., Staines, J.F. & Murphy, J. (1996) *Recreational Use of Fisheries Resources in Northern New South Wales*. Final Report, Project 57, NSW Fisheries Research Institute, Cronulla.

Steffens, W. & Winkel, M. (2002) Evaluating recreational fishing in Germany. In: *Recreational Fisheries: Ecological, Economic and Social Evaluation* (eds Pitcher, T.J. & Hollingworth, C.E.), Chapter 10, this volume. Blackwell Science, Oxford, UK.

Struna, N.L. (1996) *People of prowess: sport, leisure and labour in early Anglo-America*. University of Illinois Press, Urbana.

Treaty of Waitangi (Fisheries Claims) Settlement Act, 1992, New Zealand.

VIFTA (1997) *Submission to ban net fishing in Port Phillip Bay and Western Port*. Victorian Fishing Tackle Association, Melbourne.

WBM (1997) *Review of Scalefish Fishing Practices in Victoria's Bays and Inlets*. Fisheries Co-Management Council, Melbourne.

Chapter 3
Economic Evaluation of Recreational Fishery Policies

Murray A. Rudd, Henk Folmer and G. Cornelis van Kooten

Abstract

Recreational fisheries around the world provide humans with important economic benefits because people derive well-being from participating in the act of fishing. Many of these benefits are difficult to value, however, because they are nonmarket in nature and depend on 'free' ecological services. Other sectors of society may also depend on these public goods. It is difficult to exclude people from using public goods and there is, therefore, a tendency for them to be under-produced by the private sector. Thus, there is often a need for government policy intervention to ensure the adequate production of public ecological services and resolve conflicts over their use. Policies that affect recreational fisheries have costs and benefits, both for anglers and people in other sectors of society, that must be accounted for if social well-being is to be maximised. Economics can be used to quantify the costs and benefits of various policy options available to society and make recommendations that improve overall economic efficiency. Overall well-being (welfare) consists of the sum of 'surpluses' accruing to producers and consumers. In this chapter, we outline the principles of economic cost–benefit analysis of market and nonmarket values for recreational fisheries using examples from various jurisdictions. We also consider how economic analysis can be used to account for the transaction costs of fisheries management – costs often borne by society as a whole – for different forms of governance.

Key words: cost–benefit analysis, nonmarket valuation, policy analysis, recreational fisheries, sport fishing.

Introduction

Marine and freshwater ecosystems around the world provide humans with important recreational fisheries that generate revenue and employment, particularly in rural and developing regions. They provide economic benefits to people who derive pleasure from participating in the act of fishing, enjoying the natural environment, viewing marine wildlife, consuming the fish they catch, and/or engaging in social interactions with fishing companions. Economics is the study of how individuals and societies make choices about the use of scarce resources to maximise overall well-being. It plays a key role in efforts to evaluate the merits of policies that directly or indirectly impact recreational fisheries throughout the world. Using economics, the costs and benefits that accrue to recreational fishers and the suppliers that provide recreational fishing equipment and services can be compared under various policy options.

The healthier an ecosystem, the more that ecosystem can provide a 'flow' of fish that yields fishers increased satisfaction and well-being. Ecosystem services (see Costanza *et al.* 1997) typically have public good characteristics that provide benefits to those that use the resource but do not bear the full cost of providing healthy ecosystems. The private sector generally does not produce enough public goods from an economic efficiency or social well-being standpoint, because it is not in any economic agent's best interest to supply such goods. Individuals have an incentive to 'free-ride' by letting others bear the costs of providing the public good. In response to shortfalls in the production of public goods and conflicts between various interest groups, governments can respond with policy initiatives that prescribe or influence private behaviour by setting rules that prohibit, require, or permit specified actions designed to increase the supply of public goods.

Several types of policies have been implemented to improve recreational fisheries. One consists of putting caps on commercial fishing in order to improve the quantity and quality of stocks available for recreational fishing. Another is aimed at improving the ecological conditions for the recreational fish stocks, such as improving water quality. Both types of policies imply benefits and costs for both recreational fishers and others. In the former case, the costs may include income reductions and employment losses in commercial fisheries. In the latter case, the costs relate to such issues as water purification. A prerequisite for implementing a given policy should be that its benefits outweigh its costs. One of the methodologies that has been developed in economics to assess policies is cost–benefit analysis (CBA).

Predicting the impacts of alternative policies on overall well-being requires that policy makers consider the incentives individuals and organisations face, and how policy initiatives affect the costs and benefits associated with alternative behaviours. Economic considerations play an important role in policy design, with CBA routinely used to account for the economic costs and benefits of particular projects or policy options (Johansson 1993, van Kooten and Bulte 2000). Beyond fulfilling an accounting function, economic analysis can play an additional role in helping decision-makers understand how incentives affect different agents and lead to political mobilisation and lobbying.

The general principle of CBA is that a policy is worthwhile and should be pursued if the discounted benefits of a particular policy exceed its discounted costs. Implementing this broad principle is problematic because governments, like individuals, face budget constraints. Priority should be given to projects and programs that maximise social welfare, thus enhancing overall economic efficiency. Further, market prices do not always reflect economic values where natural resources are concerned; prices often underestimate a resource's true value or fail entirely to take the natural amenity into account. This leads to the misallocation of resources. This is the case, for example, with the systematic over-use of a marine ecosystem for commercial fishing in regions with ecosystems that are important for recreational fisheries.

The main objective of this chapter is to outline the basic principles of CBA for applications related to recreational fishery policy analysis, using examples from various jurisdictions. When considering the demand side of the economic analysis, we emphasise the need to consider nonmarket values in CBA. For both producers and consumers, it is important to differentiate between financial and economic analyses: our basic contention is that narrow financial assessments do not give an accurate accounting of the real economic benefits of recreational fishing, nor the true costs of providing the public ecosystem services needed to support recreational fisheries. There is a pressing need for economic concepts to be applied in environmental policy analyses if countries around the world are to reap the rewards of developing or maintaining vibrant recreational fisheries.

Economic policy analysis

Changes in the economic well-being of producers and consumers are the focus of economic analysis. Economic well-being includes financial measures, but is more encompassing. For example, it includes nonmarket benefits related to the camaraderie experienced while participating in recreational fishing, plus benefits from knowing that an ecosystem is protected (existence value). Measuring the nonmarket values of recreational fishing can be difficult, but this does not mean that such benefits should be ignored in policy analyses, as these values can be substantial. In Scandinavia, for instance, Kristrom and Johansson (2001) estimated that the recreational benefits from recreational salmon fishing outweigh the net benefits from commercial salmon fishing at sea. Taking account of all components of economic value can have important policy implications for recreational fisheries and the ecological systems that support them.

Economic efficiency refers to the maximisation of social welfare, which is the sum of the 'surpluses' accruing to producers and consumers. A surplus is simply the difference between total benefits and total costs, appropriately discounted to take into account the fact that surpluses accrue at different time periods. Producer surplus is simply given by the area above the supply curve and below price. Price multiplied by quantity gives total return, while the area under the supply (or marginal cost) curve for that quantity gives total (variable) cost. The difference is the producer surplus or quasi-rent (see van Kooten and Bulte 2000).

Likewise, the area under the demand (or marginal benefit) curve represents the total benefits to consumers of consuming a given quantity, while price multiplied by quantity purchased represents total cost or expenditure. The difference is referred to as consumer surplus. Consumer surplus can readily be calculated for market goods when data are available (i.e. the price of the good, the prices of substitutes, consumer income levels). This is not true, however, for amenity services such as outdoor recreation and ecosystem functions that are nonmarket in nature and not traded in established markets. Consumer surplus for recreational services can be estimated in situations where the recreational activity impacts a market transaction. There are travel cost techniques, for example, that enable the analyst to estimate a demand curve for some recreational activities (e.g. hunting, fishing, camping), and thus consumer surplus [see Duffield *et al.* 2002 for an example]. The hedonic pricing method does the same, but there are difficult statistical problems to overcome (see Freeman 1993).

Willingness to pay (WTP) for increases in the availability of an amenity or willingness to accept (WTA) compensation to forgo an increase (or decrease) in the availability of a natural resource are more appropriate concepts, compared to consumer surplus, when addressing consumer well-being arising from certain nonmarket amenities. The theoretical foundation for these concepts is illustrated with the aid of Fig. 3.1. Suppose an individual has total income M which is allocated to the purchase of market goods and services q_i ($i = 1, \ldots, n$) at prices p_i

($i = 1, \ldots, n$). Amount $q(0)$ of the public good (the environmental amenity) is initially available, so the individual is located on her indifference curve, $V(0)$, at the point where she is consuming M amount of market goods and $q(0)$ of the environmental amenity. Suppose a policy proposes to increase the availability of the environmental amenity from $q(0)$ to $q(1)$ in Fig. 3.1.

There are two candidates for the measure of consumer well-being, depending on the presumed property rights. The compensating surplus (CS) assumes the 'consumer' of the environmental amenity has the right only to $V(0)$ – that is, $q(0)$. An increase in Q to $q(1)$ will enable the consumer to reach a higher level of utility on indifference curve $V(1)$. The CS is the change in income required to restore the original level of utility, $V(0)$, but with $q(1)$ rather than $q(0)$ (see Fig. 3.1). It is the maximum amount the person is WTP for the opportunity to face $q(1)$ rather than $q(0)$.

The equivalent surplus (ES) assumes the consumer has a right to $q(1)$ in Fig. 3.1. Then ES is the minimum income that would have to be given to the consumer for her to forego the increase in the public good – the WTA compensation. In theory, WTA should be approximately equal to WTP, but in practice WTA has been shown to be substantially larger than WTP (Mansfield 1999). Of course, this has implications for CBA – benefits of a policy to enhance environmental values are higher if WTA is used rather than WTP.

Cost–benefit analysis requires that all costs and benefits be accounted for in each period over a relevant time horizon. Future costs and benefits are discounted at rate r to reflect time preferences for income (e.g. \$100 received 5 years in the future is worth less than \$100 received today; the higher the interest rate, r, the less the current worth of \$100 received in the future). The costs and benefits used in CBA are marginal values in the sense that a policy change does not have economy-wide price effects – that the project is small relative to the overall economy. If this is not true, more complicated general equilibrium analysis is re-

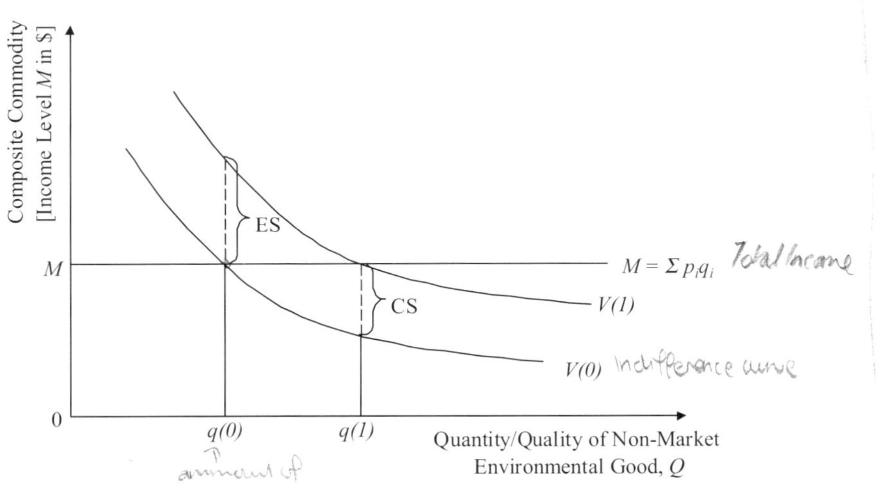

Fig. 3.1 Consumer welfare measures, compensating surplus (CS) and equivalent surplus (ES), for a nonmarket environmental good, Q. Consumption of the nonmarket good is determined by consumer income, M, the prices (p_i) of market commodities q_i, the indifference curves indicating trade-offs for market, $V(0)$, and nonmarket, $V(1)$, consumption, and property rights.

quired. Further, when distortions exist in other markets, adjustments should be made to costs and benefits so that they reflect true economic value (see van Kooten and Bulte 2000 for a further discussion of 'shadow prices'). The streams of costs (C_t) and benefits (B_t), discounted at rate r over T periods, can be summed, respectively, into the present value of costs (PVC) and benefits (PVB):

$$PVC = C_0 + \frac{C_1}{(1+r)} + \frac{C_2}{(1+r)^2} + \ldots + \frac{C_T}{(1+r)^T} = \sum_{t=0}^{T} \frac{C_t}{(1+r)^t} \quad [3.1]$$

$$PVB = B_0 + \frac{B_1}{(1+r)} + \frac{B_2}{(1+r)^2} + \ldots + \frac{B_T}{(1+r)^T} = \sum_{t=0}^{T} \frac{B_t}{(1+r)^t} \quad [3.2]$$

To address government budget constraints (not every project with PVB > PVC can be implemented) and achieve maximum economic efficiency, it is useful to rank projects by their benefit–cost ratios:

$$\frac{B}{C} = \frac{PVB}{PVC}$$

Projects ranked by their benefit–cost ratios are chosen in order until the budget is exhausted. There are a number of philosophical and pragmatic difficulties that arise when using CBA, but these are beyond the scope of this chapter. For general introductions or reviews of key issues, see, for example, Arrow *et al.* (1993), Freeman (1993), Hausman (1993), Hausman and McPherson (1996), Foster (1997), Dixon and Pagiola (1998), Pearce (1998), Weimer and Vining (1998), and van Kooten and Bulte (2000). Suffice to say that there are difficulties in (1) equating personal satisfaction with welfare and income, (2) putting economic values on public goods such as ecosystem services, (3) choosing an appropriate discount rate, and (4) addressing income redistributional effects of policies where there are identifiable losers in the policy implementation process. The last point is problematic because the generic goal in economics is Pareto efficiency – no one person can be made better off without making at least one person worse off. Pragmatically, achieving this is usually impossible so the Kaldor–Hicks compensation principle is generally invoked: if the 'winners' from a policy change gain more than the 'losers' lose, the winners could compensate the losers and still be better off. Society as a whole is left better off but, because compensation rarely occurs, some individuals are left worse off than they were prior to the policy (Hausman and MacPherson 1996).

Economic efficiency is only one criterion by which a policy or project can be judged. Other possible criteria include equity (redistribution of income according to social goals), fiscal equivalence (the beneficiaries of policies are the ones who bear the costs), administrative efficiency and accountability (of elected officials and bureaucrats), conformance to general norms and social values, and the adaptability (to changing social or ecological conditions) of institutions themselves (e.g. Ostrom *et al.* 1993, Weimer and Vining 1998). However, as Pearce (1998) points out, alternative decision rules appear to suffer as many, if not more, shortcomings as those faced by CBA.

Applying economic concepts to recreational fisheries policy analysis

Standard CBA calculates the economic costs and benefits of proposed policy alternatives for producers and consumers. Using recreational fisheries examples, we briefly illustrate the concepts that arise in undertaking research on supply-side producer surplus benefits and demand-side consumer surplus benefits. Recall that the value of a nonmarket amenity to a consumer depends on property rights and on a policy intervention to change the level of the environmental amenity being supplied. In some cases, the cost of a policy intervention can be calculated in a straightforward manner, but in other cases the costs of implementing policies that achieve the same ends can vary greatly, depending on the type of government or market organisation providing the amenity. All three components – consumer benefits, producer benefits and the costs of policy implementation – should be accounted for in economic policy analyses if social well-being is to be maximised.

The economic value of recreational fisheries to producers

Although producer surplus is theoretically easier to measure than consumer surplus, there has been very little economic research on recreational fishery service providers. Almost all work done in this field constitutes financial analyses of overall expenditures on fishing services – calculation of the price multiplied by quantity for services such as charter trips, boat rentals, accommodation, fishing supplies, and so on. Expenditures are not a surplus, and thus are not a benefit measure. Therefore, unless industry costs are quantified for specific recreational fisheries, estimates of producer surplus are not available.

Some research in the USA has gone beyond simple studies of expenditures and has assessed the operating costs of charter operators. An ongoing study of Lake Erie sport charter captains in Ohio (Lichtkoppler and Hushak 2001) examines average operating costs, cash flow and net profitability for walleye (*Stizostedion vitreum*, Percidae) charter fishing. The researchers do not explicitly calculate producer surplus but the information they collect could be used for that purpose. They found that many charter captains, particularly those that conducted less than 41 trips per year, encountered financial difficulties. This implies low producer surplus in the sector because many charter operators are likely just covering variable costs, if at all.

Another effort is currently under way by the US National Marine Fisheries Service and the Pacific States Marine Fisheries Commission (PSMFC) to remedy some of the shortcomings in the analysis of charter boat economics. New research (D. Colpo, PSFMC, personal communication) is being implemented that will survey 500 to 600 active charter boat operators on the US West Coast. Data about randomly selected day trips and annual costs will be collected from the business owners; this information can then be used to develop further insights about the financial impacts of the charter industry and economic estimates of producer surplus based on industry operating costs.

The economic value of recreational fisheries to consumers

Any particular fishing trip can be viewed as a bundle of physical and service attributes (e.g. target species size and abundance, experience of the guide, type of boat used, number of other people on the trip, accommodation, price of trip). Each attribute may have an influence on the value that a person holds for the fishing trip, but in different ways because individuals have different preferences and incomes. Individuals, including those who do not engage directly in recreational fishing activities, may also derive well-being and economic value from policies that directly or indirectly benefit recreational fisheries. For instance, fishers and nonfishers alike might experience an increase in well-being if they know policy is adopted that preserves essential fish habitat needed to protect an endangered species. The main components of economic value that ideally should be considered in CBA include (see Dixon and Pagiola 1998 for an overview):

- Direct extractive use value (e.g. fish taken as food or trophies by recreational fishers)
- Direct nonextractive use value (e.g. nonlethal catch-and-release, wildlife viewing)
- Indirect use value (e.g. the value provided by preserving key predatory fish that maintain overall ecosystem balance)
- Option and quasi-option value (i.e. the value of future direct and indirect use value and information)
- Bequest nonuse value (i.e. the value derived from knowing that future generations will be able to fish even if the person holding this value does not fish themselves); and
- Existence nonuse value (i.e. the value that people derive from knowing fish exist even if they have no plans to fish themselves)

Valuing extractive direct use for recreational fisheries

There is a substantial market demand for fish as food in some recreational fisheries. In 1999, for example, marine recreational fishers in the United States kept almost 135 million fish, about 41% of the 329 million pounds (90 000 tonnes) landed in total (NMFS 2000). For some high-quality species, such as dolphin fish (*Coryphaena hippurus,* Coryphaenidae), as much as 90% of the fish landed (1.85 million pieces) was kept by anglers, and total recreational landings exceeded those from the commercial fleet.

In Florida, anglers on multiday trips with fishing charters where dolphin were popular each paid an average of US$246 per day for the charter and $96 per day for travel and lodging (NMFS 2000). Up to 11% of Florida fishers on multiday trips also take some time off work, which costs them, on average, another $195. The opportunity cost of fishing is calculated as the foregone or shadow wage anglers could have earned if they had stayed at work; in recreational economics, opportunity cost is sometimes charged at half the wage rate in recognition of a less than one-to-one correspondence between recreational and work opportunities (Freeman 1993, pp. 451–52).

Clearly, dolphin fish landed in the sport fishery are not 'cheap' fish. The average price of frozen dolphin fillets imported into the USA during 2000, by contrast, was about $4.35 per kg (or about $2.50 per kg for whole fish, allowing for processing costs and waste). The difference between the sport and commercial fishery illustrates that other nonextractive values strongly influence the value of dolphin to sport fishers, and that a fish landed by a recreational fisher is often 'more valuable' than the same fish caught in the commercial fishery (see also Kristom and Johansson 2001). This example is based on a financial analysis of expenditures, however, and does not give an accurate estimate of economic welfare. Expenditures by anglers for hotels, restaurants, charter operators and so on provide revenue to firms supplying the recreational fishing industry; the actual surplus accruing to these suppliers depends on their cost structures and the opportunity cost of labour.

To expand on the contrast between expenditures and economic value, consider a study of the Costa Rican billfish recreational fishery during the 1993–94 season (Ditton and Grimes 1995). Total direct expenditures by foreign fishers in Costa Rica were calculated at US$3446 per trip (average length of 7 days with 4 days fishing). A total of 15 970 billfish were landed during an estimated 5219 charter trips (approximately 97% of these fish were released live and could be caught again). The financial impact on the national economy was almost $17.8m as a result of the recreational fishery. This value, however, captures neither consumer nor producer welfare. Fishers were willing to pay even higher than market prices for the high-quality fishing experience in Costa Rica: a survey indicated that anglers had a consumer surplus of $1777 per trip, or total consumer surplus of $3.99m. This consumer surplus would need to be added to earnings over and above variable costs (i.e. producer surplus or quasirent) to get total benefits. The sum of consumer and producer surplus (or total welfare) would likely be only slightly above $4.0m because producer surplus is probably small. Much of the $17.7m in revenue for suppliers would likely be used to cover the variable costs of business operations in the competitive tourism and charter sectors in Costa Rica.

Valuing nonextractive direct use for recreational fisheries

How do we put an economic value on the increased well-being that fishers derive from the fishing experience? There are several ways to infer this value: observing changes in the value of property rights; examination of actual behaviour (often travel cost expenditures) when fishing sites vary by some key attribute(s); and conducting surveys of fishers' stated preferences – that is, elicitation of WTP or WTA directly. The valuation of nonuse goods and services (i.e. bequest and existence values) can also be approached using stated preference methods (see Freeman 1993 and Shechter 2000 for general overviews of nonmarket valuation methodologies).

Where there are private property rights to provide recreational services, the value of the property right should reflect the sum of discounted future net economic benefits accruing to the owner of the property right. In Australia, for example, there is a licensing system for tourism operators in the Great Barrier Reef Marine Park (Davis and Gartside 2001). The operators pay a nominal licence fee for a 6-year permit to use the park, but the number of licences issued is limited and the licences are transferable among operators. Thus, the licence has become a valuable property right that is reportedly worth up to A$100 000 in prime tourism areas. Like commercial fishing quota, changes in the anticipated revenue stream or interest rates affect the value of the licence. Likewise, environmental degradation within the park would be expected to affect adversely the trading price for the licence. The magnitude of the change in property right value is an indicator of the economic value of environmental quality.

A more common approach to evaluate market demand for recreational fisheries is based on the travel cost method (TCM). The TCM uses surveys to elicit revealed preferences of fishers and estimate the value of recreational experiences not priced in markets. Essentially, the number of visits to a particular fishing site serves as a proxy for quantity, and costs (travel and opportunity cost of time) as a proxy for price, in the statistical analysis. Economic benefits are then calculated as an area under the appropriate estimated demand function. For example, Layman *et al.* (1996) estimated that consumer surplus for salmon fishing in the Gulkana River, Alaska was between US$15 and $104 per person per day using TCM in combination with other valuation methodologies. While TCM is commonly used to value outdoor recreation, there are difficulties in apportioning costs and benefits for multiple-purpose holiday fishing trips, and in valuing the opportunity cost of fishers' time.

Alternatively, fishers can be queried regarding their 'stated preferences' and asked directly about their willingness to pay for certain attributes; this approach is usually employed to value preservation and other attributes that leave no trace in markets. The contingent valuation method (CVM) is used to elicit WTP or WTA compensation for changes in the availability of an amenity (Hanemann 1984). An alternative is the choice experiment (Adamowicz *et al.* 1998), which seeks to value different attributes of the fishing experience by querying people directly about preferences using surveys. Choice experiments are a variant on conjoint analysis, which was developed for marketing research. Whereas CVM asks respondents whether they are willing to pay a fee to improve environmental quality, conjoint surveys ask respondents to make a choice between, or rate their relative preference for, two different product profiles. Price is one of the attributes included in each profile; this provides the basis by which to assess WTP.

Contingent valuation method has been used extensively to value environmental goods and services, including increases in the populations of fish stocks and environmental quality important for recreational fisheries. For example, CVM has recently been used to evaluate recreational fishers' WTP for five popular fishes, snapper (*Pagrus autratus,* Sparidae), kingfish (*Seriola lalandi,* Carangidae), kahawai (*Arripis trutta,* Arripidae), blue cod (*Parapercis colias,* Odacidae) and rock lobster (*Jasus edwardsii* and *Jasus verreauxi,* Palinuridae). These fish have both consumptive use and nonconsumptive use values, although snapper and blue cod are primarily valued as food fish and kingfish are valued most highly as sport fish. Table 3.1 summarises the catch and value of the five species based on a survey by the South Australian Centre for Economic Studies (Williamson 2000). Marginal WTP for kingfish, kahawai and rock lobster are based on fish caught (fishers indicated that the species were just as valuable for sport purposes as for eating), while figures for snapper and blue cod were based on fish kept (their primary value was for eating – marginal WTP for a fish released to the wild was assumed to be zero).

Table 3.1 Recreational landings and fishers' willingness to pay (WTP in US$) for catching five species of recreation fish in New Zealand (from Williamson 2000). Marginal WTP is the value of an additional animal caught (kahawai, lobster and kingfish) or kept (blue cod and snapper).

Species	Number caught (thousands)	Biomass caught (tonnes)	Total value (US$m)	Average WTP (US$/fish)	Marginal WTP (US$/fish)
Blue cod	1200	1518	13.80	11.50	0.76
Kahawai	1100	729	30.53	27.75	1.62
Snapper	4300	3229	62.35	14.50	2.70
Rock lobster	534	313	12.12	22.70	3.07
Kingfish	74	382	6.30	85.12	9.29

The use of various conjoint methods to value environmental quality is a more recent development in environmental economics and applications to recreational fisheries are limited to date. A study in New England by Roe *et al.* (1996) estimated the compensating surplus for a variety of Atlantic salmon management alternatives. They found that recreational salmon licence holders in Maine were willing to pay between $30 and $178 per day for policy initiatives that improved fishing quality by increasing run size from approximately 3000 fish to either 6000 or 10 000 fish. In the Turks and Caicos Islands, paired-comparison conjoint analysis was used to assess the impact of the abundance of spiny lobster, sea turtles and reef sharks on the willingness of dive tourists to pay for dive charters (Rudd 2001). Market share for dive charters increased significantly in simulations when divers observed more of all three types of macrofauna, demonstrating that divers held nonextractive economic value for viewing marine wildlife.

Farber and Griner (2000) used conjoint analysis in conjunction with a random utility model to estimate welfare for stream quality improvements near Pittsburgh. They found that households were willing to pay between $44 and $122 per annum for a variety of stream improvements that varied according to baseline and post-recovery conditions, and that the multiattribute context permitted the joint valuation of two substitute goods (environmental improvements on two streams).

CVM studies use the stated preferences of survey respondents to estimate measures of consumer surplus. They can be applied creatively and used to value public goods that are not traded in the marketplace. CVM surveys are not without their difficulties however and must be applied using guidelines (Hausman 1993, Arrow *et al.* 1993). While the valuation results are context-specific and there are challenges in applying the values derived in one situation to another (a practice known as 'benefits transfer'), stated preference methodologies offer the only real hope for valuing nonuse values.

Potential impacts of externalities on recreational fisheries

Negative externalities – costs within one sector that are imposed on external unrelated parties – involving recreational fishing species are common and will distort the true value of recreational fishing in society. [In economics, externalities are defined as considerations lying outside the model, unforeseen events and the like – eds.] These can arise when upland activities harm recreational fisheries (e.g. siltation from logging decreases the survival of salmon eggs in Pacific Northwest streams), commercial fishing reduces available stocks of recreational fish, or when market prices are distorted due to subsidies or other government policies (e.g. subsidising agricultural fertilisers results in excessive nutrient runoff and eutrophication of lakes used for recreational fisheries). Consider the case of recreational billfish angling in Costa Rica. Fishers are concerned about the effects of domestic longlining, a commercial fishery with revenues under US$1m, on billfish stock abundance (Ditton and Grimes 1995). Over 80% of visiting anglers surveyed indicated that they would go to other countries if billfish stocks declined due to over-fishing. The loss of surplus in the Costa Rican tourism industry might only partially be compensated for by relatively minor gains in producer surplus in the commercial longlining sector. In economic terms, the commercial longlining fishery may impose external costs on the recreational fishery, making it more expensive to find depleted fish stocks and/or reducing market demand for the Costa Rican fishing experience by making it less attractive for anglers, compared to countries with healthier stocks.

The effect of externality can be illustrated with the aid of Fig. 3.2. Let $S_{com}{}^{1}$ in Fig 3.2(a) represent the private marginal cost (supply) of fishing by the commercial sector if long-run sustainability of the stock is ignored (or discounted at a higher than socially optimal rate). Then $S_{com}{}^{0}$ represents the true costs to society. Assume a constant world price, $p_{com}{}^{*}$, for Costa Rican billfish because the small production has no impact on overall world prices. By externalising the effects of commercial fishing on stock sustainability, producers gain area $[q_{com}{}^{1} - q_{com}{}^{0}] \times p_{com}{}^{*}$ in revenue, but only area *abcd* in producer surplus. Thus, the economic benefit to the commercial longline fishery depends on the shape of the supply curve and will always be less than the increase in revenue so long as the marginal cost of catching more fish is positive.

Because the commercial fishery operates at $S_{com}{}^{1}$ rather than $S_{com}{}^{0}$, there are increased costs in the sport charter industry (e.g. charter boats must spend more time and fuel to land the same amount of fish as stocks decline). The supply curve in the sport-fishing sector shifts inwards from $S_{rec}{}^{0}$ in Fig. 3.2b to $S_{rec}{}^{1}$. The decline in billfish stocks would also cause a drop in demand for Costa Rican recreational fishing, from $D_{rec}{}^{0}$ to $D_{rec}{}^{1}$, as anglers seek out alternative destinations where stocks are more abundant and the fishing experience more rewarding. Expenditures in the sport-fishing sector would fall from $[p_{rec}{}^{0} \times q_{rec}{}^{0}]$ to $[p_{rec}{}^{1} \times q_{rec}{}^{1}]$. Note that the effect of the changes in supply and demand indicate an unambiguous decline in the angler-days of sport fishing, but the effect on the price of a billfish charter trip ($p_{rec}{}^{1}$) is ambiguous, depending on the shape and extent of the shifts in the supply and demand curves. The combined shifts leave area *rsv* as consumer surplus and *rvy* as producer surplus. The net decrease in overall social welfare is given by area $[stwx + yuwz - uvxw] = [stwx + yvxz]$, which is different from the decline in revenue. Loss in social well-being may be greater or less than the decline in revenue, depending upon the nature of supply and demand in the recreational sector.

From an economic perspective, there could be an overall gain in aggregate social welfare if the commercial fishing industry bore the full costs of billfish stock depletion (operated along $S_{com}{}^{0}$ rather than $S_{com}{}^{1}$ in Fig. 3.2a), with any resulting increase in billfish abundance more than compensated for by the increased well-being of recreational fishers and suppliers. If curtailing stock depletion was prohibitively expensive, however, and welfare in the recreational sector was only marginally enhanced by increased abundance of billfish, it could be socially optimal to increase stock depletion. The optimal level of fishing and the allocation between recreational and commercial fishing interests is a question that would require detailed empirical analysis.

Costs, benefits and incentives

It is also useful to examine changes in the well-being of various interest groups to develop an understanding of industry and interest group (actor) incentives, how particular policy proposals differentially affect actors, and how policy implementation outcomes might be affected. In Costa Rica, national longliners would be adversely affected by policy initiatives that increased billfish conservation if they already made substantial profits under the status quo. The threat of losing producer surplus could lead to lobbying if this group could mobilise

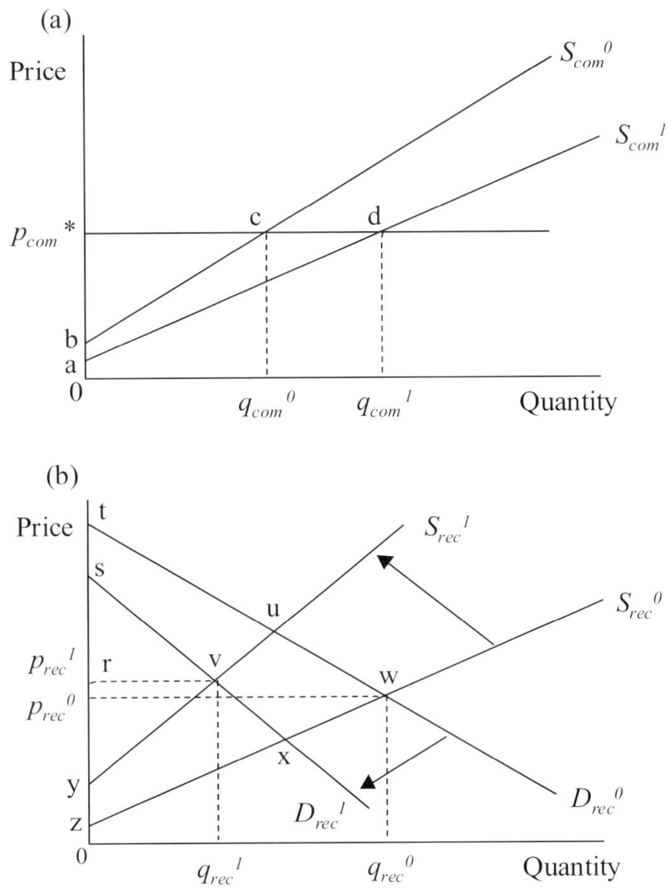

Fig. 3.2 Changes in (a) producer surplus in the commercial billfish longline fishery and (b) producer and consumer surplus in the billfish sport fishery in the face of overfishing in the commercial sector. Commercial fishing expands from q_{com}^{0}, to q_{com}^{1} when supply expands to S_{com}^{1} due to an externality. Sport fishery supply (S_{rec}^{1}) and demand (D_{rec}^{1}) contracts to q_{rec}^{1}. The net change in social welfare is given by abcd (the gain to commercial sector) less ($stwx + yvxz$) (the loss to the sport fishery sector).

politically, and there could be adverse implications for local politicians as a result. However, if there was no surplus in the industry now (boats just covered variable costs) and an increase in billfish conservation would only result in greater capitalisation and a race to capture the extra fish, there may be few incentives for fishers to organise and oppose conservation policies. Understanding economic well-being can shed light on political activities and the chances for successful policy implementation. Likewise, observing the political influence of different actors can provide strong hints at the magnitude of the surpluses those actors might enjoy (e.g. consider the political power and organisation of farmers in countries that have marketing boards or agricultural subsidisation).

Questions arise about the distributions of expenditures and economic benefits when dealing with recreational fisheries. While the 'winners' from conservation (the sport sector) in the hypothetical Costa Rican example may gain more than the 'losers' (the longliners) lose, there is no assurance that compensation will be made or that any benefits from the sport-fishing sector will flow to commercial fishers. The amount of surplus that is actually captured by Costa Rican interests in the government and the tourist industry may not be high, especially if foreigners own many of the charter boats. Politically, it may be hard to justify domestic policies where the beneficiaries of increases in overall welfare are nonresidents and the costs are borne by local citizens. Ditton and Grimes (1995) estimated a local multiplier effect of 2.0 for Costa Rica (i.e. every dollar spent by sport fishers generated two dollars of local economic activity), but Ditton and Clark (1994) estimated the billfish recreational fishery multiplier at only 1.07 in Puerto Rico. This suggests that almost all revenue that came into Puerto Rico as a result of billfish sport fishing expenditures leaked out of the economy due to Puerto Rican dependence on foreign imports [see also Kearney 2002 in relation to Australia]. In general, only a small portion of tourism revenues remains in developing countries (Gössling 1999) and, as a result, policy analysts must pay careful attention to the regional welfare effects of policy alternatives on national actors, especially in developing countries and rural coastal areas. This is an area of research where estimates derived by economic analysis may be fruitfully employed in input–output analyses of regional economic impacts (see Hamel *et al.* 2000 for an example of work proceeding in this direction). Again it is important to note that economic activity is not the same as economic benefits, or surpluses. Only if the shadow price of labour and capital are zero – there is significant unemployment, a common situation in rural areas in many developing countries – can input–output analysis be used to address benefits (van Kooten 1993).

Transaction costs and the provision of public goods

In the past, government intervention was advocated in response to market failures arising from the over-exploitation of public goods, as was the case in the fishery sector. As it became evident that the 'command and control' approach had its limits, economists advocated a move to privatisation of some public goods in the face of 'government failure' and an increased reliance on economic incentives to alter human behaviour. Now there is evidence that, when dealing with the physical and cultural reality of many renewable resource systems, neither alternative – free markets or government regulation – is robust enough to alone ensure successful management of complex economic and ecological systems (Ostrom 1990), and that local resource users need to play an important role in the management of public goods. Ostrom's empirical study of robust and durable renewable resource institutions led to the identification of eight 'design principles' that explained the success of institutions in sustaining a resource over time:

- There were clearly defined boundaries for the resource and withdrawal rights
- There was congruence between appropriation/provision rules and local conditions
- Collective choice arrangements existed, and in which local resource users participated
- There were effective monitors of resource use who were accountable to resource users
- There was a system of graduated sanctions for cheaters
- There were low-cost, local conflict resolution mechanisms
- There was recognition by government of rights of resource users to organise
- Resource management activities that require regional co-ordination were nested in a hierarchical fashion

The transaction costs of producing public goods – protecting ecosystem services necessary for sustaining recreational fisheries – are often ignored in CBA even though the role of fisheries governance is to shape behaviour so that management is conducted at minimum cost (Hanna 1999). Transaction costs need to be considered if overall social welfare is to be maximised. These include the costs of developing initial contracts between parties responsible for the production of the ecosystem services necessary for recreational fisheries (e.g. agreements to zone areas that protect essential fish habitat), the costs of information needed to manage the resource, the costs of monitoring and surveillance of 'cheaters', and the costs of imposing penalties or sanctions. These costs are usually not explicitly taken into consideration as part of a supplier's cost of production and are effectively externalised, with the costs most often borne by society as a whole via costly government management activities. Each management approach – the 'state', 'market' and 'community' – has transaction costs that will vary and depend on culture and social norms, property rights and institutional infrastructure, and the ecological nature of the fishery. The transaction costs of fisheries management can be prohibitive if done poorly and negate any surplus captured by producers and consumers. For example, Schwindt *et al.* (2000) showed that the transaction costs of management exceed the total economic benefits of the fishery for the British Columbia commercial salmon fishery. It is now widely accepted that a top-down regulatory approach will be inefficient if the policies of the government do not conform to some degree to the norms of the local resource users (Costanza *et al.* 1998). Without reasonable congruence between rules and norms, there will be widespread abuse by resource users, and government monitoring and enforcement costs will escalate (Ostrom *et al.* 1993).

Economic analysis can be used to quantify the transaction costs for different types of resource management policies and regimes. While most data available are for government-managed commercial and recreational fisheries, there are modern examples of community-based and privately managed recreational fisheries. Innovations in property rights for access to recreational fish and fishing grounds are currently being considered and it is likely there will be new forms of market-based management for recreational fisheries in the near future. Individual fishing quota for sport charter vessel operators in Alaska has, for instance, been proposed for the recreational halibut fishery. Under this plan, transferable quota could be exchanged between the commercial and recreational fisheries, allowing increased economic efficiency for the combined fishing sectors (Hamel *et al.* 2000). The North Pacific Fishery Management Council took a step in this direction in April 2001, voting to award individual catch shares to roughly 1100 halibut charter boats that operate each summer in South-east and South-central Alaska (Loy 2001). If approved, the individual quota plan could take effect in two or three years.

Government management of recreational fisheries has encountered varying levels of success around the world. In Florida, for example, licensing of local and visiting anglers has generated substantial government revenue that has been reallocated specifically towards education, research and enforcement activities in the recreational fishery. Thus, there have been generally high levels of compliance with fishing regulations and a high level of public investment in scientific research. A suite of regulatory and management tools is used in Florida, including seasonal and gear restrictions, bag and size limits, seasonal closures, and marine reserves. Government-managed recreational fisheries should be expected to be most successful in minimising the transaction costs of management when target species are widely dispersed, requiring regional management (i.e. there is congruence between institutional and ecological scale), and where citizens have input into the policy planning and implementation process.

Privately operated recreational fisheries, by contrast, are most likely to minimise transaction costs when the fishery is geographically concentrated and when private property rights can be effectively enforced using existing legal institutions. That is, unauthorised users can effectively be excluded from gaining benefits arising from the use of the recreational property. In this setting, long-term stewardship of the resource is economically sound because there are direct links between private investments in ecosystem quality and the benefits derived from selling access to the private fishing property. Leal (1996) outlines the Scottish example of salmon stream management, where privately held and transferable salmon fishing rights have existed since feudal times and are currently held in watersheds by individuals, fishing clubs, companies and nongovernmental organisations. District management boards, created by the property owners within a watershed, are responsible for monitoring, habitat improvement, restocking, and management of salmon stocks within the district. They can charge anglers for the right to fish on inland rivers and gain revenue from net fisheries at the river mouth. To accommodate growth in sport fishing demand, the Atlantic Salmon Conservation Trust (Scotland) Ltd purchased commercial fishing rights and let them go unused in order to increase the returns to anglers in the valuable inland salmon fishing streams.

The transaction costs for community-based recreational fisheries management are likely lower when relatively local in scale (Ostrom's second principle), but not easily privatised due to difficulties in excluding unauthorised resource users. Community management will result in lower monitoring costs when it is in the economic interests of community members, who are regularly on the water, to engage in monitoring and self-policing (Ostrom 1990). This tends to occur in situations when community members have a direct economic stake in the fish guiding industry and will be adversely affected by deterioration in stock abundance or habitat quality.

Consider the case of the emerging bonefish (*Albula vulpes,* Albulidae) sport fishing industry in the Turks and Caicos Islands (TCI). The fish migrate around the Caicos Bank, a shallow area of about 6000 km^2 that is also home to productive commercial queen conch and lobster fisheries. Privatisation of the fishery or fishing grounds is not possible because commercial fishers cannot be excluded from their traditional fishing grounds. Government managers have very limited information about bonefish stock size or population dynamics, and limited resources with which they could implement an effective bonefish management plan. Community management would thus seem to be a potentially viable option for the bonefish resource.

While bonefish are only worth about US$2.20 per kilogram as food in the local market (M. Rudd, personal observation), there are a number of guiding operations catering to foreign fishers who place a much higher value on the fish, which are prized for their fighting ability. Visiting fly fishers typically pay between $350 and $500 per person per day for local guiding services, all of which operate under self-imposed catch-and-release norms. There are poachers on the Caicos Bank who place gillnets at the mouths of mangrove creeks to snare bonefish. The value of the bonefish resource to the sport industry has created economic incentives for resource protection – some fishing guides regularly cut away any illegal nets that they now find (A. Danylchuk, Center for Marine Resource Studies, personal communication). The bonefish guides are actively participating in tag-and-release studies designed to improve scientific understanding of stock dynamics and the potential effects of marine protected areas on bonefish conservation. The charter companies, all of which are controlled by TCI 'belongers', also offer alternative employment opportunities for former commercial fishers in outlying rural islands. While community-based management has not been officially implemented for the TCI bonefish sport fishery, conditions are favourable for bottom-up participation in the management of the local recreational fishery. A challenge for a community-based TCI sport fishery management agency would be to limit the size of the industry, as congestion could rapidly lead to a degradation of the fishing experience (and angler WTP) even if bonefish stocks remain healthy.

Summing up the economic value of recreational fisheries for society

The overall changes in societal well-being from any proposed policy that affects recreational fisheries can, in theory, be derived by summing the appropriately discounted costs and benefits. Many narrow CBA studies in the past have only considered extractive direct use values on the demand side and narrowly defined project costs on the supply side. While this may be appropriate in some sectors, recreational fisheries depend upon ecological services that are public goods. As such, they tend to be under-produced and require either government policy intervention or collective action at the local level to increase production to socially optimal levels. This causes three main complications that must be addressed in CBA: nonextractive and nonuse values should be accounted for in order to derive all consumer welfare benefits; market distortions that indirectly affect estimates of both producer and consumer surplus in recreational fisheries should be accounted for; and the transaction costs of alternative forms of supplying public goods should be considered.

Recall from Fig. 3.1 that people are willing to pay some amount, CS, for an increase in a public good. This surplus needs to be included in CBA calculations. The costs of policy implementation need to be considered as well. Many economists consider this to be a straightforward calculation, but this may not be the case for public goods and the analyst must consider the costs for alternative forms of provision. Many policies may achieve a desired result (i.e. there can be many sufficient conditions for the policy outcome), but the costs could vary greatly depending on the social, ecological and institutional context. Hybrid forms of governance ('co-management') are possible and likely desirable for many recreational fisheries. A particular challenge in recreational fishery policy analysis will be to evaluate the efficient 'scope of government' (e.g. Hart *et al.* 1997). That is, what combination of state, market and community organisations can minimise the overall costs of producing and allocating quasi-public ecosystem services necessary for the health of recreational fisheries?

Conclusions

Economic analysis can be a useful tool for evaluating recreational fisheries policies. It can be used to assess the changes in social welfare resulting from changes in government policy (or from changes in exogenous factors such as market prices or environmental change). In addition, economic analysis can be used effectively to identify key policy actors and provide valuable insights on incentives and the patterns of behaviour that are likely to result from specific policy changes. When properly conducted – accounting for market and nonmarket values of consumers, the costs of production of firms, and the transaction costs of management – economic analysis can provide important information that can help policy makers improve economic efficiency and the effective allocation of resources to investments that benefit society as a whole.

When conducting economic analyses of recreational fisheries, it is important to consider the many nonmarket values that arise from nonconsumptive use of the fish resources and the ecosystems that sustain them. While nonconsumptive use value, indirect use value and nonuse values have been considered extensively in other fields of environmental valuation, recreational fisheries seem to be under-studied by comparison. On the supply side, there is a dire shortage of research on the cost structure of suppliers. As a result of these shortcomings, the true economic benefits to society of recreational fisheries around the world have not been adequately assessed. Given the prominence of recreational fishing as a form of leisure in many parts of the world, there is need for more detailed economic analyses of economically and socially important recreational fisheries. Finally, it is important to consider transaction costs in the economic analysis, an item often neglected in cost–benefit analysis. The costs of maintaining and managing ecosystems needed for recreational fisheries varies greatly over situations that differ in terms of ecology, social context and institutions. Only when all factors are considered will the effects of alternative policies on incentives, efficiency and social welfare be adequately understood. Without this understanding, societies will be hampered in efforts to develop effective policies and governance systems that ensure ecological and social sustainability of recreational fisheries.

References

Adamowicz, W., Boxall, P., William, M. & Louviere, J. (1998) Stated preference approaches to measuring passive use values: choice experiments versus contingent valuation. *American Journal of Agricultural Economics*, **80**, 64–75.

Arrow, K., Solow, R., Portney, P.R., Leamer, E.E., Radner, R. & Schuman, H. (1993) Advance notice of proposed rulemaking, extension of comment period and release of contingent valuation methodology report. *Federal Register*, **58**, 4601–4614.

Costanza, R., d'Arge, R., de Groot, R. *et al.* (1997) The value of the world's ecosystem services and natural capital. *Nature*, **387**, 253–260.

Costanza, R., Andrade, F., Antunes, P. *et al.* (1998) Principles of sustainable governance of the oceans. *Science,* **281**, 198–199.

Davis, D. & Gartside, D.F. (2001) Challenges for economic policy in sustainable management of marine natural resources. *Ecological Economics,* **36**, 223–236.

Ditton, R.B. & Clark, D.J. (1994) *Characteristics, attitudes, catch-and-release behaviour, and expenditures, of billfish tournament anglers in Puerto Rico.* Report prepared for The Billfish Foundation, Ft. Lauderdale, Florida. College Station, Texas: Department of Wildlife and Fisheries Sciences, Texas A&M University.

Ditton, R.B. & Grimes, S.R. (1995) *A social and economic study of the Costa Rica recreational billfish fishery*. Report prepared for The Billfish Foundation, Ft. Lauderdale, Florida. College Station, Texas: Department of Wildlife and Fisheries Sciences, Texas A&M University.

Dixon, J.A. & Pagiola, S. (1998) Economic analysis and environmental assessment. *Environmental Assessment Sourcebook Update* Number 23. World Bank, Washington, DC.

Duffield, J. W., Merritt, M.F. & Neher, C. J. (2002) Valuation and policy in Alaskan sport fisheries. In: *Recreational Fisheries: Ecological, Economic and Social Evaluation* (eds Pitcher, T.J. & Hollingworth, C.E.), Chapter 13, this volume. Blackwell Science, Oxford, UK.

Farber, S. & Griner, B. (2000) Valuing watershed quality improvements using conjoint analysis. *Ecological Economics*, **34**, 63–76.

Foster, J., (ed) (1997) *Valuing Nature: Economics, Ethics and Environment*. Routledge, London.

Freeman, A.M. (1993) *The Measurement of Environmental and Resource Values: Theory and Methods*. Resources for the Future, Washington DC.

Gössling, S. (1999) Ecotourism: a means to safeguard biodiversity and ecosystem functions. *Ecological Economics*, **29**, 303–320.

Hamel, C., Herrmann, M., Lee, S.T. & Criddle, K.R. (2000) An economic discussion of the marine sport fisheries in Central and Lower Cook Inlet. Paper presented at *IIFET 2000 Conference*, Corvallis, Oregon, July 10–15, 2000.

Hanemann, W.M. (1984) Welfare evaluations in contingent valuation experiments with discrete responses. *American Journal of Agricultural Economics*, **66**, 332–341.

Hanna, S. (1999) Strengthening governance of ocean fishery resources. *Ecological Economics*, **31**, 275–286.

Hart, O., Scleifer, A. & Vishny, R.W. (1997) The proper scope of government: theory and application to prisons. *Quarterly Journal of Economics*, **112**, 1127–1161.

Hausman, D.M. & MacPherson, M.S. (1996) *Economic Analysis and Moral Philosophy*. Cambridge University Press.

Hausman, J.A. (ed) (1993) *Contingent Valuation: A Critical Assessment*. North-Holland, Amsterdam.

Johansson, P.O. (1993) *Cost–Benefit Analysis of Environmental Change*. Cambridge University Press.

Kearney, R. E. (2002) Recreational fishing: value is in the eye of the beholder. In: *Recreational Fisheries: Ecological, Economic and Social Evaluation* (eds Pitcher, T.J. & Hollingworth, C.E.), Chapter 2, this volume. Blackwell Science, Oxford, UK.

van Kooten, G.C. (1993) *Land Resource Economics and Sustainable Development*. UBC Press, Vancouver, British Columbia.

van Kooten, G.C. & Bulte, E.H. (2000) *The Economics of Nature*. Blackwell Scientific, Oxford.

Kristrom, B. & Johansson, P. (2001) Restriktioner paa havsyrkesfiske efter lax: samhaellsekonomiska aspekter [Restrictions on commercial fishing of salmon at sea: welfare economic aspects]. Working Paper. Umeaa, Sweden: Swedish Agricultural University.

Layman, R.C., Boyce, J.R. & Criddle, K.R. (1996) Economic valuation of the Chinook salmon sport fishery of the Gulkana River, Alaska, under current and alternative management plans. *Land Economics*, **72**, 113–128.

Leal, D.R. (1996) Community-run fisheries: avoiding the 'Tragedy of the Commons'. PERC Policy Series PS-7. Bozeman, Montana: Political Economy Research Center.

Lichtkoppler, F.R. & Hushak, L. (2001) Ohio's 1998 Lake Erie Charter Fishing Industry. *Fisheries*, **26**(1), 15–23.

Loy, W. (2001) Council votes catch shares for charters: fishery proponents say halibut anglers won't see much difference. *Anchorage Daily News*, 15 April 2001.

Mansfield, C. (1999) Despairing over disparities: explaining the difference between willingness to pay and willingness to accept. *Environmental and Resource Economics*, **13**, 219–234.

NMFS (National Marine Fisheries Service) (2000) *Fisheries of the United States, 1999*. US Department of Commerce, Silver Spring, MD.

Ostrom, E. (1990) *Governing the Commons: The Evolution of Collective Action*. Cambridge University Press.

Ostrom, E., Schroeder, L. & Wynne, S. (1993) *Institutional Incentives and Sustainable Development: Infrastructure Policies in Perspective*. Westview Press, Boulder, CO.

Pearce, D. (1998) Cost–benefit analysis and environmental policy. *Oxford Review of Economic Policy*, **14**, 84–100.

Roe, B., Boyle, K.J. & Teisl, M.F. (1996) Using conjoint analysis to derive estimates of compensating variation. *Journal of Environmental Economics and Management*, **31**, 145–159.

Rudd, M.A. (2001) The nonconsumptive economic value of spiny lobster, *Panulirus argus*, in the Turks and Caicos Islands. *Environmental Conservation*, **38** (3), 226–34.

Schwindt, R., Vining, A. & Globerman, S. (2000) Net loss: a cost–benefit analysis of the Canadian Pacific salmon fishery. *Journal of Policy Analysis and Management*, **19**, 23–45.

Shechter, M. (2000) Valuing the environment. In*: Principles of Environmental and Resource Economics* (eds Folmer, H. & Landis Gabel, H.), pp. 72–103. Edward Elgar, Cheltenham, UK.

Weimer, D.L. & Vining, A.R. (1998) *Policy Analysis: Concepts and Practices*. Prentice-Hall, Upper Saddle River, NJ.

Williamson, S. (2000) The economic value of New Zealand marine recreational fishing and its use as a policy tool. Paper presented at *IIFET 2000 Conference*, Corvallis, Oregon, July 10–15, 2000.

Chapter 4
Recreational and Commercial Fishers in the Namibian Silver Kob Fishery

Ussif Rashid Sumaila

Abstract

The Namibian silver kob fishery resource is shared by recreational and commercial fishers. Recent data show that the two sectors generate N$64.23 and N$24.23 per fish (US$1.00 = N$7.96) removed from the habitat, respectively. All things being equal, one might then expect the recreational fishers to take most, if not all the fish. However, it is also reported that the two users in the fishery take about the same number of silver kob (*Argyrosomus inodorus*) per year. One of the two major goals of this chapter is to explore the reasons for these apparently conflicting data. The Ministry of Fisheries and Marine Resources of Namibia (MFMR) has recently intensified efforts to devise a management plan for this fishery. One of the crucial issues such a management plan will have to address is how best to share and allocate the harvestable part of the stock of silver kob to the two sectors of the fishery. The second goal of this chapter is to develop a model that takes into account the ecology of silver kob; the economic and social aspects of the fishery; the fishing technology employed by the two sectors; and their targeting strategies upon the silver kob stock, to study the optimal sharing of the resource.

Key words: economic, regulator's and noncooperative models, social and environmental objectives.

Introduction

The appropriate allocation of a given fishery resource between recreational and commercial users has been receiving attention in the United States for some time now (see, for example, Sutinen 1993). Indeed, since Argue *et al.* 1983, this issue is increasingly an interesting area of study in other parts of the world too (see Pitcher 1999, Kirkegaard and Gartside 1998, Hickey and Tomskin 1996). Allocation between the two sectors is likely to be more of an issue in the developing world where the recreational users are mostly from the affluent parts of the society, while the commercial users usually come from the less affluent part of the population, thereby giving such analysis interesting equity and distributional angles. The question of how best to share the resource between the two user groups is of interest because the way the resource is shared will have significant consequences on (i) the long-term sustainability of the stock, (ii) the amount of market values that can be derived from the resource, and (iii) the social and environmental objectives of the fishery.

There are three key elements of the recreational and commercial users that are important to the current analysis. First, the two sectors of the fishery appear to target different age groups of silver kob. This conclusion is derived from the silver kob catch, which shows that while both sectors remove about the same number of kob from the habitat in a year, the commercial sector's catch weighs about 50% more (see Kirchner 1998), a clear indication that this sector targets more mature kob than the recreational sector. Second, the recreational sector generates much more revenue per unit of kob removed. This second point poses an interesting research question: If the revenues from the two sectors are so much to the advantage of the recreational sector, how come the two sectors remove about the same number of kob per year? Third, the two sectors have different removal efficiencies and costs.

Putting all the above different characteristics of the sectors together in a model, we carry out an analysis to help us address the following specific questions in tune with Sumaila (1999):

- What benefits can each participant and Namibia expect under different assumptions and management scenarios?
- What are the potential effects of exploitation by commercial and recreational users on the sustainability of the fishery resource?

The next section briefly describes the Namibian silver kob fishery. Later sections present the model applied in the study, discuss data needs and calibration, and present the results and conclusions.

The silver kob fishery

We present only features of this fishery that are of direct relevance to the current paper; interested readers can refer to Kirchner (1998) and the references therein for more details. Two species of kob are found in Namibian waters, silver kob (*Argyrosomus inodorus*) and westcoast dusty kob (*Argyrosomus coronus*). The silver kob fishery consists of a recreational fishery (rock-and-surf angling and ski boats) [in South Africa 'ski boats' are 4–8 m vessels with semidisplacement hulls and outboard engines, launched off slipways and beaches, as used originally for water skiing – eds], and commercial (ski boat and line fish boat) fishery. The line fishery [rod-line-and-baited-hook – eds] in Namibia is multispecies, multisector. Silver kob is exploited and sold locally as the 'catch of the day' by commercial fishers.

The Namibian coast is well known for its rock and surfing angling opportunities. Each year thousands of anglers visit the coast in hope of good line fish catches. There is usually a huge influx of tourists to the Namibian coast during Christmas and Easter holidays by both Namibian and South African anglers. The latter usually stay in rented accommodation and generate income for local businesses as well as foreign currency for Namibia.

In 1998, there were 14 registered Namibian line fish boats. A line boat is between 15 and 21 m long and carries between 9 and 15 fishers. The fishing gear is a line with a hook and each fisher usually operates two lines. The duration of a fishing trip is between 5 and 8 days, and on average one or two trips per month is conducted by each boat. Between 2 and 12 tonnes of beheaded and gutted silver kob, kept on ice, are usually landed with each trip. Currently there is no quota on catches of line fish boats. In other words, the fishery is unmanaged, a point we make use of when we model the fishery.

As at 1998, nine ski boats are licensed and have the right of commercial exploitation. In addition, approximately 25 ski boats operate infrequently as recreational fisheries boats throughout the year and reach a peak activity during December and January every year. A ski boat is approximately 5–6 m long and can carry between four and six fishers. Fishing is carried out with boat, rods and reels and normally only one rod per fisher is used. Historically, neither commercial nor recreational ski boat catches were recorded. However, in 1994, the MFMR introduced a daily catch log system for commercial ski boats. A number of regulations have been imposed on the recreational fishers by the MFMR. For instance, anglers are limited to 30 fish per day; 60 per vehicle if two or more anglers are in the vehicle, and no trading of fish. [Descriptions of another African line fishery are provided by Griffiths & Lamberth 2002 – eds.]

The model

As in Sumaila (1999), we develop a noncooperative and regulator's model for the fishery. The two players in the model are defined as the recreational and commercial fishers. To be able to do this, we assumed that both the commercial and recreational fishers are organised into two distinct groups, each acting as a unit for management purposes. Even though this is not the case on the ground at the moment, this arrangement is a possibility in the near future because members of each group have much in common. Hence, it will not be too difficult to align their interests, and work together as a group. What is more, a paper like this one could serve as a catalyst for such developments.

We apply Nash's noncooperative equilibrium concept to determine likely solutions in a nonmanaged silver kob fishery. The idea of a Nash equilibrium can easily be expressed in terms of benefit functions (Binmore 1982). A pair of strategies (H^*_c, H^*_r) is a Nash noncooperative equilibrium in a two-player game if H^*_c is a best reply to H^*_r and H^*_r is simultaneously a best reply to H^*_c. This is the same as requiring that the inequalities:

$$Bc\ (H^*c,\ H^*r) \geq Bc\ (Hc,\ H^*r) \qquad [4.1]$$
$$Br\ (H^*c,\ H^*r) \geq Br\ (H^*c,\ Hr)$$

hold for all strategies H_c and H_r available to the two players, respectively. In the context of the present chapter, these denote harvest (removal) rates of the commercial and recreational users, respectively. B_c and B_r denote the net private benefits from the use of silver kob by the commercial and recreational users, respectively.

Given the above, the noncooperative management problem facing commercial users can be stated as (see Flåm 1993 and Sumaila 1997):

$$\max_{Hc} \sum_{t=1}^{T} \left[\rho_{c,t-1} B_{c,t} \right] \qquad [4.2]$$

subject to the relevant constraints, in particular, those related to the natural stock dynamics of silver kob.

Similarly, the noncooperative management problem facing recreational users can be stated as:

$$\max_{H_r} \sum_{t=1}^{t} \left[\rho_{r,t-1} B_{r,t} \right] \qquad [4.3]$$

subject to the relevant constraints. The two user groups are assumed to pursue their stated objectives unilaterally under the assumption that their rival will do the same.

Under a managed fishery regime or the regulator's model, it is assumed that a body, which may be a government authority, a community-based management body or sole owner of the resource, regulates the fishery. This body or regulator is assumed to be concerned with maximising total benefits from the use of the resource. Elements of the overall benefit include private economic benefits, social benefits, e.g. the need to preserve regional settlement, and environmental benefits not traded in the market, e.g. benefits derived from keeping the diversity of the marine habitat. The regulator faces two broad challenges. First, the optimal harvest for the two groups of users in each period must be determined to ensure that the stated objectives are met. Second, the regulator has to put in place a management plan to implement the desired harvests. This model is a framework that can be used to provide quantitative results to help with the first task.

In addition to B_c and B_r, let us introduce two additional benefit functions denoted by B_e and B_s to capture environmental and society-wide benefits, respectively. The private benefits are assumed to depend positively on own harvest (H_c and H_r for the commercial and recreational users, respectively) and negatively on the other user's harvest of silver kob. Social benefits may depend in some way on the harvest taken by one of the groups, say commercial users, for social, cultural or other reasons. Environmental benefits, on the other hand, depend on the amount of standing stock of fish, $(N - H_c - H_r)$, where N is the total biomass at a given time. Alternatively, environmental benefits can be modelled to depend on the total harvest taken, that is, $(H_c + H_r)$. Formally, we have,

$$B_c(H_c); B_r(H_r); B_e(N - H_c - H_r) \text{ and } B_s(\theta_{s,c} H_c + \theta_{s,r} H_r) \qquad [4.4]$$

where $\theta_{s,r}$ and $\theta_{s,c}$ are the social parameters attached to the harvest of the recreational and commercial sectors, respectively, and

$$\begin{aligned} &\partial B_c / \partial H_c > 0; \partial B_c / \partial H_r < 0; \\ &\partial B_r / \partial H_r > 0; \partial B_r / \partial H_c < 0 \\ &\partial B_e / \partial (N - H_c - H_r) > 0 \text{ or } \partial B_e / \partial (H_c + H_r) < 0 \end{aligned} \qquad [4.5]$$

The regulator's problem is to maximise total net benefits B_t through the choice of $H_{c,t}$ and $H_{r,t}$ for $t = 1..T$, where T is the last (terminal) period. That is:

$$\max_{Hr,Hc} \sum_{t=1}^{T} [B_t] \rho_{t-1} \qquad [4.6]$$

subject to the relevant constraints.

In the above equation,

$$B_t = B_{c,t} + B_{r,t} + B_{e,t} + B_{s,t} \tag{4.7}$$

$$\rho_{t-1} = \frac{1}{(1+r)^{t-1}},\ \rho_0 = 1,\ t = 1,\ldots,T. \tag{4.8}$$

The parameter ρ is the discount factor, and r is the discount rate; N_t is the standing biomass in period t.

Since an important goal of this paper is to provide quantitative results for the silver kob fishery, we proceed in the next section to define specific functional forms for the equations of the model for this fishery.

Specific functional forms for the silver kob fishery

The fishery operates under the natural stock constraint of silver kob. Let this stock constraint be described by the following Beverton and Holt stock recruitment function:

$$\begin{aligned} &n_{0,t} \le f(B_{t-1}), \\ &n_{a,t} + h_{a,t} \le \xi_{a-1} n_{a-1,t-1}, \qquad \text{for} \quad 0 < a < A \\ &n_{A,t} + h_{A,t} \le \xi_A n_{A,t-1} + \xi_{A-1} n_{A-1,t-1}, \end{aligned} \tag{4.9}$$

where $n_{a,t}$ is the biomass of kob in numbers;

$$f(B_{t-1}) = \frac{\alpha B_{t-1}}{1+\gamma B_{t-1}} \tag{4.10}$$

is the Beverton–Holt recruitment function (in this model, recruitment refers to the number of age zero fish that enter the habitat in each fishing period);

$$B_{t-1} = \sum_a p_a w_{sa} n_{a,t-1} \tag{4.11}$$

represents the postcatch biomass in numbers; p_a is the proportion of mature fish of age a; w_{sa} is the weight at spawning of fish of age a; α and γ are constant parameters $f'(0)$ is the number of recruits per unit weight of biomass "at zero" or the population level); S_a is the natural survival rate of fish of age a; and $h_{a,t}$ denotes the combined harvest of fish of age a, in fishing season t by all agents.

Let's define the private benefit functions mentioned above by the following specific functional forms:

$$B_c(H_c) := p_c H_c(N, e_c) - C_c(e_c) = p_c q_c N e_c - k_c e_c^{0.01} \tag{4.12}$$

$$B_r(H_r) := p_r H_r(N, e_r) - C_r(e_r) = p_r q_r N e_r - k_r e_r^{0.01} \tag{4.13}$$

where q is the catchability coefficient, p is price in the case of commercial fishers or value in the case of recreational fishers, k is unit cost of effort, and e is the effort level. The exponent 0.01 is used to ensure convergence in the model by making the objective function concave (see Flåm 1993 and Sumaila 1997). Equations (4.12) and (4.13) stipulate that the net benefits to the two users are given by the difference between total revenue and total cost.

Model data

Except where it is stated otherwise, the data reported in this section are either taken from Kirchner (1998) or calculated from information taken from there. The biological parameters α and γ are given values of 1.5 and 1, respectively, to produce a normalised recruitment of 1, which represents about half a million of recruits. The normalised initial number of the different age groups of silver kob in the model were assumed, for age groups 0,1,...,10 to be [1, 0.79, 0.62, 0.49, 0.39, 0.31, 0.24, 0.19, 0.15, 0.12, 0.09]. This gives a total initial biomass of about 14 000 tonnes, which is well within the range of estimated biomass of silver kob in Namibian waters. Applying data reported in Kirchner (1998), we determine the proportion of mature silver kob at ages 1,..., 10 to be [0.25, 0.75, 1, 1, 1, 1, 1, 1, 1, 1, 1]. Natural survival rate of silver cod is taken to be 0.79. The age dependent weight of silver kob, W_{age}, is calculated from the equation (Kirchner 1998):

$$W_{age} = 11.38 * (1 - e^{-0.15*age})^{2.53} \qquad [4.14]$$

The environmental parameter, $\beta > 0$, if the social planner cares about the environment. The social parameters, $\theta_{s,c}$ and $\theta_{s,r}$ also take values greater than or equal to zero. Setting $\theta_r = 0$ and $\theta_c > 0$ implies that benefits to the commercial group are valued more than those to the recreational group by the social planner (Lopez *et al.* 1994). For runs of the regulator's model in this paper, β is set to 50, $\theta_{s,c} = 5$ and $\theta_{s,r}$ is equal to 0. These values give reasonable quantitative outputs from the model. It should be noted that sensitivity analysis shows that variations from these values do not change the results of the model qualitatively.

The catchability coefficients are different for the two groups. Using data in Kirchner (1998), and the method presented in Sumaila (1997), we computed the age dependent catchability coefficients of [0, 0.002, 0.003, 0.003, 0.003, 0.02, 0.06, 0.06, 0.06, 0.05, 0.05] and [0, 0.003, 0.003, 0.003, 0.002, 0.001, 0, 0.001, 0.004, 0.006, 0.006] for the commercial and recreational fisheries, respectively. The cost parameters for the two players were difficult to determine, so an initial assumption is made and several sensitivity analyses carried out to determine values for k, which lead to an outcome close to what happens in the fishery currently. Price per fish of kob removed by the commercial and recreational sectors of the fishery are N$24.3 and N$64.3 (US$1.00 = N$7.96), respectively. A discount factor of 0.935 is assumed for both the commercial and recreational fishers.

Results

Same catchability and equal costs scenario

We use the noncooperative game model in a dynamic sense to help us capture the current situation in the fishery. To do this, we first assume that both of the two fisheries have the same catchability coefficients for kob, and face the same costs of removing a unit weight of fish. We run the model under this scenario to help isolate benefits that actually accrue to recreational fishers as against benefits that are usually perceived to be accruing to them. In addition, this run will help us explain why even though the relative revenues from the two sectors are so different, they harvest about the same number of kob. Table 4.1 below presents the average (i) annual standing biomass, (ii) harvest of kob and (iii) discounted economic benefits to the two groups when both cost and catchability are assumed to be equal to them.

The main point to note here is that the recreational fishers remove almost three times more kob under this scenario, and derive a lot more benefits than the commercial sector. When most policy makers and some researchers discuss the allocation of fish to recreational users *vis-à-vis* the commercial sector, these are the benefits they see. But in reality this is not what actually turns up because the two fisheries do not face the same costs, and usually have different catchability coefficients. This is why, in reality, the commercial users in the fishery take a lot more fish than they take under this scenario.

The equal costs scenario

Here, we plug in the catchability coefficients calculated for the two fisheries, but maintain the equal cost hypothesis. In this way, we are able to isolate the impacts of catchability and costs separately. Table 4.2 is similar to Table 4.1. We see from this table that the relative benefit to the commercial sector has increased significantly from only 13% to about 41% compared to the previous scenario. Similarly, the commercial sector now removes far more of the harvest than in the previous scenario. Both of these results indicate that the difference in catchability between the two sectors has significant impacts on the potential outcomes of the game.

The noncooperative scenario that captures the current state of the fishery

We conduct sensitivity analysis on the cost parameters in the model until the relationship between the catches in weight of the two fisheries are in the ratio of approximately 1.5 in

Table 4.1 Average annual benefits (N$m), harvest (tonnes) and standing biomass (tonnes) when cost and catchability are similar.

	Benefits	Harvest	Biomass
Commercial	7.22	620	6355
Recreational	54.66	1655	
Total	61.88	2275	6355

Table 4.2 Average annual benefits (N$m), harvest (tonnes) and standing biomass (tonnes) when costs are equal.

	Benefits	Harvest	Biomass
Commercial	13.04	1105	6765
Recreational	31.84	990	
Total	44.88	2095	6765

Table 4.3 Average annual benefits (N$m), harvest (tonnes) and standing biomass (tonnes) when cost of fishing is calibrated to capture current state of fishery.

	Benefits	Harvest	Biomass
Commercial	13.76	1165	7015
Recreational	2.54	810	
Total	16.3	1975	7015

favour of the commercial fishers. The costs parameters that are used to achieve this are then applied to run the regulator's model. The results presented in Table 4.3 show that harvest in weight by the commercial sector is about 1.5 times harvest by the recreational sector, as reported for the fishery. The recreational to commercial cost ratio per unit of fishing that produces this outcome is 20. This ratio may appear large but if one considers this to be a comparison of the opportunity cost of the time of recreational fishers compared to those of commercial fishers in Namibia, it suddenly becomes realistic.

Before we proceed to present results from the regulator's model, we display in Fig. 4.1 a comparison of the benefits to the recreational fishers versus those to Namibia under the three noncooperative scenarios discussed above. We see from the figure that both Namibia and the anglers get many benefits under scenarios with equal costs and/or similar catchability coefficients. However, the story is very different in the more realistic noncooperative model. In this case, Namibia reaps a lot but the anglers do not. This is because the anglers incur costs in time and otherwise. It is important to be aware of this when designing management and allocation policies in fisheries involving commercial and recreational fishers. The benefits that affect anglers' behaviour are the ones they receive (net of their costs) not what the larger society, in this case, Namibia realises.

The regulator's model results

Using the model parameter values in the above noncooperative scenario, we run the regulator's model, and compare the outcome with those under noncooperation. The average com-

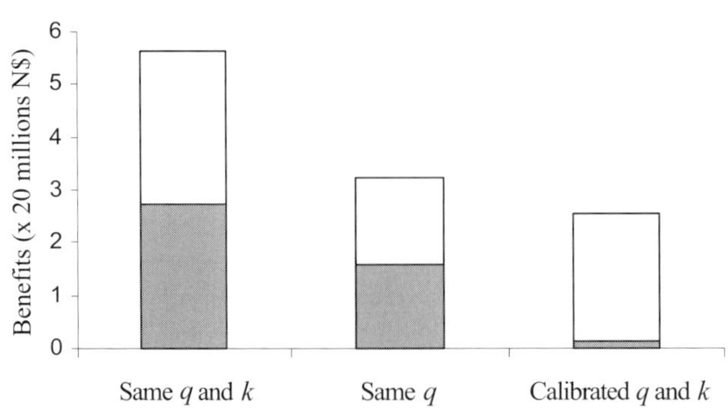

Fig. 4.1 Compares discounted benefits to recreational fishers (shaded) with those accruing to Namibia (open). q = catchability coefficient and k = unit cost of employing fishing effort.

Table 4.4 Average annual benefits (N$m), harvest (tonnes) and standing biomass (tonnes) for the regulator's and noncooperative models.

	Regulator's			Noncooperative		
	Benefits	Harvest	Biomass	Benefits	Harvest	Biomass
Commercial	10.76	805	7500	13.76	1165	7015
Recreational	12.88	1220		2.54	810	
Total	23.64	2025	7500	16.3	1975	7015

bined annual standing biomass, harvest levels, and discounted economic benefits to the two groups are given in Table 4.4. We see from this table that the regulator's model produces good outcomes relative to the noncooperative: it generates about 45% more benefits than in the case of the noncooperative scenario. The interesting point is that the regulator's scenario achieves this improvement in benefits while also improving the standing biomass of kob by about 6% compared to under noncooperation.

The regulator's model is able to deliver these win-win results (higher average biomass and higher economic benefits) because of two main reasons. First, this model does a much better job at allocating the harvestable part of the kob stock between the two sectors: while the commercial sector removes only about 27% of silver kob in the regulator's model, it removes around 39% in the noncooperative model. Second, the regulator's model uses less fishing effort overall.

Concluding remarks

This chapter has presented a model to help determine both the optimal total harvest of Namibian silver kob, and the allocation of this to the commercial and recreational users of the resource. The results of the analysis show that it is not enough to look at the revenue generated by a unit of fish removed from the habitat to determine how best to allocate the resources. The costs incurred by the parties, especially the costs to recreational fishers, have to be taken into account. In addition, it is important to remember that recreational and commercial users face different types of catchability constraints. In many instances, commercial users, because of their superior knowledge of the fishery, are able to enjoy great advantages in terms of their fishing selectivity, which partly compensates for their lower revenues per unit of fish removed. Finally, our model predicts that taking the main features of the Namibian silver kob fishery (including social and environmental concerns) into account, about 2000 tonnes of kob should be removed from the habitat, with the commercial sector removing about 800 tonnes, whilst the recreational sector takes about 1200 tonnes. This will leave on average about 7500 tonnes of standing biomass through time.

References

Argue, A.W., Hilborn, R., Peterman, R.M., Staley, M.J., & Walters, C.J. (1983) Strait of Georgia Chinook and Coho Fishery. *Canadian Bulletin of Fisheries and Aquatic Sciences*. Fisheries and Oceans, Bulletin 211, Ottawa.

Binmore, K. (1982) *Fun and games: a text on game theory*. Chancellor Press, London.

Flåm, S.D. (1993) Path to constrained Nash equilibria. *Applied Mathematics and Optimisation* **27**, 275–289.

Griffiths, M.H. & Lamberth, S.J. (2002) Evaluating a marine sport fishery in South Africa. In: *Recreational Fisheries: Ecological, Economic and Social Evaluation* (eds Pitcher, T.J. & Hollingworth, C.E.), Chapter 16, this volume. Blackwell Science, Oxford, UK.

Hickey, P. & Tomskin, H. (1996) *The Social, Economic and Management Aspects of Recreational Fisheries*. Fishing News Books, Blackwell Science, Oxford.

Kirchner, C.H. (1998) *Population dynamics and stock assessment of the exploited silver kob (Argyrosomus inodorus) in Namibian waters*. PhD thesis, University of Port Elizabeth, South Africa.

Kirkegaard, I.R. & Gartside, D.F. (1998) Performance indicators for management of marine recreational fisheries. *Marine Policy*, Vol. 22, No. 4–5, pp. 413–422.

Lopez, R., Shah, F.A. & Altobello, M.A. (1994) Amenity benefits and the optimal allocation of land. *Land Economics*, **70**, 53–62.

Pitcher, T.J. (ed) (1999) Evaluating the benefits of recreational fisheries. *Fisheries Centre Research Reports* 7(2).

Sumaila, U.R. (1997) Cooperative and non-cooperative exploitation of the Arcto-Norwegian cod stock in the Barents Sea. *Environmental and Resource Economics,* **10**, 147–165.

Sumaila, U.R. (1999) An economic model of recreational and commercial fishers. In: *Evaluating the benefits of recreational fisheries* (ed Pitcher T.J.) *Fisheries Centre Research Reports* 7(2), pp. 30–35.

Sutinen, J. (1993) Recreational and commercial fisheries allocation with costly enforcement. *American Journal of Agricultural Economics*, **75**(5), 1183–1187.

Chapter 5

Projecting Recreational Fishing Participation

Eric M. Thunberg and J. Walter Milon

Abstract

This chapter describes the methods used to develop projections of future numbers of marine recreational anglers in two separate surveys conducted in the North-eastern and South-eastern Atlantic coastal states of the USA. Similar data were collected in each survey, and both used the same survey method, yet differences in sample frame required different approaches to estimate recreational fishing probabilities. In the North-east a predictive statistical model was developed while in the South-east population proportions were used to estimate participation probabilities. The basic methods of applying these two approaches are described and compared. In each case, projections of future angler participation were based on projections of population and demographic change developed by the US Bureau of the Census. Limitations of reliance on demographic change as the primary basis for making participation projections are noted and alternative survey design possibilities are discussed. The results of the two studies suggest that fishery managers in the North-east and South-eastern regions of the USA will be faced with a growing and diverse recreational fishing constituency. Long-term planning will be required to deal with increased demands for both public and private recreational fishing services.

Key words: census, demographics, participation, projections, recreational fishing.

Introduction

> I think there is a world market for about five computers.
>
> Thomas J. Watson, Chairman of the Board, IBM, 1943

> 50 years hence … we shall escape the absurdity of growing a whole chicken in order to eat the breast or wing, by growing these parts separately under a suitable medium.
>
> Winston Churchill, 1932

> A severe Depression like that of 1920–21 is outside the range of probability.
>
> Harvard Economic Society, 1929

> Where a calculator on the ENIAC is equipped with 18 000 vacuum tubes and weighs

> 30 tons, computers in the future may have only 1000 vacuum tubes and weigh perhaps only 1.5 tons.
>
> *Popular Mechanics*, 1949

With predictions like these it seems rather audacious even to be considering projecting recreational fishing participation. Nevertheless, none of these predictions was considered outrageous at the time they were made as they do in hindsight. Such is the business of making projections. Predictions may look good in the short run but many look pretty foolish in the long run. So why do we make them? Making projections is a fundamental part of planning. Proactive decision making requires some capacity to anticipate events or processes that need to be addressed before small problems become large ones. In this chapter we provide an overview of two alternative procedures for projecting potential changes in recreational fishing participation in terms of numbers of recreational anglers. Note that projecting numbers of anglers is only one aspect of recreational fishing. Projections of numbers of fishing trips or changes in recreational expenditures are also important planning considerations. Nevertheless, given space limitations we focus on projecting recreational fishing participation. With this information fishery managers may be better able to anticipate the direction and magnitude of growth in recreational fishing participation to be better able to anticipate potential conflicts among resource users. Similarly, if managers can anticipate the changing composition of the recreational fishing public, then they may be better able to design programs and develop the infrastructure to accommodate this change.

A variety of factors influence participation in recreational fishing. Constraints or barriers to participation (Jackson 1988) can be grouped into five major categories: (1) lack of interest; (2) lack of time; (3) lack of money; (4) lack of facilities; and (5) lack of skill (Searle and Jackson 1985, Kay and Jackson 1991). Social and cultural constraints such as age, gender, and income also affect recreational fishing participation (Aas 1995). Although economic and demographic factors are generally insufficient by themselves to explain why individuals make the choices they do, simple relationships have been developed between demographic variables and recreational participation to project future participation in recreational fishing (Loomis and Ditton 1988). Murdock *et al.* (1992) projected recreational fishing participation to the year 2050 based on projected national changes in population growth, age structure, minority populations, and household composition. Edwards (1989) developed predictive models to project marine recreational fishing for coastal states to the year 2025. Similarly, Milon and Thunberg (1993) developed predictive models to project participation rates of Florida resident recreational anglers through 2010.

In this chapter we illustrate two procedures for projecting recreational fishing participation. In each case, a common set of demographic variables was collected through a random telephone survey [see Lyle *et al.* 2002 for an Australian example – eds] designed to obtain information on participation of saltwater anglers in different regions of the USA. The North-east region was comprised of North Atlantic and Mid-Atlantic coastal states from Maine to North Carolina and the South-east region included coastal states along the South Atlantic and Gulf of Mexico from South Carolina to Louisiana. Detailed technical reports for each method are available in PDF format at the following web address www.st.nmfs.gov/st1/econ/pubs.html. The North-east region study is described in Thunberg *et al.* Differences in both sample frame and sample sizes led to different methods for estimating recreational fishing participation probabilities. In the following section aspects common to each approach are described. These include a common survey format and survey instrument as well as common procedures for projecting participation in saltwater recreational fishing based on population projections provided by the US Bureau of the Census. Next, the differences between the two methods in estimating participation probabilities are described. The advantages and disadvantages of both approaches are then discussed in a concluding section. Note that the basic procedures described herein are representative of two broad classes of analytical methods that have been used to estimate recreational fishing participation in our own work and by others. Our intent is to offer general guidance in using these methods via illustration with actual applications, not to suggest that these methods should be used in all circumstances.

Recreational fishing participation survey

Participation data were collected as a component of the National Marine Fisheries Service (NMFS) Marine Recreational Fishing Statistics Survey (MRFSS). A detailed description of the MRFSS can be found at the web address www.st.nmfs.gov/st1/recreational/index.html [and in Gentner and Lowther 2002 – eds]. The primary purpose of the MRFSS is to provide annual estimates of marine recreational catch and effort through a combination of a household telephone survey (to estimate fishing effort) and a field intercept survey (to estimate catch). As such, the participation survey was administered as an add-on to the base MRFSS. Since the telephone portion of the MRFSS is a population survey implemented using a stratified random design it was selected as the vehicle for the participation surveys.

The MRFSS telephone survey is administered to residents of coastal counties (generally defined as counties within 25 or 50 miles of ocean coastline). Strata are the coastal counties and sample size for each stratum is proportional to the square root of the county population. The survey is conducted in six waves beginning with wave 1 (January/February) and ending with wave 6 (November/December). Interviewing is conducted during a two-week period beginning the last week of the wave and continuing into the first week of the next wave.

The MRFSS household survey is designed to gather information for all fishing trips taken within the two-month wave. For each trip, detailed data are gathered on fishing mode (shore, party or charter, and private or rental boat) and primary fishing location (estuary, bay, sound, and distance from shore). Since the primary purpose of the MRFSS is to gather information on fishing trips, data are not normally collected on individuals or households that have not fished during a given wave and demographic or economic data are not collected in the base survey.

The participation survey in both regions was administered in the following manner. In accordance with MRFSS survey protocol described above, the individual answering the telephone was asked whether any member of the household had taken a saltwater recreational fishing trip at any time within the past two months. With a negative response, the MRFSS interview is normally terminated. However, for the participation survey, each respondent was queried as to whether he/she had never fished; not fished in past 12 months; or had fished at least once in past 12 months but not during the past two months. The interview was continued for all individuals in the latter two categories to collect data on age, ethnicity, education, gender, income, and employment status. In the North-east region demographic data for a sample of individuals that had never fished were also collected. Data for individuals that had never fished were not collected in the South-east due to budgetary constraints. For individuals that had fished in the past two months the standard MRFSS data for every fishing trip taken in the past two months were collected. Demographic data were also collected for all 'two-month' anglers in each region.

In the North-east region 55 553 contacts were made to households with a working telephone while a total of 157 637 households were contacted in the South-east region (Table 5.1). Since different sampling rates were assigned to each of the participation categories described above, the participation survey was not initiated with every household that was contacted. Based on the specified sampling rates, a total of 11 060 and 25 975 interviews were initiated in the North-east and South-east respectively. Both surveys achieved relatively high completion rates calculated as the proportion of completed to initiated surveys. The cost of the base MRFSS household survey in the North-east and South-east regions was approximately US$225 000 and $450 000 respectively; the additional cost of the participation survey was $26 000 and $20 000 respectively.

Population projections

The US Bureau of the Census has produced USA population projections by age, gender and race in five-year intervals from 1995 to 2025 (Campbell 1996). These projections are provided on a state-by-state basis, but are not broken out by coastal and noncoastal counties. Since the participation survey sampling frames were limited to coastal counties, prorating the survey-based results to an entire coastal state required certain assumptions. In the North-east region, the census projections were prorated by the ratio of coastal county population to total state population by (demographic groupings) using 1990 census data. This procedure assumes constant proportional population growth by demographic grouping in coastal counties and noncoastal counties over time.

In the South-east region, the demographic profiles of coastal and noncoastal residents were assumed to be the same. For states such as Florida, this assumption is not a problem since

Table 5.1 Summary of sample size for participation survey in the North-east and South-eastern regions of the USA.

	North-east	South-east
Households screened	53 553	157 637
Participants		
2-Month	4 168	13 610
12-Month	2 590	12 365
Nonparticipants*		
Never fished	44 714	110 762
Fished but not in past 12 months	2 081	19 970
Interviews initiated	11 060	25 975
Completed interviews	8 621	21 121
Per cent completed	77.9	81.3

* Demographic data for nonparticipants was collected in the North-east but not in the South-east.

all counties are classified as coastal. In most other states in the South-east region (Georgia is the exception), coastal participants greatly outnumber noncoastal participants. Therefore, using the demographic profile of coastal participants as a proxy for nonparticipants should not cause a serious problem.

Once the sample-based participation probabilities were estimated, projections of recreational fishing participation were produced by multiplying the participation probabilities by the projected population size. Since participation probabilities were estimated by demographic component, the population projections of recreational fishing participation can provide not only an estimate of the total number of participants but may also indicate changes in the underlying demographic composition of participants.

Estimating participation probabilities

The difference in sampling frames between the North-east and South-east region has important implications for how participation probabilities may be calculated. With a population survey (as conducted in the North-east), participation probabilities can be estimated using statistical methods from the sample data or by using a method based on population proportions similar to that described in Loomis and Ditton (1988) hereafter referred to as the population proportions method. Where only a segment of the population is surveyed (as in the South-east region), estimation of participation probabilities is limited to the population proportions method.

Statistical probability models: the North-east region survey

In the North-east region survey, participation was modelled as a binary process where a value of 1 was assigned to a respondent that had fished in the past 12 months and a value of 0 was assigned otherwise. Binary processes are typically modelled using either probit or logit regression. Both approaches are based on cumulative probability distributions (the cumulative normal for the probit, and the cumulative logistic for the logit) which assures that predicted values cannot exceed the 0 to 1 interval (Pyndick and Rubinfeld 1981). Logistic regression was selected in this study because of its computational simplicity. Note that sample data must also be weighted according to the statistical design of the stratified random survey (Cochran 1977). This meant that sample data had to be weighted by strata, different sample sizes across waves, and the different sampling rate assigned to individuals that never fished (a detailed description of the statistical weighting procedures is provided in Thunberg *et al.* 1999). The demographic data (age, income, education, ethnicity, and gender) collected in the add-on to the MRFSS household telephone survey were used as explanatory variables. Household income and education were treated as continuous variables, while a series of dummy variables were constructed to represent ethnicity and gender–age group effects. Based on previous research, fishing participation for men and women was found to change with age, with participation increasing early in life and decreasing late in life (Milon and Thunberg 1993). To reflect this participation pattern, combinations of gender–age group dummy variables were constructed. Specifically, a total of five dummy variables were defined (males 16–24, males 65+, females 16–14, females 25–64, and females 65+) using males age 24–64 as the base group. Males

24–64 was selected as the base because in this particular instance, there was no statistically significant difference in male participation probability among these four census cohorts. An additional dummy variable (REGION) was included to account for regional differences (Maine to New York as compared to New Jersey to Virginia) in participation probability.

For nearly all of the demographic variables, the estimated coefficients from the regression analysis were statistically significant at the 5% level of significance (Table 5.2). Positive coefficients indicate that participation probability increases as the variable increases and vice versa for negative coefficients. For example, the probability of being a recreational fishing participant increases with income, but decreases with educational status.

The probability that an individual (P_i) with a given set of demographic characteristics (X_i) would be a saltwater recreational fishing participant was calculated by:

$$P_i = \frac{1}{1+e^{-(\alpha+\beta X_i)}} \qquad [5.1]$$

where α is the intercept and β is the estimated coefficient for the dependent variable from the logistic regression. Setting income and education equal to their estimated population median values and systematically applying the estimated coefficients expected participation probabilities for each demographic subgrouping were derived (Table 5.3).

Table 5.2 Coefficient estimates for North-east region saltwater recreational fishing participation model.

Variable	Estimated coefficient	Standard error
Intercept	-1.2012*	0.0979
Household income	0.1139*	0.0200
Education	-0.1217*	0.0212
Ethnicity	-0.7882*	0.0925
Female 16–24	-2.2180*	0.2325
Female 25–64	-1.5325*	0.0818
Female 65+	-3.0222*	0.2880
Male 16–24	-0.1853	0.1146
Male 65+	-0.3971*	0.1312
Region	0.2155*	0.0624

* Statistically significant at the 5% level or greater for Chi square with one degree of freedom.

Table 5.3 Predicted probability of participation by region and gender age-group.

States and age group	Females	Males
ME to NY		
16 to 24	0.023	0.153
25 to 64	0.045	0.179
65+	0.010	0.128
NJ to VA		
16 to 24	0.029	0.183
25 to 64	0.055	0.213
65+	0.013	0.154

Table 5.4 1990 North-east region coastal county population by region and gender age-group (millions).

Age group	Females	Males
ME to NY		
16 to 24	1.143	1.408
25 to 64	5.976	5.497
65+	1.745	1.085
NJ to VA		
16 to 24	0.966	1.010
25 to 64	4.158	3.956
65+	1.092	0.714

Based on these probabilities, males age 25 to 64 were most likely to participate in saltwater recreational fishing, while females over the age of 64 were least likely to participate. The expected number of participants in any given subgroup can be calculated by the product of the predicted participation probability (Table 5.3) and the total population size of the subgroup (Table 5.4). Summing products across all demographic subgroups gives an estimate of 3.1 million saltwater recreational anglers in the North-east region in 1994 (the year in which the survey was conducted).

Population proportion methods: the South-east region survey

Although population proportion methods may use either parametric (Edwards 1989) or non-parametric (Loomis and Ditton 1988) approaches, both approaches require an independent estimate of total recreational fishing participation from which the participation proportions can be calculated. Participation probabilities for the South-east region were estimated using a nonparametric approach developed by Loomis and Ditton (1988). Participation proportions were derived for various age/gender/ethnic groups within each state in the South-east region. These demographic variables (age/gender/ethnicity) were selected because: (a) previous research (Edwards 1989, Milon and Thunberg 1993, Aas 1995 and Thunberg *et al.* 1999) indicated that they were consistent predictors of marine recreational fishing participation; (b) these variables were included in the South-eastern add-on survey; and (c) US Census Bureau projections were available for these demographic groups.

Mathematically, the sample participation proportion was defined as:

$$P_{ij} = (\text{Participants in sample cohort}_{ij}) / (\text{Participants in the sample for state j}) \qquad [5.2]$$

where i (i = 1,…,n) represents age/gender/ethnic cohort groupings and j represents a state within the South-east region. The sample participation proportion for each state was then multiplied by the MRFSS estimate of the number of participants in the year of the add-on survey to determine the total number of participants (PT) in each cohort group:

$$PT_{ij} = P_{ij} \times \text{Number of MRFSS Resident Participants in state j} \qquad [5.3]$$

The sample information was then extrapolated to the population of each state by computing population participation proportions (PR) as:

$$PR_{ij} = PT_{ij} / Population_{ij} \qquad [5.4]$$

In theory, the resulting participation proportions are equivalent to the participation probabilities in Table 5.3.

Projecting recreational fishing participation

Although different approaches were used to estimate participation probabilities in the North-east and South-east regions, the methods used to project numbers of participants to the year 2025 were identical. The future number of participants in each demographic cohort FPT_{ij} was calculated by applying the participation probability for each group to the US Census Bureau population projection for that group:

$$FPT_{ijt} = PR_{ij} \times \text{Projected Population}_{ijt} \qquad [5.5]$$

where t denotes projection year. By summing across all FPT_{ijt} the total number of participants in each state was then estimated (Table 5.5). Given this estimate of total participants, the proportion of anglers to the total population, also known as the participation rate, can be calculated by dividing the total number of participants in each year by the total population projection for that year.

The participation projections suggest that the number of recreational fishing participants will continue to increase in both the North-east and South-eastern regions of the USA. Mis-

Table·5.5 Projected number of saltwater recreational fishing participants (thousands) by state (2000–2025).

State	Projection year					
	2000	2005	2010	2015	2020	2025
Connecticut	277.4	282.4	290.4	297.2	302.6	307.1
Delaware	74.2	78.0	80.1	81.1	81.7	81.7
Massachusetts	396.5	406.6	416.1	422.7	426.9	429.4
Maryland	424.7	440.4	455.6	467.5	476.9	484.5
Maine	92.0	95.4	98.8	101.4	103.1	103.8
New Hampshire	67.9	72.1	75.1	77.2	78.4	78.9
New Jersey	747.8	768.8	792.6	811.6	828.0	841.6
New York	828.0	836.3	854.6	869.1	880.5	889.7
Rhode Island	84.8	86.8	89.6	91.8	93.5	94.8
Virginia	290.7	305.2	318.7	329.2	337.5	344.0
North-east region	3284.1	3372.0	3471.6	3548.9	3609.0	3655.5
Regional participation rate %	10.95	10.88	10.79	10.67	10.54	10.38
North Carolina	546.1	588.9	622.9	644.6	660.1	667.4
South Carolina	239.3	253.4	267.4	277.6	258.4	290.3
Georgia	104.9	112.5	118.8	123.7	128.0	131.4
Florida	2582.5	2805.3	3024.0	3194.5	3337.9	3443.8
Alabama	185.7	195.5	204.9	211.3	215.6	217.6
Mississippi	146.1	151.9	156.4	158.6	159.7	159.6
Louisiana	527.6	536.2	549.6	559.2	566.1	569.9
South-east region	4332.2	4643.6	4944.0	5170.0	5352.8	5479.9
Regional participation rate %	11.56	11.97	11.99	12.00	11.93	11.64

sissippi is the only state in both regions that was projected to experience a decline in the number of saltwater fishing participants. The participation rate for the North-east region was projected to decline in every year from 2000 to 2025. By contrast, the participation rate in the South-east region is projected to increase to 2015 and decline thereafter. In 2015 the majority of the US 'Baby Boom' generation will have reached age 65 or greater. After 2015 the overall rate of growth in participation in marine recreational fishing declines as this dominant population cohort moves past 65 years of age.

The projection results indicate three dominant trends. First, while participation rates are projected to decline after 2015, the overall number of recreational fishing participants is likely to increase. This means that fishery managers will have a larger recreational constituent base to deal with. These constituents are likely to increase demand for recreational fishing services from both the private and public sector. A larger fishing population may result in increased fishing mortality on recreational stocks. Increased fishing pressure may require implementation of more restrictive management measures or may require greater emphasis on catch-and-release fishing [see chapter by Policansky 2002 – eds]. The second dominant trend is the ageing of the recreational fishing public. This trend may require increased infrastructure development to accommodate the physical needs of an older population that may include such things as wheelchair access or installation of handrails at fishing sites or providing written materials in larger print. The third dominant trend is the changing ethnic composition of the fishing population. Potential differences in species targeting preferences as well as cultural differences in motivations for fishing will need to be considered as fishery managers seek to design effective management programs. Considered together, these three trends indicate that fishery managers will need to be responsive to a larger and more diverse fishing public throughout the US Atlantic and Gulf of Mexico coastal states.

Conclusions

In this chapter we have described two alternative methods of providing projections of recreational fishing participation. Either of these methods will be feasible for countries where a periodic population census is carried out. While it would have been possible to produce our own population projections the fact that the US Census Bureau already does this simplified our task enormously. In addition to the census information, each of the studies required collection of a rather basic set of demographic variables. These sample data were used to produce population estimates of participation probabilities.

Of the two methods described to estimate participation probabilities, both yield valid estimates of recreational fishing participation, but the statistical approach (as compared to the population proportions method) has greater flexibility for several reasons. First, use of a probability model allows direct estimation of participation probabilities from which total participation can be estimated. No independent estimation of total participation is required. Second, participation probabilities are derived from sample observations from an entire region. Such pooled data may provide better predictions since more information is included from the sampled population. Third, the statistical approach can be applied to lower sample sizes. Pooling data assures that there are no empty cells in a given cohort. For example, due to sample size limitations the population proportion method could not have been applied in the North-east region without combining data across multiple states. Fourth, the statistical approach offers the opportunity to develop alternative model specifications and to test hypotheses about the effects of specific combinations of demographic variables on recreational fishing participation. Lastly, and perhaps most important, the statistical approach allows for the incorporation of factors other than demographic variables. The principal advantage of the population proportions method is that it does not require information from individuals that have never fished. Given that nonanglers may be a relatively large proportion of the total population, then not having to survey this group will reduce overall survey costs.

Projecting participation based solely on demographic change has its constraints. All of the participation estimates provided in this study assume that the factors that influenced participation in 1994 will not change (an assumption common to similar studies: Edwards 1989, Milon and Thunberg 1993, Murdock *et al.* 1992, Loomis and Ditton 1988). These factors are not likely to remain constant nor is participation merely a function of demographics. Individual attitudes, experiences, social norms, and opportunity determine whether an individual will engage in any given recreational activity. The extent to which demographics are correlated with these decisions is not static. For example, changing gender norms may lead to increased participation rates among women relative to current and past levels of female participation. Similarly, lifestyle changes among older individuals may result in higher participation rates among this segment of the population.

Within the context of current studies the ability to examine the potential effect of these social changes would be limited to varying the participation probabilities for particular segments of the population to develop a range estimate of participation. A broader approach would be to design a survey specifically to examine recreational fishing participation (i.e. the fact that the primary purpose of the MRFSS was not to collect social and economic data limited the range of data elements that could be collected). Such a survey could be designed to elicit the full range of reasons why people do or do not fish. The implications for changing specific constraints or barriers to participation identified by Jackson (1988) could be examined by developing a statistical probability model and constructing a variety of different scenarios. For example, recreational fishing participation has been shown to be positively related to whether an individual was taken fishing as a child (Milon and Thunberg 1993). The effect on future participation of a campaign to promote recreational fishing in children could be examined by computing the change in participation probability as the proportion of people that have been fishing as a child increases. In a similar manner, the effects of changing per capita income on participation could also be examined. Any number of other scenarios or possibilities could be developed depending on survey design and specific interests of the researcher.

Projections of future participation cannot be used to predict how many people will actually participate in any given year. Interannual differences in participation are likely to depend on fluctuations in short run economic, climatic, and resource conditions. However, the projections are likely to provide reasonable estimates of longer-term trends in recreational fishing participation.

Lest we get too carried away, it is important to be mindful of the limitations of making projections. We began with a note of caution on the perils of projecting and we end with a reminder of the same.

> Predicting is usually very difficult, especially when it is about the future.
>
> attributed to Yogi Berra

Acknowledgements

The authors appreciated the comments from Dr Phil Logan and Dr Fred Serchuk of the North-east Fisheries Science Center as well of those of two anonymous reviewers in improving the quality of the presentation. The authors also thank the Marine Recreational Fishing Statistics Survey staff and Maury Osborn in particular for her assistance in making the participation surveys possible, as well as her continued diligence and support in facilitating social science data collection.

References

Aas, Ø. (1995) Constraints on sportfishing and effect of management actions to increase participation rates in fishing. *North American Journal of Fisheries Management*, **15**, 631–638.

Campbell, P.R. (1996) *Population projections for states by age, sex, race, and hispanic origin: 1995 to 2025*, U.S. Bureau of the Census, Population Division, PPL-47 pp. 1–49.

Cochran, W.G. (1977) *Sampling Techniques*. pp. 1–417. John Wiley and Sons, New York.

Edwards, S.F. (1989) Forecasts of in-state participation in marine recreational fishing. *Transactions of the American Fisheries Society*, **118**, 564–572.

Jackson, E.L. (1988) Leisure constraints: a survey of past research. *Leisure Sciences*, **10**, 203–215.

Kay, T. & Jackson, G. (1991) Leisure despite constraint: the impact of leisure constraints on leisure participation. *Journal of Leisure Research*, **23**(4), 301–313.

Loomis, D.K. & Ditton, R.B. (1988) Technique for projecting the future growth and distribution of marine recreational fishing demand. *North American Journal of Fisheries Management*, **8**, 259–263.

Lyle, J.M., Coleman, A.P.M., West, L., Campbell, D. & Henry, G.W. (2002) New large-scale survey methods for evaluating sport fisheries. In: *Recreational Fisheries: Ecological, Economic and Social Evaluation* (eds Pitcher, T.J. & Hollingworth, C.E.), Chapter 15, this volume. Blackwell Science, Oxford, UK.

Milon, J.W. (2000) Current and Future Participation in Marine Recreational Fishing in the South-east U.S. Region. U.S. Department of Commerce, *NOAA Technical Memorandum* No. NMFS-F/SPO-44. pp. 1–44.

Milon, J.W. & Thunberg, E.M. (1993) A regional analysis of current and future Florida resident participation in marine recreational fishing. Gainesville, Florida: *Florida Sea Grant College Program*; SGR-112. pp. 1–96.

Murdock, S.H., Backman, K., Ditton, R.B., Hoque, Md. N. & Ellis, D. (1992) Demographic change in the United States in the 1990's and the Twenty-first Century: implications for fisheries management. *Fisheries*, **17**(2), 6–13.

Policansky, D. (2002) Catch-and-release recreational fishing: a historical perspective. In: *Recreational Fisheries: Ecological, Economic and Social Evaluation* (eds Pitcher, T.J. & Hollingworth, C.E.), Chapter 6, this volume. Blackwell Science, Oxford, UK.

Pyndick, R.S. & Rubinfeld, D.L. (1981) *Econometric Models and Economic Forecasts*. pp. 1–630. McGraw-Hill, New York, NY.

Searle, M.S. & Jackson, E.L. (1985) Socioeconomic variations in perceived barriers to recreation participation among would-be participants. *Leisure Sciences*, **7**(2), 227–249.

Thunberg, E., Steinback, S., Gray, G., Gautum, A. & Osborn, M. (1999) Volume III: Summary report of methods and descriptive statistics for the 1994 North-east region marine recreational fishing participation survey. U.S. Department of Commerce, *NOAA Technical Memorandum* No. NMFS-F/SPO-39. pp. 1–48.

Chapter 6

Catch-and-Release Recreational Fishing: a Historical Perspective

David Policansky

Abstract

Catch-and-release (C&R) fishing has evolved and become increasingly widespread among recreational anglers in North America and elsewhere. Initially it was mainly a byproduct of regulations limiting the size, number, or species of fish that could be retained, but it has become widely adopted as a specific management regulation and as a voluntary approach to conservation. It has an important role as one of the tools available to fishery managers. Even though it usually promotes the conservation of a fish population, it does have adverse effects, mainly hooking mortality but also more subtle ones such as effects on growth or reproduction. The evolution of C&R has been associated with changes in angler motivation and management issues. Crowding of fishing venues is receiving increasing attention from management agencies and even a state legislature. Catch-and-release also is becoming controversial on ethical and moral grounds. The motivations and ethics of anglers as well as society at large will continue to evolve, and fishery management must therefore adapt as well.

Key words: angler crowding, angler motivations, catch-and-release, cruelty of fishing, historical perspective, recreational angling.

Introduction

Fishing is an old human activity that involves typical common-pool resources. For traditional (consumptive) fishing, in which the intent is to keep the fish to eat or sell, the commons resource is the fish. If an individual subtracts one fish, one less fish is available to others. Fishery management has been profoundly dependent on the commons aspect of consumptive fishing. However, catch-and-release (C&R) fishing has changed the face of recreational fishing and its management. Not only has it changed the effects of fishing on the fish, it has changed the way anglers and management agencies view fishing by changing the incentives and disincentives associated with it. Now the common-pool resource in short supply often is space and related intangibles, such as the quality of the experience (Policansky 2001). In this chapter, I review the evolution of C&R in recreational fishing and its effects on changing incentives, disincentives, and management practices. I also discuss some more philosophical issues related to C&R fishing. My focus is on North America.

What are recreational fishing and catch-and-release (C&R)?

Recreational fishing is fishing primarily for recreation or enjoyment as opposed to fishing whose main purpose is the production of food or other products. Other kinds of fishing, e.g. commercial or subsistence fishing, often bring pleasure to their practitioners, but their primary purpose is to get food or other products for consumption, sale, or trade. Although it can be difficult sometimes to classify an activity unambiguously as recreational fishing (which can also be an important source of food), recreational fishing meets psychological, environmental, social and other nonutilitarian needs (Fedler and Ditton 1994).

Catch-and-release fishing is fishing in which fish are returned alive to the water. It is a relative term. Some few unwanted fish must have been released alive from the earliest days of fishing. More importantly, limits on the size, number, or species that may be retained at a particular time and place compel at least some C&R if the limits are observed. *Voluntary* C&R is the release of fish alive even though those fish could legally be retained. *Regulatory* C&R is required by or as a consequence of regulations. *Total* C&R, in which all fish are released alive, can be voluntary or regulatory (although hooking mortality, described later, makes total C&R more of a theoretical ideal than a practical reality).

A brief history of recreational fishing and C&R

Recreational fishing as understood today is at least several centuries old, and probably much older (McDonald 1963, Schullery 1987). The early fifteenth century *A Treatyse of Fysshinge with an Angle* described what was clearly nonprofessional, i.e. recreational, fishing with a rod and line in Britain. The document, most often seen in its first printed form of 1496 (Berners 1496) but known as well from a midcentury manuscript and probably written about 1420 (McDonald 1963), is attributed to Dame Juliana Berners, almost surely a pseudonym (Blades 1881, McDonald 1963, Schullery 1987, 1999). The following excerpt makes clear that the purposes and pleasures of early fifteenth century fishing were similar to those of early twenty-first century fishing: 'For [the angler] cannot lose more than a line or a hook, of which he can have plenty of his own making, or of other men's making, as this simple treatise will teach him; so then his loss is no grievance. And he can have no other grievances, unless some fish breaks away from him when he is on his hook, in the landing of that same fish, or in any case, he does not catch him. This is no great hardship, for if he fails with one, he cannot fail with another, if he does as this treatise which follows will instruct him – unless there are no fish in the water where he is angling. And yet, at the very least, he will have his wholesome and merry walk at his own ease, and also many a sweet breath of various plants and flowers that will make him right hungry and put his body in good condition. He will hear the melodies of the harmony of birds. He will also see the young swans or cygnets following their brood swans, the ducks, the coots, the herons, and many other birds with their broods, which seems to me better than all the noise of hounds and the blasts of horns and other amusements that falconers and hunters can provide, or the sports that fowlers make. And if the angler catches the fish with difficulty, then there is no man merrier than he is in his spirits.' (Translation by S. Kuhn of a mid-century manuscript of the Treatyse, in McDonald 1963.)

Although the Treatyse emphasised the sporting aspects of fishing, the use of fish as food also mattered, and it provided descriptions of their eating quality. Even if obtaining food was not the main object of the exercise, the fish often were eaten. In the fifteenth century, even in Britain, fish were plentiful and bag limits were not in force (although there was already a closed season for salmon by 1295 – Hickley 1998). For that reason, C&R would have been unnecessary, and was not mentioned in the Treatyse, although it did advise anglers not to be greedy and take too many fish, because 'That could easily be the occasion of destroying your own sport and other men's also.'

The essence of sport in sportfishing was at first clearly related to obtaining food, and although some fish must have been released, C&R was not part of most American anglers' consciousness well into the nineteenth century. Herbert (1851), writing about North America under the pseudonym of Frank Forester, wrote 'Again, it is true that no sportsman captures that, which captured, is worthless; and that to be game, whether bird, beast, or fish, is to be eatable … It is remarkable, however, that all those fish which are the most game, the boldest, the strongest, the bravest, and the most obstinate, are invariably the finest also for culinary purposes, and the most highly appreciated by the gourmet on the board, as well as by the fisher in the river or mere.' This is in clear contrast to the view, established already by the beginning of the twentieth century and one that provides a basis for C&R, that inedible or at best barely edible fish could be considered game fish worthy of angling. For example, Gregg (1902) described ladyfish (*Elops saurus*) as 'one of the gamest fish that swims,' despite rating it as 'barely edible.'

Gradually, increasing popularity of recreational fishing coupled with industrial, residential, and agricultural conversion and alteration of the land and watersheds led to depletion of native fish populations. That depletion resulted in more and more need for bag and size limits and other protective measures and hence regulatory C&R. In England, during Queen Elizabeth's reign (1558–1603) the following minimum size-limits were in law: 10 inches for pike (*Esox lucius*), 16 inches for salmon (*Salmo salar*), 8 inches for trout (*S. trutta*), and 12 inches for barbel (*Barbus barbus*) (Wright 1858). During the reign of George II (1727–1760), there were also size-limits for roach and it was illegal to take, possess, or sell any undersized fish or fish caught out of season (Wright 1858).

Wright wrote 'It is unsportsmanlike to take any fish when full of spawn, for then man becomes the wholesale destroyer of fish by tens, or even hundreds of thousands …'. This passage implies that he released roe-laden fish. He also advocated the release of undersized roach (*Rutilus rutilus*), for which Britain's Parliament had set a size limit of 8 inches. Wheeley (1897), for example, fished extensively in the River Thames for barbel, of which he wrote 'As food, I can only consider barbel worthless; I have twice tried to eat them, but have abandoned the attempt …'. Thus, it seems that he probably released the fish (although some people fed them to pigs (Maxwell 1904)). He also described his pleasure that gross-weight tournaments were disappearing, because 'gross-weight fishing tends to deplete the stock of fish in any water; when the practice is indulged in, the competitors keep every fish that is sizeable in order that the weight may tell up.' In other words, the tournaments discouraged the release of fish, although C&R tournaments were becoming more popular in Britain at around this time (Hughes 1999).

On the River Thames, by the 1890s bylaw No. 7 required that 'No person shall fish for pike with any device or tackle that does not admit of the pike taken therewith being returned to the water without injury' (Wheeley 1897). Those limits implicitly required C&R (and bylaw No. 7 explicitly did), for if an angler catches a fish that is too large or too small, or is out of season, or is of a species that is totally protected, the law required it to be released alive and unharmed.

Not all English writers saw regulatory C&R as a solution to the depletion of fish populations. Cholmondeley-Pennell (1893) recognised that the recent increase in angling pressure would lead to the reduction of fish populations, but saw pisciculture as the answer. Indeed, enhancement has been considered the main method of providing angling (and even commercial fishing) opportunities in the face of overexploitation of freshwater fishes down to the present day (e.g. Kron 1995, NRC 1996, Ross and Loomis 2001).

When Europeans arrived in North America in the seventeenth century, fish were plentiful on this continent as well (Nielsen 1999). Although fishing pressures increased faster in the east than in the west, the abundance of fishes in earlier days was unfathomable to modern anglers and biologists. For example, in Oregon's Willamette River, the Multnomah Rod and Gun Club was deploring the decline in trout (probably rainbows, *Oncorhynchus mykiss*) fishing in 1893 and wanted to compensate for it by introducing largemouth bass (*Micropterus salmoides*). Describing the fishing, Lampman (1946) wrote 'you may judge of how poor the fishing was hereabouts in 1893 when you are told that not more than a decade later, in 1904, it had deteriorated until the lawful daily bag limit was a beggarly 125 fish, with no closed season!' On one day in the early 1880s, H.P. Wells (1885) and a friend caught about 150 trout *apiece* in Maine's Magolloway River. They practised C&R, releasing all but about 20 fish that were badly injured. Wells also reported another trip, to a lake, when he killed a 4-pound male brook trout (*Salvelinus fontinalis*), while releasing females and smaller males. Wells did not discuss C&R explicitly, and so it is not clear whether he released fish he did not want for conservation purposes, but it seems likely.

Voluntary C&R also was practised – at least occasionally – in saltwater, as well. Gregg (1902) described an outing he was on in 1892 where his party 'trolled for Bluefish [*Pomatomus saltatrix*], Spanish Mackerel [*Somberomorus maculatus*], and Kingfish [*S. cavalla*], catching a large number, returning all to the water except a few for the table.' He rated all of those species as excellent table fare, and so it seems clear that they were released for conservation purposes. Gregg also made a revealing confession, 'which may read like a too late repentance for past errors or iniquities.' He confessed that in the 1850s and 1860s, he observed angling expeditions during which hundreds of large-mouth bass and pickerel (*Esox niger*) were 'buried on the shore near the hotel many days in succession.' He also caught far more fish than he could use on one of those expeditions, although he kept only the bass and pickerel, releasing the others. In addition, on a trip in the 1880s, he caught more than 200 bass, which he gave away. He justified the action by saying the fish were used, but nonetheless, he 'promise[d] never to do it again.' The stories are important because they illustrate the importance of distinguishing the evolution of ethical standards in one individual from the evolution of ethical standards in the angling community or even in society at large. Both occur, but the evolution of ethical standards in an individual can occur faster than in the general angling community. By 1902, Gregg's standards for releasing fish appear to have evolved to a state that anglers in general would not reach until the last third of the twentieth century, and even then some individuals had not adopted those standards.

In the United States, where the sporting tradition derived from the British sporting tradition in the nineteenth century (Reiger 1975), sportsmen urged conservation. Indeed, H.W. Herbert ('Frank Forester'), one of the early North American advocates of the sporting tradition in angling, came to North America from England (Reiger 1975). R.B. Roosevelt (1884) wrote that he 'was one of the first to press on the State and National Governments the importance of establishing fishery commissions' and that he was appointed to the New York State Fishery Commission when it was created in 1867. However, it would appear that he expected the fishery commission to increase fish populations rather than restrict fishing, for he wrote as well that recent improvements in travel in Canada and the Provinces had been accompanied by 'fishing restrictions, license fees, and government interference, which more than counterbalance the advantages.' Sportsmen's urgings for conservation and even restrictions became more widespread by the 1870s (Reiger 1975). Explicit advocacy of C&R fishing soon followed. Schullery (1987) described an article in a 1919 issue of *Outlook* magazine whose author (Harold Pulsifer) explained that barbless hooks allowed him to release unharmed fish he didn't want to keep. Catch-and-release fishing as a desirable general approach entered the popular literature in the late 1930s, notably through the writings of Roderick Haig-Brown and Lee Wulff. Haig-Brown (1939) wrote of an angler's 'killing less than ten fish because he hooks, plays and returns to the water unharmed all fish in excess of that number; after all, his pleasure is not in killing, and sportsmen have used this way out of the difficulty for a great number of years.' Wulff (1939) provided a complete and clear argument for C&R fishing. He pointed out that as a result of environmental degradation and increased fishing pressure, 'the day of the great catches is passing ... There is a growing tendency among anglers to release their fish, returning them to the water in order that they may furnish sport again for a brother angler. Gamefish are too valuable to be caught only once ... [S]eemingly, angling is reaching a new high plane when a fisherman can spend a day on the lake or stream, catching fish and returning them to the water again, unharmed, to come home empty handed. That angler keeps no trophy to show his fellow men as proof of his prowess but contents himself with the pleasures of a day well spent in the surroundings he loves. He has fished for sport and not for glory. Upon him and those who follow his leadership the future of angling depends.' The last sentence of his book returns to the message: 'The fish you release is your gift to another angler and, remember, it may have been someone's similar gift to you.' To Wulff, C&R was more than a practical management tool, it was a point of honour and sportsmanship. Gradually, C&R fishing has increased – especially in inland waters and especially for salmonids – and by the end of the twentieth century it had become an indispensable tool for managing recreational fisheries (e.g. Barnhart and Roelofs 1977, 1987), a mantra for many in the United States (e.g. McIntyre 2000, Talbot 2001), and widely used elsewhere, e.g. in South Africa (Row 1993), Europe (Hickley and Tompkins 1998), and New Zealand (Carbines 1999).

The spread of restrictive fishing regulations, which sometimes implicitly and sometimes explicitly required C&R, changed the way many anglers spent their time and money. Certain prestige or premium destinations – places famous for large or abundant wild fish and attractive surroundings – became more and more popular, especially as there was now a way for people to fish without depleting the fish populations (Barnhart and Roelofs 1977, 1987). As a result, these destinations have become increasingly crowded, mainly by people who have travelled from far away to visit them. This tendency has been amplified by prosperous economies, increasing leisure (Schullery 1987), and increasingly easy and inexpensive means of travelling long distances.

Much of the growth of C&R and crowding of premium destinations are features of fisheries that target salmonids. There is a long history of exalting the status of salmonids over that of other sport fish, but only salmon (*Salmo* and *Oncorhynchus*), trout and char (*Salmo*, *Oncorhynchus*, and *Salvelinus*), and grayling (*Thymallus*), and not North American whitefish and ciscoes, or British powan and vendace (*Coregonus*). In Britain, fish have long been divided into 'sporting fish' – salmon, trout, and grayling – and 'coarse fish' (all others) (Maxwell 1904). In North America, the named division of freshwater sport fish into coarse and game fish is much less common than in Britain but, especially in the northern regions, many anglers' attitudes are similar to those of their British counterparts. In addition to the historical snobbery associated with fishing for salmonids, there is perhaps a practical reason for much of the focus of C&R to be on these fishes. They generally live in cold water, and cold waters tend to be less productive than warmer waters (Noble and Jones 1999). Also, warm- and cool-water streams often are less heavily fished than cold-water ones, largely because they have fewer ramps, streamside paths, and other access points (Noble and Jones 1999, Rabeni and Jacobson 1999), and so C&R is probably more critical in cold-water fisheries than in warm-water ones.

The tendency of premier fishing locales to become crowded has received much attention on electronic fishing bulletin boards, fishing magazines, the literature of fishing, and elsewhere, but it has not received much attention from fishery managers, especially in the United States, until very recently. The essence and nuances of the problem were captured well by Canadian writer Mallory Burton (2000):

> He reached into a vest pocket, extracting two spools. 'I never made it out your way for the steelhead, but I always wanted to. Is the Bulkley as good as they say?' 'Probably. Used to be. I saw a fistfight on the Bulkley two years ago, and I disappeared before they could nab me for a witness. Haven't been back. There are other places to go, not as famous. Too many people on the Bulkley.'

The thoughtful attempts of the British Columbia Ministry of Environment, Lands and Parks to deal with crowding on the Bulkley River are discussed in a later section.

Catch-and-release as a management tool

Catch-and-release is but one of a set of management tools available to fishery managers, and must be viewed in that context. Indeed, C&R regulations would be difficult to implement in the absence of any other regulations, and as mentioned above, often derive from them. Important management tools include licenses and permits, size limits, creel or bag limits, seasonal restrictions and closures, area restrictions and closures, and gear restrictions (Noble and Jones 1999). Those regulations all have the potential to reduce total fishing mortality, although they are of varying effectiveness (Radomski *et al.* 2001, Cook *et al.* 2001); all of them can influence the sizes and kinds of fish that are caught. Every state of the United States and every province in Canada has most of the above regulations, if not all of them. Catch-and-release regulations are widespread in North America but not universal, and those jurisdictions that use them generally do so on a few specific waters or occasionally throughout the jurisdiction or range for a particular species (e.g. seasonal C&R regulations for smallmouth bass and striped bass in Maryland). An example of the latter is the Canadian federal legislation enacted in the 1980s requiring anglers to release all Atlantic salmon larger than 63 cm fork-length (Wilkie *et al.* 1996).

As resource managers and anglers increasingly recognised current and foresaw future pressures on recreational fisheries, restrictions and even explicit C&R increased in popularity. Albert Hazzard in Michigan was one of the main moving forces in bringing C&R to recreational fishery management (Schullery 1987). Today it is a mainstay (Barnhart and Roelofs 1977, 1987), as exemplified by a recent guide book that describes only C&R waters in Maryland (Gelso and Coburn 2000). It also is very much part of many anglers' consciousness as 'the right thing to do.' For example, Slipke *et al.* (1998) attributed a doubling of catch rates of smallmouth bass (*Micropterus dolomieu*) in the Shoal Reach of the Tennessee River in Alabama to voluntary C&R rather than explicit fishing regulations. Finally, it has been called 'the liar's friend,' because 'all trout grow when you let them go' (Hughes 1999).

Catch-and-release has a number of associated management issues, of which hooking mortality is perhaps the most significant (e.g. Clark 1983, Barnhart and Roelofs 1987, Muoneke and Childress 1994, Diodati and Richards 1996, Cox-Rogers *et al.* 1999, Wilde *et al.* 2000), but also including effects on growth, reproduction, and population characteristics. Usually the effects of C&R are compared with consumptive fisheries, and are viewed positively; however, it is important also to consider C&R as an alternative to not fishing at all, and then its effects are usually negative from the point of view of natural resources.

The most obvious negative effect of C&R is hooking mortality. Hooking mortality is the number of deaths that result from hooking, playing, handling, and releasing a fish as a percentage of the total number of fish hooked and released. It usually is measured by keeping captured fish and observing them for various periods and comparing the number of deaths with those observed in a control group of similar fish. (But Schill *et al.* (1986) estimated hooking mortality by counting dead fish visually and estimating their efficiency at finding dead fish, and by using data on the number of fish hooked from surveys.) Most of the mortality observed is so-called short-term mortality, i.e. it occurs within 24 hours, but unfortunately there is little consistency in the literature in the use of the terms short-term or immediate as opposed to delayed mortality. For example, Wilkie *et al.* referred to mortality occurring within 12 hours as 'delayed,' while Nelson (1998) referred to mortality occurring up to 3 days after hooking as 'short-term.' Nonetheless, even when fish are held for up to 5 days, most mortality occurs within the first 24 hours.

Many factors make hooking mortality quite difficult to estimate accurately (Cox-Rogers *et al.* 1999). Some of the factors that affect it are fishing-gear type, angler experience, water temperature, and hook size. High water temperatures appear to be the most consistent and significant factor that increases hooking mortality. In general, hooking mortality is usually reported to be low for freshwater salmonids and centrarchids, often considerably less than 5% (Barnhart and Roelofs 1987). Perhaps the extreme is the report of Schill *et al.* (1986), who estimated that hooking mortality for cutthroat trout (*Oncorhynchus clarki*) in Yellowstone National Park was 0.3% per capture. Total mortality from C&R was 3% because on average, each trout was caught an estimated 9.7 times during a 7-week period in July and August of 1981. In the common snook (*Centropomus undecimalis*), Taylor *et al.* (2001) reported hooking mortality of only 2.1%, even though 90% (421) of the 470 fish were caught on live bait. For largemouth bass in California tournaments, Lee (1987) reported hooking mortalities of 2.3% and 1.7% in 1985 and 1986, respectively. Low hooking mortalities for other species also are reported in other papers in Barnhart and Roelofs (1987).

However, occasionally hooking mortality can be much higher (Muoneke and Childress 1994). Cox-Rogers *et al.* (1999) reported hooking mortalities of up to 30% for young chinook salmon (*Oncorhynchus tshawytscha*), and Muoneke and Childress (1994) and others reported hooking mortalities of up to 33% for channel catfish (*Ictalurus punctatus*), 77% for black crappie (*Pomoxis nigromaculatus*), 88% for bluegill (*Lepomis macrochirus*), 45% for red drum (*Sciaenops ocellatus*), and 56% for spotted seatrout (*Cynoscion nebulosus*). Hooking mortalities often are high for striped bass (*Morone saxatilis*) – up to 67% in fresh water – and strongly correlated with temperature (Wilde *et al.* 2000). Even in salt water, mortality was as high as 26% in a recent study (Diodati and Richards 1996), although overall mortality was much lower (9%) than for many freshwater studies. All the mortality occurred within the first 24 hours, even though all captured fish were held for 48 hours or more. In Atlantic salmon, hooking mortality of fish in water of around 22°C was 40% as compared to zero at 6°C (Wilkie *et al.* 1996).

Despite the wide range of hooking mortalities reported, the general conclusion – borne out by experience – is that C&R causes low enough hooking mortality to be an effective management tool in most circumstances (e.g. Barnhart and Roelofs 1977, 1987, Griffith 1999, Noble and Jones 1999). It provides the basis for many fishing tournaments in North America and Britain, some of which are very lucrative (Hughes 1999; see Lyons *et al.* 2002 for more details – eds). The studies, and related ones on the effects of extreme exercise (e.g. Wilkie *et al.* 1997), also provide guidance on when C&R might be less effective, as for example at high water temperatures for striped bass (Wilde *et al.* 2000) or for Atlantic salmon (Wilkie *et al.* 1996, 1997).

In addition to hooking mortality, C&R can have other effects on fishes. For example, Lewynsky and Bjornn (1987) showed that C&R affected the social structure of cutthroat trout (*Oncorhynchus clarki*) populations by causing the dominant fish to be replaced in its preferred site in a pool. Sublethal effects on reproduction and growth also have been reported (Muoneke and Childress 1994) as well as on physiology (e.g. Wilkie *et al.* 1997). Finally, fish in C&R waters learn to avoid fishing gear and otherwise change their behaviour (Lewynsky and Bjornn 1987); this is widely recognised by anglers, who often refer to heavily fished areas by names such as 'the PhD Pool.' Interestingly, this learning is adaptive but not evolutionary. Only if there is significant and selective fishing mortality will evolutionary (i.e. genetic) changes in response to fishing occur (Policansky 1993a, 1993b). In general, these effects are not sufficient to counteract the effectiveness of C&R as a management tool, although there probably are some circumstances under which it might be less effective, especially high water temperatures.

Two regulations often associated with C&R are the requirements to use flies only and to use barbless hooks. Catch-and-release regulations are complicated by the opinion of many anglers – often not supported by research – that the use of artificial flies and especially barbless hooks produces lower hooking mortality than other methods (Schullery 1987, Schill and Scarpella 1997). These views are further influenced by long-held and passionate opinions about the intrinsic superiority of fly fishing over other fishing techniques (Schullery 1987).

Roosevelt (1884), for example, wrote, 'There is but one mode of taking [trout] – namely, with the fly; although it is said that poachers and pot hunters capture them with worms, minnows, nets, and even with their own roe. These villainies are not at present punished with death nor even imprisonment for life; but our legislature is looking into the matter, and there is no telling how soon such statutes may be passed.' Although few if any fly anglers use such words today, even when their tongues are in their cheeks, the basic sentiment of fly fishing's intrinsic superiority remains commonplace. As a result, it is difficult to know exactly why the many special-regulation waters – in which fly fishing only is permitted and with various degrees, often total, of C&R – were established. While bait fishing sometimes makes it impossible to release a fish alive, sometimes it and other techniques, even those involving treble hooks, do not, and yet fly fishing is most often associated with C&R, especially for salmonids. The same is true of requirements for barbless hooks, despite the general lack of scientific support for their effectiveness in reducing hooking mortality (e.g. Schill and Scarpella 1997). (Barbless hooks do, however, reduce the severity of injuries that result from accidental hooking of anglers, which should be sufficient reason to use them.)

The association of C&R with flies-only and barbless-hook regulations will continue to challenge those who study the motivations and rewards of C&R as well as its effectiveness, but clearly those hook regulations arise from issues that are largely social rather than biological (Schill and Scarpella 1997, Radomski *et al.* 2001).

Despite the physiological effects and mortalities caused by C&R, it is generally accepted as an important, even critical, aspect of modern recreational fishery management (Clark 1983, Barnhart and Roelofs 1987, Muoneke and Childress 1994). Various studies, that of Schill *et al.* (1986) in particular, suggest that the population effect of C&R fishing can be low (but see below for striped bass and Atlantic salmon). Many state fish and game agencies have pamphlets or web pages that advocate C&R. Many states have explicit total C&R regulations for some waters. In most cases, the mortality is low enough that no-take C&R allows for high fishing effort without the need for enhancement (stocking). Even when the cumulative mortality of multiple captures is taken into account (e.g. Clark 1983, Schill *et al.* 1986, Barnhart and Roelofs 1987), total C&R usually permits much higher fishing effort than all other regulatory regimes except for those in which size limits are so high and bag limits so low that they approach total C&R. Indeed, in one case the 'recreational fishery of the Ancien Canal Charleroi-Bruxelles [in Belgium] leads to obvious under-exploitation of the roach stocks because the main angling activity is [voluntary] catch-and-release,' but predatory fish were not released (Gerard 1998). However, for striped bass in estuaries and along the US east coast, C&R has been questioned because perhaps 2.5–3 million fish are caught and released each year (Diodati and Richards 1996). By some estimates, this could reflect a higher fishing mortality than that caused by the landings of commercially caught striped bass. Diodati and Richards estimated that the overall mortality rate from C&R was 9%, which would mean that 234 000 fish were killed out of the estimated 2.6 million caught and released by recreational anglers per year between 1989 and 1992. Given the considerable uncertainties in the estimates of landings of recreationally caught fish, let alone those released, it seems that Diodati and Richards were right to recommend the development of models and data for estimating the effects of recreational angling on species such as striped bass. Maine closed the 2000 angling season for Atlantic salmon year-round (including C&R) because wild salmon populations are so depleted.

Attributes of the resource that affect management

In traditional consumptive fisheries, the resource being managed is the fish. Fish usually are publicly owned or not owned at all (Bader 1998). The degree of definition of property-rights has long been recognised a basic problem in understanding the exploitation of a common-pool resource that fishing is. Although Graham (1935) and Gordon (1954) are usually credited with establishing the economic theory of fishing as a common-pool resource issue, Gilbert and O'Malley (1920) clearly understood and described the problem. They made clear that the problems of protecting the salmon resource could not be understood and managed without considering basic economics. As long as there was open access to the fishery and competition for every fish – the classic common-pool resource problem – no individual could accomplish anything towards conservation. They wrote 'Whatever the individual [salmon] canner spared for spawning purposes, his competitors would thankfully accept and place in cans. He is powerless to conserve either the public interest, or even – should he be sufficiently enlightened to see it – his own private interest, by listening to counsel of moderation.' This has been the basic problem of consumptive fisheries, both recreational and commercial, when fishing pressure is enough to reduce the fish populations.

The simplest way that fishery resources have been protected from overexploitation has been to limit the total catch allowed. This is the basis for the management of many commercial and recreational fisheries. However, simply limiting the total catch, while it may well protect the resource, often results in social disruption and economic waste. In general, protecting common-pool resources from overexploitation in an economically and socially rational way requires finding a way to limit access to the fishery (i.e. exclude many or even most potential users) and creating incentives for users to invest in the fishery instead of overexploiting it (Ostrom *et al.* 1999). An increasingly widely used, although controversial, method of creating incentives for users to invest in a fishery – particularly commercial fisheries – is the establishment of individual fishing quotas or ITQs (see NRC 1999b for an extensive discussion).

However, the recent advent of C&R fishing provides a major difference between recreational and commercial fisheries. In addition, many species of recreational importance, especially inland species, are maintained by extensive stocking programs. Thus protecting the resource from overexploitation is not always the central issue that it is in commercial fisheries. With recreational fisheries – especially C&R fisheries – the limiting resources are access to the fishery and space to fish. Interestingly, this common-pool resource problem has seldom been considered by recreational fishery managers in the United States until very recently. As an example, in a recent survey that asked the heads of freshwater fisheries-management agencies to list important management issues, space and angler crowding were not mentioned (Ross and Loomis 1999). Yet crowding is clearly a problem, and it is recognised sometimes. Merritt and Criddle (1993) studied the very crowded and sometime contentious chinook salmon fishery on Alaska's Kenai River, and they did record crowding as a significant problem. On North Carolina's Outer Banks, the beaches are crowded with off-road vehicles in the fall. In some places there are three rows of vehicles for hundreds of yards along the beach. Fishing magazines and electronic bulletin boards, especially those devoted to fly fishing, have frequent discussions about the overcrowding of popular or publicised waters, which are made even more popular by C&R. If the total amount of recreational angling (number of anglers multiplied by mean time spent fishing) continues to increase, and if accessible fishing habitat continues to decrease, the crowding problem will become increasingly serious. Thus I consider it in some detail here.

Excluding everyone from a fishery is possible, although poaching remains problematic, but excluding only some people while allowing others to fish is more difficult (Bader 1998, Policansky 2001). The two major elements of solving commons problems have long been used by private individuals or groups for angling. They band together and form an angling club to buy the land containing the fishable water. This allows them to limit access and it vests them in the property, so they have an interest in protecting (investing in) the resource.

How can a public management agency achieve this? In the United States, there are many difficulties in limiting access to publicly owned land and causing people to invest in it. I briefly review some of the difficulties, and then discuss Montana's and British Columbia's approaches to the problem. My aim is to provide a basis for thoughtful consideration of methods of addressing angler overcrowding (and the resultant adverse environmental effects), and to help relate them to C&R angling, not to recommend adoption of the Canadian or any other particular plan.

The basis in US law for a state to exclude only the residents of other states from access to its fishing waters is controversial, although it clearly is permissible to charge nonresidents more for the access (Bader 1998). The Privileges and Immunities clause of the United States Constitution (Article 4 section two) says 'The Citizens of each State shall be entitled to all Privileges and Immunities of Citizens in the several States.' The 14th Amendment re-emphasises that by guaranteeing that the privileges and immunities granted by any state apply to the citizens of all states. The constitutionality of different treatment of in-state and out-of-state fishers has been considered by the courts. In Alaska (*Carlson* v *Alaska*, 798 P. 2d 1269 (1990)), the court said that '[C]ommercial fishing is certainly a sufficiently important economic activity to be protected through the Privileges and Immunities Clause, *although sportfishing and hunting are not*.' [Emphasis added.] As described below, when one includes guides and outfitters, sportfishing becomes a significant industry. Combined with individual recreational fishing, it far outweighs commercial fishing in economic terms in many areas. In some places, recreational fishing almost surely is the largest contributor to the local economy.

In the United States, direct limitation of access to public angling waters is very uncommon. Access to some national parks with good angling is limited by limiting the number of cars that can enter, and some public rivers in the United States have limits on the numbers of boats or rafts that can be on the river at any one time. Permits are required for some wilderness areas, and they are limited in number. Limits on entry for recreational angling to public waters to maintain or improve the quality of the angling itself is very uncommon. For example, Colorado's 1999 fishing regulations explicitly provided for limiting access of anglers to some waters, but only when there is damage to the riparian environment caused by too many people, or in case of dangerous water conditions.

A few exceptions involve streams in the south-eastern United States. Some streams in Stone Mountain Park in North Carolina have a 'beat' system: permits are issued on a daily basis until a limit is reached, and thereafter no additional anglers are allowed on the streams for the specified period (North Carolina Department of Parks and Recreation, no date). Dukes Creek and its tributaries within the state-owned Smithgall Woods-Dukes Creek Conservation

Area in Georgia are open to fishing year-round under a permit system according to Georgia's 1999 Fishing Regulations, and some public waters on Long Island, New York also have limited entry (Bergen 2000). A limited number of permits can be obtained by advanced registration. Recently, in its 2000 session, Montana's legislature considered a bill (HB 228) that would have restricted out-of-state anglers on the Beaverhead and Big Hole rivers by requiring special nonresident licenses for those rivers. The licenses would have been limited and available through a lottery system. Although the bill did not pass, the Montana Fish, Wildlife, and Parks Commission decided on a plan to limit commercially guided float trips and out-of-state anglers in boats to certain days and specific sections of the Beaverhead and Big Hole rivers, reserving stretches of each river for use by Montanans on weekends (Anez 2001). The regulations will be in effect for 2 years, during which time fish and game officials will develop permanent management plans for the rivers and study the overall problem of overcrowded rivers in Montana (Anez 2001.) The plan has generated some controversy and is likely to receive attention from other state agencies. With luck, people will take advantage of the opportunity for a thoughtful study of the overcrowding problem, and how to balance that problem against the economic and political benefits (and disbenefits) of angling tourism in Montana.

The well-developed British Columbia Bulkley River Angler Use Plan (BRAUP) is instructive (Policansky 2001). The plan developed from a report by the Fisheries Branch of the BC Ministry of Environment, Lands and Parks (MELP) (MELP 1996) that concluded that 'Angler crowding on rivers is the greatest concern of anglers and angler groups. It is most common on the premier steelhead rivers because the choice places to fish are limited. Thus, it became necessary to classify waters to maintain a measure of quality angling and to reduce crowding.' The BRAUP is explicitly intended to reduce angler crowding and to enhance the quality of the angling experience. Without C&R, which is common for steelhead fishing in the region (personal communication, Bob Hooton, BC Ministry of Environment, Lands and Parks, Vancouver, February 2001), angler crowding would have been less severe, because the fish would have been exploited more and less available to anglers.

The plan recommends that the Bulkley River be managed to meet four objectives: (1) to provide a quality angling experience for all anglers, especially BC residents; (2) to protect the opportunities for BC residents to participate in the fishery; (3) to provide for a stable and economically sound guiding industry; and (4) to contribute to the economic diversification of local communities, including Houston, Telkwa, Smithers, Moricetown, and Hazelton, BC, by enhancing tourism and business opportunities in those communities. This remarkably candid and clear statement of objectives by itself distinguishes the BRAUP from many other planning documents. It recommends management of the river such that the total number of rod-days (i.e. one angler fishing for one day) be limited to 10 500. Of those, 1504 should be guided and 8996 unguided. The number of guides is limited to seven, and there is no restriction on who their clients may be. The plan recommends that 7056 (68%) of the nonguided days be allocated to BC residents, 540 (6%) to residents of other parts of Canada, and 1400 (26%) to residents of other countries. The plan further recommends that only Canadians be allowed to fish the Bulkley River on weekends during the peak fishing season (September 1 – October 31) unless they are guided. Furthermore, the plan provides a working definition of a quality angling experience: 'An experience whereby the angler has a reasonable expectation of catching a fish during the course of an angling day plus a reasonable opportunity during the course of that day to arrive at a section of water which is not already occupied by another angler' (MELP 1998). That definition provides an objective measure for the success of the plan, even if perhaps it is not quantitative.

Despite the thoughtfulness and candour of the plan, and despite its explicit attention to the space problem in recreational fishing, it raises interesting questions. How is this plan to be achieved? The plan acknowledges that 'the best way to ensure compliance with the proposed allocation of rod-days is to implement a limited entry system for all classes of anglers (including both residents and non-residents of BC).' However, there is no mechanism currently available to limit entry (Reid White, Skeena Region Fish and Wildlife Manager, pers. comm., 1999).

Another question concerns the very high rod limit in the plan described above (10 500), which exceeds the average annual number of rod-days over the past 30 years (7602) by 38% and exceeds the 1997 number of rod-days (3983) by 163%. Those limits are so high that the problem of angler overcrowding that was the basis for the plan will not be fully solved by the plan, because its proposals do not become active until the overcrowding becomes much worse than it currently is. Perhaps this is due to a common problem associated with stakeholder groups, namely the difficulty they can have in arriving at difficult solutions. Another possibility is that guided trips, with mainly nonresident clients, are perceived by residents as contributing disproportionately to crowding. Since the number of guided rod-days is fixed, the higher the total number of rod-days allowed, the smaller the proportion of guided rod-days will appear to be, no matter how many people actually fish the river. It would be advantageous to any group of users to minimise their apparent contribution to the crowding problem. But perhaps a different approach will be taken. British Columbia Fisheries has recently been considering managing crowding through 'smart licenses' and other methods that might eliminate reliance on rod-days (BC Interagency Review Team 2000).

The Dean River in British Columbia is also managed under the broad auspices of British Columbia's angling guide policies. Canadian residents with a valid fishing license can fish it during the season, and the limited number of guides can fill their rod allocations with whoever pays them to fish. Non-Canadians must apply through a lottery system for access to fish the river, and fewer permits are granted than the number of applicants. Thus the river has a limited-entry system for non-Canadians. This is in accordance with the objective in the Bulkley River plan of 'adopting strategies which ensure that BC residents are the last group of anglers to be restricted in the event that restrictions become necessary in order to meet the objective of preserving a quality angling experience on the Bulkley River' (MELP 1998). This program appears to be meeting its objectives (Leggett 2000).

Additional questions arise in considering these plans as models for use elsewhere. It is clear that in the United States there are political and perhaps economic consequences of instituting such a plan, although Montana's experience is too new to draw any conclusions from it. It is at least possible that there will be lawsuits on whether it is legally permissible to exclude or even limit nonresidents of Montana from fishing there. Even Canada has not implemented a system that actually discriminates against nonresidents of a province. And it is not clear how willingly US anglers would accept being excluded from fishing that is supported by their own taxes, and so it is not clear how successful or widespread limited-entry schemes for recreational angling might become (e.g. Bergen 2000, Berry 2000), and whether reciprocal exclusions are likely to occur.

Finally, it essential to understand the role that guides, outfitters, and charter-captains play in recreational fishing if its economic and social aspects are to be understood. Even in Wells's day, well-to-do anglers made use of 'professionals' who located fish for them, rowed their boats, advised them, and obtained bait. Many modern recreational fisheries have a very heavy involvement of guides, in particular those fisheries that require equipment and specialised skills. Two examples are recreational fisheries for steelhead (*Oncorhynchus mykiss*) in British Columbia and for bonefish (*Albula vulpes*) in Florida. Those are almost entirely C&R fisheries in which success is made much more likely if one has expert knowledge and a boat. As a result, nonresidents of the areas are unlikely to be successful unless they hire guides. In those areas, guide and outfitter associations have considerable economic and political influence; in some places in Canada, it is required that nonresidents hire guides in order to obtain fishing permits. In Maryland's Chesapeake Bay, a specified portion of the allowable catch of striped bass is allocated to charter boats, separate from the recreational and commercial allotments. Guiding, chartering, and outfitting represent a well-capitalised and influential industry. That industry, which provides a valued service, influences regulations, politics, and angler attitudes. It needs to be taken into account in any analyses of recreational fishing.

Future trends

It is risky to extrapolate from the present to guess at future trends, especially when recreational fishing involves so many social, economic, political, ethical, and cultural factors. Furthermore, those factors are changeable and flexible. For example, it seems clear that an angler's expectations of a place can affect her or his perception of the quality of the experience (e.g. Ross and Loomis 2001). A ten-fish day might be very disappointing for a bass angler in a stocked lake; a two-fish day might represent the fishing-trip of a lifetime for a salmon angler in Nova Scotia or a bonefish angler in the Florida Keys. Even a zero-fish day could be rated a success if the scenery, food, wildlife viewing, and company were excellent.

Yet one fairly recent development seems worth mentioning, and that is the idea that C&R fishing represents an inappropriate use of or unacceptable cruelty to fish for no utilitarian purpose (Fig. 6.1; see Aas *et al.* 2002 for a discussion of European views). At least some Native American and First Nation peoples in North America consider that sport angling is disrespectful to the animal if it is not used for food (P. Cochran, Alaska Native Science Commission, Anchorage, Alaska, personal communication, 2001; Yukon Renewable Resources 1998). The idea that fishing is cruel has a long history (Schullery 1987). Some writers mention it oddly, almost as if it did not apply to them. Thus Wulff (1939) wrote that anglers who play their fish too long are showing 'insensibility to the sufferings of their quarry.' He even devoted most of a page to describing what it would feel like for a person to be hooked by some sharp object attached to an invisible but ineluctable force, dragging the person toward some unidentified doom. Yet he did not explicitly write about the cruelty of angling as an ethical issue. Roosevelt (1884) wrote of fly fishing for brook trout: 'How splendid is the sport, to deftly throw the long line and small fly with the pliant single-handed rod, and with eye and nerve on the strain, to watch the loveliest darling of the wave, the spotted naiad, dart from her mossy bed, leap high into the air, carrying the strange deception in her mouth, and turning in her flight, plunge back to her crystal home, with the cruel hook driven into her lips by a

Fig. 6.1 DILBERT reprinted by permission of United Features Syndicate, Inc.

skilful turn of the angler's wrist ...'. Roosevelt reported that he repeated this cruelty 70 times ('I caught seventy trout') during one three-hour period, but the cruelty did not dissuade him from fishing, either.

It has dissuaded others. Luce (1959) clearly expressed the argument that C&R fishing is cruel:

> 'To hook trout and put them back into the water, unless they are too small to keep and quite uninjured, is to inflict pain, however small the amount, unnecessarily, and it therefore comes under the definition of cruelty. It is a mild form of cruelty; but it is cruelty; it involves the infliction of pain without the hunter's justification for doing so ... It may look paradoxical at first sight to hold that it is cruel to hook and release trout, and not cruel to hook and kill them; but such is the case. Cruelty is largely in the mind and motive, and it is just here that the primary object of angling become important for both theory and practice. *The primary objective of justifiable angling is to catch fish for food; there are various pleasures incidental to angling; but they cannot justify the infliction of pain or death.*' [Emphasis in original.]

For Luce, catching fish for food was justifiable and not cruel. De Leeuw (1996) took the argument further. He argued – questionably, in my opinion – that fish caught by recreational fishing suffer more than those caught by commercial fishing, and so even recreational fishing with aim of keeping the fish for food is unjustifiably cruel. In Germany, the Animal Protection Act (*Tierschutzgesetz*) makes it illegal to harm an animal for an inappropriate (nonutilitarian) use, which effectively makes voluntary C&R fishing illegal (Berg and Rösch 1998). Kerasote (1997), McIntyre (2000), and Talbot (2001) have discussed C&R recently in terms of its being wantonly cruel, and Talbot advocated a 'new' fishing ethic of catching only enough fish to eat and then stopping. Given the ease for most people in North America and Europe of obtaining food without fishing, the distinction between fishing for food and for sport in the twenty-first century is not entirely clear, except that some people who fish for food stop fishing when they have enough fish to eat. Catch-and-release anglers, probably believing that they are not affecting the fish population adversely, often catch and release dozens of fish or even more. In any event, the issue is unlikely go away soon. Some animal-rights organisations, notably People for the Ethical Treatment of Animals (PETA), campaign against fishing and hunting.

Perhaps the most thoughtful discussion of this and other aspects of fishing is Hersey's (1987). In a conversation between a fisher (F) and a stranger (S), he writes:

> 'S: But I must remind you – forgive me if I step on your toes – I have to remind you of my real distaste for what I think of as the fishing mystique: all that notion of the elegance and nobility of a blood sport. F: Try to remember that we are going after food – that we are, in a way, exploring our place in the systems of life in the universe. I grant you that our place, when we think we've found it, isn't always comfortable. Sometimes it's awful, but often it's also awesome and very beautiful. It's what we have and must live with, at any rate.'

Conclusions

Like commercial fishing, recreational fishing is an important activity in North America and elsewhere, involving millions of people and billions of dollars. Although it has many features in common with commercial fishing, changes in people's economic status, lifestyles, and their perceptions of the environment have led to significant changes. Perhaps the largest of these is the increase in C&R fishing. Its advent has allowed more anglers to fish harder without the need for efforts to enhance the fish populations. It has resulted in differing motivations, rewards, and economic benefits and disbenefits. It also appears to have led to angler crowding in some premier destinations, and as a result, more jurisdictions are considering or even implementing restrictive angling regimes. A few are limiting nonresidents more than they limit residents.

Understanding recreational fishing also requires an understanding of the guiding, chartering, and outfitting industry. It requires a recognition that angler attitudes and motivations continue to evolve and comprise many factors, most not requiring the retention of fish (Aas and Ditton 1998, Cook *et al.* 2001, Radomski *et al.* 2001, Calvert this volume). As Hersey's (1987) fisher said, 'Fishing is complicated.'

Acknowledgements

I thank Paul Schullery for his thoughtful and elegant writings about fly fishing and for helping me decipher history's code. The Virtual Flyshop electronic bulletin board (www.flyshop.com) is a treasure trove of valuable information and helpful contributors. Although there are too many to acknowledge them all, Jeff Serena, John Mundinger, Hermann Schibli, and Preston Singletary in particular were helpful with information. Bob Hooton and Art Tautz of the British Columbia government were enormously helpful in providing documents and insight related to that province's angling plans. Terry Roelofs provided valuable literature citations. Chip Rizzotto of Montana's Paradise Valley gave me information, insight, and encouragement as I tried to understand Montana's political process. The thoughtful comments of an anonymous reviewer and the editors of this volume were enormously helpful.

References

Aas, Ø., Thailing, C. & Ditton, R.B. (2002) Controversy over catch-and-release recreational fishing in Europe. In: *Recreational Fisheries: Ecological, Economic and Social Evaluation* (eds Pitcher, T.J. & Hollingworth, C.E.), Chapter 7, this volume. Blackwell Science, Oxford, UK.

Aas, Ø. & Ditton, R. (1998) Human dimensions perspective on recreational fisheries management: implications for Europe. In: *Recreational Fisheries: Social, Economic and Management Aspects* (eds Hickley, P. & Tompkins, H.) Chapter 22 pp. 153–164. Fishing News Books, Oxford, UK.

Anez, B. (2001) To Answer Overcrowding Problem, FWP Will Limit Use of Beaverhead, Big Hole Rivers. *The Missoulian*, 2/10/2001.

Bader, H. (1998) Who Has the Legal Right to Fish? Constitutional and Common Law in Alaska Fisheries Management. *Marine Advisory Bulletin No. 49*. University of Alaska Sea Grant College Program, Fairbanks.

Barnhart, R. & Roelofs, T. (eds) (1977) *Catch-and-Release Fishing As a Management Tool. A National Sport Fishing Symposium*. California Cooperative Fishery Research Unit, Humboldt State University, Arcata, CA.

Barnhart, R. & Roelofs, T. (eds) (1987) *Catch-and-Release Fishing: A Decade of Experience. A National Sport Fishing Symposium*. California Cooperative Fishery Research Unit, Humboldt State University, Arcata, CA.

B.C. (British Columbia) Interagency Review Team. (2000) *Discussion paper: Sustaining A Quality Angling Experience in BC: Summary of Findings and Potential Directions for the Future*. BC Fisheries, Victoria, BC.

Berg, R. & Rösch, R. (1998) Animal welfare and angling in Baden-Württemburg, Germany. In: *Recreational Fisheries: Social, Economic and Management Aspects* (eds Hickley, P. & Tompkins, H.), Chapter 14 pp. 88–92. Fishing News Books, Oxford, UK.

Bergen, G. (2000) Limited entry to public trout water: The pros and cons. In: *Wild Trout VII: Management in the New Millennium: Are We Ready?* (eds Schill, D., Moore, S., Byorth, P. & Hamre, B.), pp. 213–214. Symposium held at Old Faithful Inn, Yellowstone National Park, October 1–4, 2000. National Park Service (and others), Yellowstone National Park, Wyoming.

Berners, Dame Juliana (1496) *A Treatyse of Fysshinge With an Angle*. Wynkyn De Worde, St. Albans, England. (Facsimile reprint 1880, Elliott Stock, London.)

Berry, D.K. (2000) Limited entry – no demand way! In: *Wild Trout VII: Management in the New Millennium: Are We Ready?* (eds Schill, D. Moore, S. Byorth, P. & Hamre, B.), pp. 211–212. Symposium held at Old Faithful Inn, Yellowstone National Park, October 1–4, 2000. National Park Service (and others), Yellowstone National Park, Wyoming.

Blades, W. (1881) Introduction to facsimile reprint of Berners, Dame Juliana [1486]. *The Boke of St. Albans*. Elliott Stock, London.

Burton, M. (2000) *Green River Virgins and Other Passionate Anglers*. The Lyons Press, New York.

Calvert, B. (2002) The Importance of Angler Motivations in Sport Fishery Management. In: *Recreational Fisheries: Ecological, Economic and Social Evaluation* (eds Pitcher, T.J. & Hollingworth, C.E.), Chapter 9, this volume. Blackwell Science, Oxford, UK.

Carbines, G.D. (1999) Large hooks reduce catch-and-release mortality of blue cod (*Parapercis colias*) in New Zealand. *North American Journal of Fisheries Management*, **19,** 992–998.

Cholmondeley-Pennell, H. (1893) *Fishing*. 5th edn, Badminton Library of Sports and Pastimes. Longmans Green & Co, London.

Clark, R.D., Jr. (1983) Potential effects of voluntary catch-and-release of fish on recreational fisheries. *North American Journal of Fisheries Management*, **3**, 306–313.

Cook, M.F., Goeman, T.J., Radomski, P.J., Younk, J.A. & Jacobson, P.C. (2001) Creel limits in Minnesota: A proposal for change. *Fisheries*, **26**(5), 19–26.

Cox-Rogers, S., Gjernes, T. & Fast, E. (1999) A Review of Hooking Mortality Rates for Marine Recreational Coho and Chinook Salmon Fisheries in British Columbia. *Canadian Stock Assessment Secretariat Research Document 99/127*. Ottawa, Ontario.

De Leeuw, A.D. (1996) Contemplating the interests of fish: The angler's challenge. *Environmental Ethics*, **18**, 373–390.

Diodati, P. & Richards, R.A. (1996) Mortality of striped bass hooked and released in salt water. *Transactions of the American Fisheries Society*, **125,** 300–307.

Fedler, A.J. & Ditton, R.B. (1994) Understanding angler motivations in fisheries management. *Fisheries*, **19**(4), 6–13.

Gelso, C. & Coburn, L. (2000) *Guide to Maryland Trout Streams: The Catch-and-Release Streams*. K & D Limited, Inc, Woodbine, MD.

Gerard, P. (1998) The impact of angling on the population dynamics of roach (*Rutilus rutilus*) in a former ship canal. In: *Recreational Fisheries: Social, Economic and Management Aspects* (eds Hickley, P. & Tompkins, H.), Chapter 32 pp. 235–244. Fishing News Books, Oxford, UK.

Gilbert, C.H. & O'Malley, H. (1920) Special Investigation of the Salmon Fishery in Central and Western Alaska. In: *Alaska Fisheries and Fur Seal Industries in 1919*, pp. 143–160, US Bureau of Fisheries, Washington, DC.

Gordon, H.S. (1954) The economic theory of common property resources: The fishery. *Journal of Political Economy*, **62**, 124–142.

Graham, M. (1935) Modern theory of exploiting a fishery, and its application to North Sea trawling. *Journal du Conseil*, **13**, 264–274.

Gregg, W. H. (1902) (Assisted by Captain J. Gardner) *Where, When, and How to Catch Fish on the East Coast of Florida*. The Matthews-Northrup Works, Buffalo and New York.

Griffith, J.S. (1999) Coldwater streams. In: *Inland Fisheries Management in North America* (eds Kohler, C.& Hubert, W.), 2nd edn, Chapter 18 pp. 481–504. American Fisheries Society, Bethesda, MD.

Haig-Brown, R. (1939) *The Western Angler*. Derrydale Press, New York.

Herbert, H.W. (1851) *Frank Forester's Fish and Fishing of the United States and British Provinces of North America*. Springer and Townsend, New York.

Hersey, J. (1987) *Blues*. Alfred A. Knopf, Inc, New York.

Hickley, P. (1998) Comments concerning a code of conduct of good pratice for recreational fishing. In: *Recreational Fisheries: Social, Economic, and Management Aspects* (eds Hickley, P.& Tompkins, H.), Chapter 38 pp. 299–304. Fishing News Books (Blackwell Science Ltd.). Oxford, UK.

Hickley, P. & Tompkins, H. (eds) (1998) *Recreational Fisheries: Social, Economic, and Management Aspects*. Fishing News Books (Blackwell Science Ltd.). Oxford, UK.

Hughes, R. (1999) *A Jerk on One End: Reflections of a Mediocre Fisher*. Ballantine, New York.

Kerasote, T. (1997) Catch and Deny. *Orion* Winter 1997, 24–27.

Kron, T. (1995) Prince William Sound salmon enhancement programs and considerations relative to wild stocks. In: *Interactions Between Cultured Species and Naturally Occurring Species in the Environment: Proceedings of the Twenty-Second U.S.-Japan Aquaculture Panel Symposium*, Homer, Alaska, August 21–22, 1993 University of Alaska, Fairbanks. Alaska Sea Grant College Program Report AK-SG-95–03. (eds Collie, M.R. & McVey, J.P.), pp. 49–53.

Lampman, B.-H. (1946) *The Coming of the Pond Fishes*. Binfords and Mort, Portland, Oregon.

Lee, D.P. (1987) Mortality of tournament caught and released black bass in California. In: *Catch-And-Release Fishing: A Decade of Experience. A National Sport Fishing Symposium.* California Cooperative Fishery Research Unit (eds Barnhart, R. & Roelofs, T.), pp. 207–215. Humboldt State University, Arcata, CA.

Leggett, J. (2000) British Columbia's Dean River – limited entry: A management tool for maintaining a quality angling experience. In: *Wild Trout VII: Management in the New Millennium: Are We Ready?* Symposium held at Old Faithful Inn, Yellowstone National Park, October 1–4, 2000. (eds Schill, D., Moore, S., Byorth, P. & Hamre, B.), pp. 205–207. National Park Service (and others), Yellowstone National Park, Wyoming.

Lewynsky, V.A. & Bjornn, T.C. (1987) Response of cutthroat and rainbow trout to experimental catch-and-release fishing. In: *Catch-and-Release Fishing: A Decade of Experience. A National Sport Fishing Symposium.* California Cooperative Fishery Research Unit (eds Barnhart, R. & Roelofs, T.), pp. 16–32 . Humboldt State University, Arcata, CA.

Luce, A.A. (1959) *Fishing and Thinking.* Hodder and Stoughton, London.

Lyons, J., Hickley, P. & Gledhill, S. (2002) An Evaluation of Recreational Fishing in England and Wales. In: *Recreational Fisheries: Ecological, Economic and Social Evaluation* (eds Pitcher, T.J. & Hollingworth, C.E.), Chapter 12, this volume. Blackwell Science, Oxford, UK.

Maxwell, Sir Herbert (1904) *British Fresh-Water Fishes.* Hutchinson & Co, London.

McDonald, J. (assisted by Kuhn, S. &Webster, D., and the editors of *Sports Illustrated*) (1963) *The Origins of Angling and a New Printing of The Treatise of Fishing with an Angle Attributed to Dame Juliana Berners.* Doubleday and Company, Inc, Garden City, NY.

McIntyre, T. (2000) The insufferable sport. *Field and Stream*, March 2000, 32–35.

MELP (British Columbia Ministry of Environment, Lands and Parks) (1996) *Angling Guide Report and Policy Review Implementation: Working Towards Improvement.* British Columbia Ministry of Environment, Lands and Parks, Fisheries Branch, Victoria.

MELP (British Columbia Ministry of Environment, Lands and Parks). (1998) Angling Use Plan Bulkley River: Approval. Skeena Region, Ministry of Environment, Lands and Parks, Smithers, BC. Available at http://www.elp.gov.bc.ca/ske/bulk_aup/bul_aup.html. Last accessed May 2001.

MELP (British Columbia Ministry of Environment, Lands and Parks). (1999) Angling Use Plan Letter to Stakeholders, February 10 1999. Skeena Region, Ministry of Environment, Lands and Parks, Smithers, BC. Available at http://www.elp.gov.bc.ca/ske/bulk_aup/rdletter.html. Last accessed May 2001.

Merritt, M.F. & Criddle, K.R. (1993) Evaluation of the analytic hierarchy process for aiding management decisions in recreational fisheries: A case study of the chinook salmon fishery in the Kenai River, Alaska. In: *Proceedings of the International Symposium on Management Strategies for Exploited Fish Populations.* Alaska Sea Grant College Program Report 93–02, University of Alaska Fairbanks. (eds Kruse, G., Eggers, D.M., Marasco, R.J., Pautzke, C. & Quinn, T.J., II). pp. 683–703.

Muoneke, M.I. & Childress, W.M. (1994) Hooking mortality: A review for recreational fisheries. *Reviews in Fisheries Science*, **2**, 123–156.

National Research Council (NRC) (1996) *Upstream: Salmon and Society in the Pacific Northwest.* National Academy Press, Washington DC.

National Research Council (NRC) (1999a) *Sustaining Marine Fisheries.* National Academy Press, Washington, DC.

National Research Council (NRC) (1999b) *Sharing the Fish: Toward a National Policy on Individual Fishing Quotas.* National Academy Press, Washington, DC.

Nelson, K. (1998) Catch-and-release mortality of striped bass in the Roanoke River, North Carolina. *North American Journal of Fisheries Management*, **18**, 25–30.

Nielsen, L.A. (1999) History of inland fisheries management in North America. In: *Inland Fisheries Management in North America* (eds Kohler, C. & Hubert, W.), 2nd edn, Chapter 1 pp. 3–30. American Fisheries Society, Bethesda, MD.

Noble, R.L. & Jones, W.T. (1999) Managing fisheries with regulations. In: *Inland Fisheries Management in North America* (eds Kohler, C. & Hubert, W.). 2nd edn, Chapter 17 pp. 455–477. American Fisheries Society, Bethesda, MD.

North Carolina Division of Parks & Recreation (no date) *Stone Mountain State Park. Division of Parks and Recreation*, North Carolina Department of Environment and Natural Resources, Durham.

Ostrom, E., Burger, J., Field, C.B., Norgaard, R. & Policansky, D. (1999) Revisiting the commons: Local lessons, global challenges. *Science*, **284**, 278–282.

Policansky, D. (1993a) Fishing as a cause of evolution in fishes. In: *The Exploitation of Evolving Resources*. Lecture Notes in Biomathematics Volume 99 (eds Stokes, T.K., McGlade, J.M. & Law, R.), pp. 2–18. Springer-Verlag, Berlin.

Policansky, D. (1993b) Evolution and management of exploited fish populations. In: *Management Strategies for Exploited Fish Populations*. AK-SG-93–02. (eds Kruse, G., Eggers, D.M., Marasco, R.J., Pautzke, C. & Quinn, T.J.) pp. 651–664. Alaska Sea Grant College Program, Fairbanks, Alaska.

Policansky, D. (2001) Recreational and commercial fisheries. In: *Protecting the Commons: A Framework for Resource Management in the Americas* (eds Burger, J., Ostrom, E., Norgaard, R.B., Policansky, D. & Goldstein, B.), Chapter 7 pp. 161–173. Island Press, Washington DC.

Rabeni, C.F. & Jacobson, R.B. (1999) Warmwater streams. In: *Inland Fisheries Management in North America* (eds Kohler, C. & Hubert, W.), 2nd edn, Chapter 19 pp. 505–528. American Fisheries Society, Bethesda, MD.

Radomski, P.J., Grant, G.C., Jacobson, P.C. & Cook, M.F. (2001) Visions for recreational fishing regulations. *Fisheries*, **26**(5), 7–18.

Reiger, J. (1975) *American Sportsmen and the Origins of Conservation*. Winchester Press, New York.

Roosevelt, R.B. (1884) *The Game Fish of the Northern States and British Provinces*. Orange Judd Company, New York.

Ross, M.R. & Loomis, D.K. (1999) State management of freshwater fisheries resources: Its organizational structure, funding, and programmatic emphases. *Fisheries*, **24**(7), 8–14.

Ross, M.R. & Loomis, D.K. (2001) Put-and-take fisheries: Investigating catch and retention assumptions. *Fisheries*, **26**(2), 13–18.

Row, J. (1993) Tag and release in the tournament environment. In: *Fish, Fishers and Fisheries. Proceedings of the Second South African Marine Linefish Symposium*, Durban, 23–24 October 1992. Oceanographic Research Institute Special Publication No. 2 (eds Beckley, L.E. & van der Elst, R.P.), pp. 191–195. Oceanographic Research Institute, Marine Parade, Durban, South Africa.

Schill, D.J. & Scarpella, R.L. (1997) Barbed hook restrictions in catch-and-release trout fisheries: A social issue. *North American Journal of Fisheries Management*, **17**, 873–881.

Schill, D.J., Griffith, J.S. & Gresswell, R.E. (1986) Hooking mortality of cutthroat trout in a catch-and-release segment of the Yellowstone River, Yellowstone National Park. *North American Journal of Fisheries Management*. **6**, 226–232.

Schullery, P. (1987) *American Fly Fishing: A History*. Lyons and Burford, New York.

Schullery, P. (1999) *Royal Coachman*. Simon and Schuster, New York.

Slipke, J.W., Maceina, M.J., Travnichek, V.H. & Weathers, K.C. (1998) Effects of a 356-mm minimum length limit on the population characteristics and sport fishery of smallmouth bass in the Shoals Reach of the Tennessee River, Alabama. *North American Journal of Fisheries Management*, **18**, 76–84.

Talbot, R. (2001) A conservation ethic that can help preserve our fisheries. *Fly Fisher*, **32** (4), 14–16, 18 (May 2001).

Taylor, R.G., Whittington, J.A. & Haymans, D.E. (2001) Catch-and-release mortality of common snook in Florida. *North American Journal of Fisheries Management*, **21**, 70–75.

Wells, H.P. (1885) *Fly-Rods and Fly-Tackle*. Harper Brothers, New York.

Wheeley, C.H. (1897) *Coarse Fish*. George Routledge and Sons, London.

Wilde, G.R., Muoneke, M.I., Bettoli, P.W., Nelson, K.L. & Hysmith, B.T. (2000) Bait and temperature effects on striped bass hooking mortality. *North American Journal of Fisheries Management*, **20**, 810–815.

Wilkie, M.P., Davidson, K., Brobbel, M.A. *et al.* (1996) Physiology and survival of wild Atlantic salmon following angling in warm summer waters. *Transactions of the American Fisheries Society*, **125**, 572–580.

Wilkie, M.P., Brobbel, M.A., Davidson, K.G., Forsyth, L. & Tufts, B.L. (1997) Influences of temperature on the postexercise physiology of Atlantic salmon (*Salmo salar*). *Canadian Journal of Fisheries and Aquatic Sciences*, **54**, 503–511.

Wright, W. (1858) *Fishes and Fishing. Artificial Breeding of Fish, Anatomy of Their Senses, Their Loves, Passions, and Intellects. With Illustrative Facts*. Thomas Cautley Newby, Publisher. London.

Wulff, L. (1939) *Lee Wulff's Handbook of Freshwater Fishing*. F.A. Stokes, New York.

Yukon Renewable Resources (1998) *Sport Fishing Regulations summary, 1998–1999*. Yukon Renewable Resources, Whitehorse, Yukon Territory, Canada.

Chapter 7
Controversy over Catch-and-Release Recreational Fishing in Europe

Øystein Aas, Carol E. Thailing and Robert B. Ditton

Abstract

This paper discusses various perspectives on catch-and-release in Europe. The discourse on catch-and-release is multifaceted, with much controversy among anglers, biologists and fishery managers. The different views by fishery and nation have implications for all because the European Union (EU) now makes decisions in this field. This paper analyses the debate by identifying multiple perspectives towards catch-and-release. Two dichotomous positions are as follows.

- For some in Europe, catch-and-release is an unethical and reprehensible fishing practice. They see fishing solely as a means of catching fish and when the angler's catch goals or legal limit is attained, there should be no more fishing. These groups may regard catch-and-release as 'playing with fish for no good reason.' There are concerns about animal welfare issues, about catch-related mortality and physiological injury to fish, and whether solely catch-and-release fishing will be allowed in the future.
- For others in Europe, catch-and-release fishing is both an ethical and a conservative approach to resource utilisation. They argue this approach is consistent with the fishery management goal of sustainability in situations where many fish populations are under pressure. They argue that, for sustainability, catch-and-release has to be preferable to catch-and-kill.

The implications of these two opposing views and other views from elsewhere, for instance North America, are discussed. Content analysis techniques (Weber 1990) were used to clarify the discourse in Europe. Case examples of government actions both pro and con are presented.

Key words: catch-and-release, conflict, content analysis, discourse, recreational fishing.

Introduction

The term 'catch-and-release fishing', defined as angling where all fish caught have to be released, can have several meanings for anglers, conservationists, managers, scientists, and politicians. It can mean a harvest regulation, a management strategy, an angling philosophy, and for some specialist anglers, a strict religion. The concept and the angling behaviour it implies have significant implications for several important elements of recreational fishing, and for tourism fishing businesses. Catch-and-release can influence stock characters important to stock recruitment and stock composition such as average size, age, and abundance. This again, influences catch probabilities and fishing quality through number and size of fish caught. Catch-and-release inevitably also implies various angler motivations, attitudes, and outcomes, both psychological and physical. It is well documented that anglers vary in their preferences for catching fish and keeping fish, both in North America and Europe (Fedler and Ditton 1986, Aas and Kaltenborn 1996, Aas *et al.* 2000). Different angler groups seem to differ in their view on catch-and-release regulation (Teisl *et al.* 1996). More specialised anglers (Bryan 1977), fly anglers and conservation organisations with their roots in the angling public seem to be the strongest supporters. The catch-and-release practice is also presented as an ethical issue, both from people who support and oppose it (Bielak 1986, Berg and Rötsch 1998).

The concept has roots on both sides of the Atlantic. The European tradition is largely tied to coarse angling, mostly for cyprinids (Lowerson 1989), while the emergence of catch-and-release in coldwater fisheries is a more recent, primarily North American, idea. Here, it originated in the mid-1950s (Barnhart 1989), and has since spread to European game fisheries for trout and salmon. Scandinavian recreational fishing is historically strongly influenced by British game angling traditions. During the 1970s, the influence of North American angling ideas and trends on Europe increased strongly. Catch-and-release in Europe is consequently now practised in salmonid fishing (game angling), and within the coarse fishing tradition for cyprinids, where the tradition is probably longer (Lowerson 1989) and where many species have limited or no value as food. The growth of catch-and-release in Europe can also be judged as an example of modernisation and globalisation. Through generally improved communication, and more travel and tourism, the world is getting tied closer together. Resources are redefined, and practices spread quickly across continents.

Catch-and-release is to a large degree practised on a voluntary basis (NASCO undated). So far, catch-and-release is very limited in European saltwater angling. However, the increased growth in tourism angling in areas with limited resources, and where commercial fisheries are under a strict quota regime, makes the issue of angling quotas and catch-and-release regulations more relevant than only a few years ago.

The rapid growth in catch-and-release among European anglers has led to a loud and contentious debate or discourse pro and con catch-and-release. Participants have been different angler groups and organisations, management agencies, scientists, conservation groups of different kinds, politicians and government representatives, as well as international organisations. Among professional groups, veterinarians have played a significant role in addition to fish biologists. The debate includes all the aspects mentioned above: biological effects on fish stocks, hooking mortality issues, physiological stress, angling behaviour, angling motivation and management issues regarding compliance/noncompliance. In addition, the issue of animal welfare has been prominent (i.e. Berg and Rötsch 1998, Hickley 1998).

It is here imperative to underline that there are large differences within Europe regarding fish resources, management challenges, angling participation, angling practices, legislation and traditions that naturally influence the relevance of and direction of the discussion about catch-and-release between regions and countries. For instance, in northern countries recreational fishing includes activities with a strong harvest focus, similar to subsistence fishing in other countries and regions, even if it takes place in a typical leisure context (Aas and Skurdal 1996). In southern areas, participation in angling is low compared to north European countries, with differences in both species composition and resource challenges.

In this chapter we will take a closer look at the catch-and-release discourse in Europe. Instead of a study of the volume of catch-and-release fishing in Europe, angler attitudes towards it, and under what circumstances they fish catch-and-release (a Europe-wide angler survey would be required to achieve this!), we have a perspective from NGOs and government agencies on where they would like anglers to be on this subject. We utilise content analysis to systemise arguments used in available papers discussing catch-and-release from different European countries. Finally, the findings will be discussed in relation to different traditions and challenges in Europe, the situation in North America, and in relation to the development of common regulations within the European Union.

Methods

Content analysis, as described by Weber (1990), was used to identify the underlying meanings attributed to catch-and-release fishing as reported throughout Europe. The sample texts for analysis were drawn from available literature using general current periodicals from the internet-based OVID database as well as OVID Social Science Journals, and the Cambridge Scientific Abstracts. These databases cover not only natural science journals but also social science journals, and therefore the population of written materials sampled included technical and scientific writings by academics, as well as editorial and opinion pieces. Thus, the population sampled was not restricted to any particular group of journals. This was further extended by including European contacts identified by either their personal involvement in or listing on the American Fisheries Society Committee on the Human Dimensions of Recreational Fisheries, the European Inland Fisheries Advisory Commission (EIFAC), and the Food and Agriculture Organization of the United Nations (FAO). The first organisation was chosen to highlight social scientists, while the latter two were used to sample fisheries managers and scientists in the European community as a whole.

There are clear limitations to the sampled literature worth noting. While every effort was made to sample all readily available literature on this subject, there is likely much 'grey literature' in Europe and the United States that was not identified or sampled. Also, because we were limited to literature available in English, documents in other languages were not analysed, with a few exceptions where articles were translated. A future effort whereby personnel in all European nations are solicited for information is needed to yield additional useful information on the subject of catch-and-release fishing. Additionally, while every effort was made to sample the breadth of literature in the time available, a more comprehensive, funded research effort is required to be confident with the relative distribution of themes developed in this exploratory paper. Due to time, language, resource, and funding constraints, this effort relied heavily on literature available in English from several European countries where our sources were located: Finland, Norway, Scotland, Germany, Belgium, the Czech Republic, Northern Ireland, and England.

A broad search was conducted for 'fishing' to include those countries where catch-and-release is and is not practised. Using all of the literature identified, basic content analysis was conducted to identify the perceptions of the activity of fishing in general as well as those toward the practice of catch-and-release. The initial neutral search for the term 'fishing' provides a broad, unbiased sample of the general activity. Modifiers such as 'recreation', 'sport', 'subsistence', 'welfare', and 'catch and release', which attach meanings to the term, were added subsequently to narrow the search. These terms were not chosen randomly, but were subsequently identified as the most emphasised themes and terminology in the literature after the data were gathered (Marshall and Rossman 1999).

To maintain a unified perspective of the text, entire documents were analysed instead of isolated sections. Text was analysed to determine which key words were present and how those words were being used throughout the text (Weber 1990). Values were not attributed based solely on word frequency counts because of the various contexts in which words such as 'recreational', 'fishing', 'angling', etc., were presented as well as the multiple meanings of these words (Weber 1990). Synonyms, words similar in meaning or representing similar ideas, were grouped categorically. For example, words such as 'recreation', 'recreational' and 'leisure' were grouped together as representing a similar idea of enjoyment and relaxation. Additionally, these groupings were also placed into categories of positive, negative, and neutral perceptions of fishing and catch-and-release fishing.

Individual words and phrases were analysed and related words representing similar ideas were grouped into categories. However, terms listed under the same umbrella heading can only represent similar ideas or interests, not necessarily interchangeable words. The themes identified were as follows (Table 7.1): catch-and-release, subsistence, mortality, welfare (for the animal), recreation, resource (fish as a resource), management, regulatory, licence/permit, requirements (in relation to other categories), sustainability/conservation, sport/game, respect, keeping (in reference to only keeping the fish), voluntary (in relation to other categories), skills/training, individual (characteristics of the individual fish).

In addition to the limitations of the sample, there are also limitations to this type of content analysis. First, because one word may be used differently even within the same text and might have multiple meanings, extracting only single words may create a false unity (Weber 1990). Therefore, the context of the word usage was important. Secondly, the amount of inferential reasoning used in this type of analysis entails a great deal of interpretation by the researcher (Marshall and Rossman 1999). Coinciding with this level of interpretation, synonyms need to be recognised as only similar words, not exactly interchangeable words. The fact that the methods involved human analysis and not a computer program did not hinder this effort how-

Table 7.1 Prominent themes identified in the reviewed texts.

Catch-and-Release	Opposition
Subsistence	Support
Mortality (and survival)	Requirements
Welfare	Skills/Training
Recreation	Sustainability/Conservation
Ethics	Sport/Game
Resource	Individual (fish as an)
Management	Keeping (not releasing)
Regulation	Voluntary
License	Respect

ever, because the same problems are typically encountered with content analysis computer programs (Weber 1990).

Results

Certain keywords were identified as indicating relevant themes throughout the reviewed literature. Themes related to fishing as a 'subsistence' activity were found in literature from Finland, Norway, and Germany. However, terminology related to catch-and-release fishing as an activity (such as 'catch and release,' 'releasing,' and 'returning') was found in the literature from Scotland, Belgium, Northern Ireland, and England as well as Finland, Norway, and Germany. Therefore, isolation of any one term or theme does not provide an accurate picture of the activity within individual countries; themes do not appear to be mutually exclusive. It is the combination of terminology and themes that help provide a clearer understanding of the meanings attached to catch-and-release fishing in Europe.

Certain thematic categories emerged within individual countries. An overview is presented in Table 7.2. The prevalence of words relating to individual fish 'mortality' (not to be confused with the category of 'sustainability/conservation' of fish populations) was evident in the reviewed literature. Although it might be inferred initially that mortality coincides with 'welfare' (the welfare of individual fish), this is not necessarily the case. Mortality in this case refers to the life/death/survival of individual fish and welfare refers to the 'well-being' of a living fish. 'Mortality' concerns were present in countries regardless of their support or opposition to the activity of catch-and-release fishing and therefore, alone, are not good indicator terms of catch-and-release perceptions. Where 'mortality' concerns could be used either to support or to oppose catch-and-release, its context in terms of the welfare theme, as well as other themes such as recreation (fishing for enjoyment, relaxation, leisure, etc.), subsistence (fishing for food), respect (for the fish), and management (management of fish as a 'resource' in some cases, fish as an animal in another) could help clarify the intended meaning.

Mortality was a theme present throughout the literature in Finland, Norway, Germany, Belgium, and United Kingdom. In Finnish literature, the term 'welfare' was also one of the most prevalent themes, roughly equivalent in frequency to that of 'mortality.' In a 1998 *Ilta-Sanomat* article, the minister responsible for fishery management challenged catch-and-release fishing and its consistency with the Animal Protection Act of Finland. He questioned whether it exposed the fish to 'unnecessary suffering, pain and agony' (Manninen 1998).

Table 7.2 Six most prominent themes found in the reviewed literature.

Finland	Norway	Scotland	Belgium	Czech Republic	Northern Ireland	England and Wales	Germany
Welfare	Recreation	Conservation	Recreation	Regulations	Mortality	Conservation	Welfare
Mortality	Subsistence	Regulatory	Resource	Licence	Recreation	Management	Regulatory
Management	Welfare	Management	Management	Keeping	Conservation	Regulatory	Licence
Regulatory	Mortality	Resource	Conservation	Recreation	Resource	Mortality	Skills/ Training
Resource	Ethics	Sport/Game	Licence	Sport/Game	Management	Resource	Mortality
Conservation	Management	Mortality	Regulatory	Conservation	Regulatory	Licence	Recreation

Further, there were more references to the 'subsistence' theme than to the 'recreation' theme in the Finnish literature. In some instances, catch-and-release was viewed as 'playing with food.' While the themes of 'respect' and 'ethics' were used by those opposing catch-and-release, they were also used by supporters of catch and release, who claimed it was a 'correct practice by many environment-conscious fishers' (Vainio 1998 – website).

In Norway, the theme of 'welfare' occurred with more frequency than mortality. However, terminology regarding 'recreation' as well as 'subsistence' occurred more frequently than both mortality and welfare, with fishing as a recreational activity being a more prevalent theme than subsistence fishing. Groups like the Norwegian Advisory Board on Animal Welfare Issues insisted that catch-and-release shifted the motive of fishing from its original purpose of harvesting food to that of purely sport or entertainment (Norwegian Advisory Board on Animal Welfare Issues 1998). With catch-and-release as a fairly new approach in Norway, rebuttals to the subsistence argument abound in the literature. Those in support of catch-and-release counter the subsistence theme with arguments that the main purpose for all recreational angling was the experience, 'not because it [was] necessary to [human] survival' (Storaas and Punsvik 1998). In the Scottish literature reviewed, 'mortality' concerns were mentioned more often than the welfare theme. However, the 'management' theme of fish as a 'resource' as well as the 'sustainability/conservation' theme (referring to reproduction of populations, sustainability of the fishing activity, conserving the resource/fish populations) dominated the literature reviewed from this country. Carefully releasing salmon (if that was the method chosen), according to guidelines to promote the survival of the fish, was identified as a way of making a real contribution to conservation in Scotland.

The limited available literature reviewed from Belgium and the Czech Republic resulted in some different patterns regarding themes. In Belgium, little reference was made to mortality and even less to the welfare theme. Instead, the literature revealed a strong theme of fishing as 'recreation' with emphasis on the experience rather than fishing as a 'subsistence' activity. The strong theme of recreation was accompanied by an equal presence of the 'management' theme, as well as fish as a 'resource' terminology and the 'sustainability/conservation' theme. In the literature reviewed, anglers financed the management and conservation of the fisheries resources through their purchase of fishing permits. However, while there was mention of and support for the practice of 'catch-and-release,' there was also mention of and support for the practice of 'keeping' the fish within the 'recreation' theme (this idea differs from strict subsistence, where subsistence was a by-product of recreational fishing, rather than fishing for the purpose of subsistence).

In the Czech Republic, the mortality and welfare themes were again subordinate to that of the 'recreation' theme. As in Scotland, the 'regulatory' (including legislation, guidelines, and rules) theme was notable in the literature from the Czech Republic. Coinciding with the regulatory terminology was also a strong reference to the requirement for licences and permits. The literature emphasised angler requirements to observe regulations set by the Anglers' Board such as closed seasons and requirements for equipment, specialised training, and licences. Additionally, statistical catch report forms, issued at the time of the licence, were to be returned when the permit expired, or anglers risked not obtaining permits for the following year. Management, sustainability/conservation, and resource themes were present, but not in as a great a proportion as in Belgium. While there was no direct reference to the subsistence theme, the 'keeping' the fish theme (with example terminology such as 'fish caught', fish 'taken into possession', and fish 'retained') frequented the literature. Equivalent in frequency to this theme was the recreation theme, which referenced increased leisure time coinciding with an increased demand for the 'sport' as well as an expanding 'recreational sector'.

In the catch-and-release literature reviewed from Northern Ireland, 'mortality' and 'recreation' terminology represented the most prominent themes. In comparison to the 'welfare' theme, 'mortality' terminology was proportionately greater by roughly three to one. In reference to the 'mortality' theme, catch-and-release was supported as a means of reducing mortality if the correct methods were used. The themes of 'sustainability/conservation' and 'resource' maintained a strong presence, just slightly less frequent than the 'mortality' and 'recreation' themes. Catch-and-release fishing was suggested as a contribution in a situation with reduced salmon stocks: 'At a time when salmon stocks are under threat, there is an increasing recognition of the importance of conserving as many fish as possible by fishery scientists, and anglers' (Ulster Angling Federation 1998 – quoted from website). Following the above themes were 'management' and 'regulatory' themes with use of additional terminology such as 'requirements' for maintaining future generations of fish stocks. The 'welfare' theme was mentioned least in the literature reviewed from Northern Ireland; it appeared approximately one-third less frequently than 'mortality,' 'recreation,' 'sustainability/conservation,' and 'resource' terminology.

In the literature reviewed from the United Kingdom, references to the activity of catch-and-release prevailed. It was through the most predominant theme of 'sustainability/conservation', especially in reference to salmon stocks, that support for catch-and-release emerged. Certain examples were found which portrayed anglers as the 'last predators' to inhibit the spawning of salmon; catch-and-release was seen as the most effective solution (Charles Teace Campaign for Catch and Release 1999 website). The themes appearing next were the 'management' and 'regulatory' themes, which occurred with approximately equal frequency. This included seasonal openings and closings as well as emphasis on size and catch requirements. The 'mortality' theme occurred next in frequency and was followed by the theme of fish as a 'resource' and the terminology of 'licences' as 'requirements.' The literature reviewed from the United Kingdom emphasised angler requirements to purchase not only permits from proprietors of particular fisheries, but also licences from the State. The revenue is collected by the National Rivers Authority and used to manage general as well as area specific fisheries. The themes of 'recreation,' 'keeping' the fish, and 'welfare' appeared only sparingly in the literature reviewed.

The situation was different in Germany, where the predominant or most frequent theme was animal 'welfare.' The 'mortality,' 'recreation,' and 'sustainability/conservation' themes referencing catch-and-release fishing were all subordinate to that of 'welfare.' Welfare concerns included alleviating pain and suffering as well as anaesthetising the fish and killing it immediately. The 'regulatory' theme was also a strong voice with many references to the German Code of Practice. Strict words such as 'forbid' and 'not allowed' were also used in direct reference to catch-and-release as well as live baiting and competitive fishing tournaments. Included in the regulatory theme in Germany, as well as some of the other countries reviewed, was the reference to licence and permit 'requirements'. In Germany, the 'regulatory' category included requirements for the 'skills and training' theme. For Germans, this means classes as well as exams before the issuance of a 'required' 'licence' [see also Steffens and Winkel 2002 for more detail of German angling regulations – eds]. Although not as strict as Germany, other countries such as Finland, Norway, Scotland, England and Wales all made references to licence requirements and the skills needed to practise catch-and-release fishing.

Whereas the German literature used prohibitory language in direct reference to catch-and-release, other countries used optional terminology such as 'voluntary' to characterise catch-and-release fishing. Especially in Great Britain, the 'voluntary' aspect of catch-and-release fishing was emphasised. The literature implied that catch-and-release was a method that fishers would want to 'voluntarily undertake' to help 'boost fish stocks,' but it was up to the angler's judgement (Charles Teace Campaign for Catch and Release 1999 – website). For example, although it was inferred that frequent fishers would probably choose to practise catch-and-release, no one would expect a first-time fisher to release his first fish. Literature from Scotland, Belgium, the Czech Republic, and Northern Ireland also directly used the word 'voluntary' when discussing catch-and-release fishing. Another important aspect of the literature was the use of terminology relating to the theme of fishing as a 'sport/game'. This terminology was used in Finland, Norway, Scotland, the Czech Republic, Northern Ireland, and in England and Wales. In this context, 'sport/game' was not seen as antiwelfare or encouraging cruelty, but more as a challenge and/or inclusion of fish as a 'game' animal (such as in hunting). The 'sport/game' theme was not found in the reviewed literature from Belgium or Germany. While use of the 'resource' theme predominated in the literature, reference to fish as 'individuals' (including direct references to fish as animals and usually coinciding with welfare concerns) was found in the literature of Finland, Norway, Scotland, and Germany.

Discussion

The European discourse about catch-and-release is characterised by complex and multifaceted arguments, with arguments from most themes of concern used both to support and oppose this practice. There are likely groups both pro and con catch-and-release in most countries studied. In some countries the practice of catch-and-release seems to be gaining increasing support, primarily in northern countries with cold-water game fisheries, i.e. the British Isles, Norway and Sweden. In other areas, the best example being parts of Germany, a catch-and-release fishery on fish above minimum size is prohibited.

The discussion about catch-and-release has definitely led to a fundamental and philosophical debate about the meanings and ideas of angling and/or recreational fishing, among anglers, bureaucrats, conservationists and scientists in many countries.

From the content analysis, there seem to be some systematic differences worth noting. In east-central mainland Europe, the focus seems to be strongly on formal regulations related to fishing and treatment of the catch. This is not unexpected on the basis of the strong bureaucratic and regulatory traditions in former Eastern Europe countries. In Great Britain the focus is strong on the conservation and management aspects related to catch-and-release, i.e. the conservation effect and eventual mortality problems. Subsistence aspects seem to be strongest in northern countries like Norway and Finland, as a part of their rural culture.

Another difference that emerged was that between representatives for authorities on one side and angler organisations on the other. Authorities were generally more sceptical to catch-

and-release compared to angler organisations. International organisations or bodies were often neutral to catch-and-release, often underlining the voluntary aspect and the need to take into account different cultures and traditions in different countries, in combination with varying biological needs and effects of catch-and-release (NASCO undated).

The growth in common rules and regulations within the European Union in many areas, including veterinary issues and animal welfare, raises the question of whether the Union will establish regulations regarding catch-and-release that are obligatory for member countries. The generally strong position of Germany within the Union has led some angling groups to fear a total prohibition of catch-and-release. Looking closer at the EU countries in this analysis illustrates very different views on the issue between countries like Germany, Belgium, Finland and the United Kingdom. In addition, the NASCO intergovernmental salmon conservation group with a strong EU representation has a pragmatic and conservation-oriented view on catch-and-release. This suggests that an EU decision prohibiting catch-and-release in general is not likely in the near future.

The different perspectives on catch-and-release in the European countries studied may be attributable to cultural differences in the conduct and meaning of recreational fishing in these countries (Haworth 1983, Aas and Kaltenborn 1995), or to differences in values or views from anglers on one side and managers on the other. Although hook and line appears to be the predominant but not exclusive gear used by anglers in Europe, fishing appears to differ in terms of the mix of extrinsic and intrinsic motivations involved. For example, in those countries (e.g., Norway and Finland) where fishing for many is mainly about harvesting fish, catch-and-release and its challenge, sport, and adventure benefits are viewed by some groups as antithetical. Other groups often view these perspectives as complementary, and want to combine the opportunity to harvest fish with catch-and-release in certain situations. In other countries (e.g., Scotland, Belgium, Czech Republic) where fishing is about harvest as well as the challenge and sport involved in hooking and landing fish, catch-and-release is more likely to be seen as an appropriate solution for anglers for lessening their impacts on fish populations where angler harvest can affect fish populations. Another possible explanation is the generally strong involvement from the veterinary profession in this discussion in several countries (for instance, Norway and Germany). Animal welfare seems to be a more prominent theme among veterinarians compared to fish biologists. However, their perspective is strongly influenced by the farming and husbandry context (Storaas and Punsvik 1998).

Except in those areas where catch-and-release is currently illegal (certain states in Germany) or for those fisheries where the main focus is on harvest, it would appear that, as the meaning of recreational fishing changes over time in response to various conditions, so too can we expect more or less emphasis on the practice of catch-and-release.

Whereas many recreational fishing NGOs encourage the practice of catch-and-release in the United States, less than 10% of all US anglers are affiliated to fishing organisations. These groups have been successful in promoting catch-and-release as the norm among billfish, trout and black bass anglers, for example, but overall catch-and-release is still practised by a minority of anglers in the USA [Policansky 2002 gives an account of catch-and-release in North America – eds].

As mentioned earlier, further analysis of European literature as well as additional European contacts are needed to understand more completely the perceptions of catch-and-release fishing in Europe and the relative distribution and number of people supporting the different views and attitudes. This will entail a more in-depth, funded study with individuals and agencies from those countries as well as possible angler surveys within those countries. This study would require translators from various countries to increase the breadth of the material capable of being evaluated. Additionally, multiple coders in the content analysis technique (Ritter *et al.* 1992) are necessary to improve the reliability of the categories identified and the conclusions drawn from such a study.

References

Aas, Ø. & Kaltenborn, B.P. (1995) Consumptive orientation of anglers in Engerdal, Norway. *Environmental Management* **19**, 751–761.

Aas, Ø. & Skurdal, J. (1996) Resident and non-resident fishing in a rural area in Norway: sport and subsistence – conflicts or coexistence? *Nordic Journal of Freshwater Research,* **72**, 45–51.

Aas, Ø., Haider, W. & Hunt, L. (2000) Angler responses to potential harvest regulations in a Norwegian sport fishery: a conjoint based choice modeling approach. *North American Journal of Fisheries Management,* **20**, 940–950.

Affre, P. (1985) Fishing in southern Europe especially France. In: *World Angling Resources and Challenges* (*Proceedings of the First World Angling Conference.* Cap d'Agde, France September 12–18, 1984. Compiled under the direction of Harry, E.K. –President).

Atlantic Salmon – A case against catch-and-release (anonymous manuscript).

Barnhardt, R.A. (1989) Symposium review: Catch-and-release fishing, a decade of experience. *North American Journal of Fisheries Management,* **9**,74–80.

Berg, R. & Rosch, R. (1998) Animal welfare and angling in Baden-Wurttemberg, Germany. In: *Recreational Fisheries: Social, Economic and Management Aspects* (Proceedings from the 19th session of the European Inland Fisheries Advisory Commission) (eds Hickley, P. & Tompkins, H.), pp. 88–92. Food and Agriculture Organization of the United Nations. Rome, Italy.

Bielak, A. (1988) Promoting catch and release: the ultimate low consumption fishing technique. In: *Recreational Fisheries Conference Proceedings 1986.* pp. 57–67. Canadian Special Publication of Fisheries and Aquatic Sciences 97. DFO Canada 1988.

Bninska, M. (1998) Summary report of the symposium topic session on the interactions between fisheries and outside influences. In: *Recreational Fisheries: Social, Economic and Management Aspects* (Proceedings from the 19th session of the European Inland Fisheries Advisory Commission) (eds Hickley, P. & Tompkins, H.), Chapter 10. Food and Agriculture Organization of the United Nations. Rome, Italy.

Bryan, H. (1977) Leisure value system and recreational specialization: The case of trout fishers. *Journal of Leisure Research,* **9**, 174–187.

Bult, T.P. (National Institute for Coastal and Marine Management/RIKZ) Middelburg, The Netherlands. Personal E-mail.

Caffrey, J.M., Allison, J.P., Donnelly, R.E., Tierney, D.M. & Monahan, C. (1998) Development and management of recreational angling in Irish canals. In: *Recreational Fisheries: Social, Economic and Management Aspects* (Proceedings from the 19th session of the European Inland Fisheries Advisory Commission) (eds Hickley, P. & Tompkins, H.), Chapter 29. Food and Agriculture Organization of the United Nations. Rome, Italy.

Charles Teace Campaign for Catch and Release (1999) The Atlantic salmon catch-and-release campaign. http://www.gamefishing.co.uk/catch-and-release/(last cited on 1 May 2001).

Cowx, I.G. (1998) Aquatic resource planning for resolution of fisheries management issues. In: *Recreational Fisheries: Social, Economic and Management Aspects* (Proceedings from the 19th session of the European Inland Fisheries Advisory Commission) (eds Hickley, P. & Tompkins, H.), Chapter 16. Food and Agriculture Organization of the United Nations. Rome, Italy.

Currie, B. (1999) Gamefishing in Scotland: An overview of Scottish gamefishing. http://www.gamefishing.co.uk/fishings/scotland/overview.htm (last cited on 2 May 2001).

CyberTrout.Com (1997) Fly fishing in Germany. http://www.cybertrout.com/germany.htm (last cited on 10 May 2001).

District boards and close season for salmon and trout. http://www.wsi.walshsimmons.co.uk/wtf/content/2b.html (Last cited on 10 May 2001).

Ellis, J.W. (1998) Trends in recreational angling: the British Waterways experience. In: *Recreational Fisheries: Social, Economic and Management Aspects* (eds Hickley, P. & Tompkins, H.), Chapter 11. Food and Agriculture Organisation of the United Nations. Rome, Italy.

Fedler, A. & Ditton, R.B. (1986) A framework for understanding the consumptive orientation of recreational fishers. *Environmental Management* **10**, 221–227.

Fischer, Johanne. Deputy Director: EuroGOOS Secretariat, Southampton Oceanography Centre. Personal E-mail.

Fishing in England and Wales: NRA structure, close seasons, license duties etc. http://www.where-to-fish.com/etc/la.html (last cited on 10 May 2001).

Fishing the Kenai National Wildlife Refuge – Combat fishing the Kenai and Russian rivers. http://www.r7.fws.gov/nwr/kenai/fishing.html (Last cited 5 January 2000).

Food and Agriculture Organization Fisheries Department (FAO Fisheries Circular No. 911) (1996) *Fisheries and aquaculture in Europe: Situation and outlook in 1996*. FAO, Rome, Italy.

Food and Agriculture Organization of the United Nations (FAO Fisheries Circular No. 942) (1999) *Review of the state of world fishery resources; inland fisheries*. FAO, Rome, Italy.

Frank, V., Lejeune, A. & Herman, D. (1998) Belgium. Recreational fisheries survey in the Liege province of Belgium. In: *Recreational Fisheries: Social, Economic and Management Aspects* (Proceedings from the 19th session of the European Inland Fisheries Advisory Commission) (eds Hickley, P. & Tompkins, H.), Chapter 5. Food and Agriculture Organization of the United Nations. Rome, Italy.

Fururly, J.G. (1997) Is it wrong to show pictures of dead fish on the Internet? http://home.sol.no/~jangf/salmon/deadfish.htm (last cited on 10 May 2001).

Haworth, J.T. (1983) Satisfaction statements and the study of angling in the United Kingdom. *Leisure Sciences,* **5**, 181–196.

Hickley, P. (1998) Comments concerning a code of good practice for recreational fishing. In: *Recreational Fisheries: Social, Economic and Management Aspects* (eds Hickley, P. & H. Tompkins), pp. 299–304. Fishing News Books, Blackwell Scientific/FAO/EIFAC.

Hickley, P. & Tompkins, H. (eds) (1998) *Recreational Fisheries: Social, Economic and Management Aspects* (Proceedings from the 19th session of the European Inland Fisheries Advisory Commission). Food and Agriculture Organization of the United Nations. Rome, Italy.

Hickley, P. (1998) Comments concerning a code of good practice for recreational fishing. In: *Recreational Fisheries: Social, Economic and Management Aspects* (Proceedings from the 19th session of the European Inland Fisheries Advisory Commission) (eds Hickley, P. & Tompkins, H.), Chapter 38. Food and Agriculture Organization of the United Nations. Rome, Italy. http://home.sol.no/~jangf/salmon/debate.htm (last cited on 10 May 2001)

Hunter, B. (1995) Personal letter to Orri Vigfusson.

Lowerson, J. (1989) Angling. In: *Sport in Britain – a social history* (ed Mason, T.), pp. 12–44. Cambridge University Press.

Manninen, T. (translated by Anna-Liisa Toivonen) (1998) Is fish releasing animal torture? *Ilta-Sanomat*, April 18, 1998, 1–2.

Marshall, C. & Rossman G. (1999) Designing qualitative research. 3rd edn Sage Publications, Inc, California.

NASCO *Guidelines on catch and release*. (No date) Brochure published by the North Atlantic Salmon Conservation Organization. (An inter-governmental commission consisting of USA, Canada, Denmark (on behalf of Faroe Islands and Greenland), European Union, Iceland, Norway and Russia).

North Atlantic Salmon Conservation Organization. Pamphlet: *Guidelines on catch and release*. Edinburgh, Scotland.

Norwegian Advisory Board on Animal Welfare Issues. (1998) *Consideration*: ANGLING – Catch-and-release practice. Unpublished report.

Pierrou, Ulf. Fisheries Director, Laensstyrelsen i Dalarnas laen, Falun, Sweden. Personal E-mail.

Policansky, D. (2002) Catch-and-release recreational fishing: a historical perspective. In: *Recreational Fisheries: Ecological, Economic and Social Evaluation* (eds Pitcher, T.J. & Hollingworth, C.E.), Chapter 6, this volume. Blackwell Science, Oxford, UK.

Raat, Alex Dr. Head Research Department: Organization Improvement Inland Fisheries. Personal E-mail.

Ritter, C., Ditton, R. & Riechers, R. (1992) Constraints to sport fishing: Implications for fisheries management. *Fisheries*, **17**(4), 16–19.

Spitler, R.J. (1998) The animal rights movement and fisheries: They're Heeeeere! *Fisheries*, **23**(1), 21–22.

Steffens, W. & Winkel, M. (2002) Evaluating recreational fishing in Germany. In: *Recreational Fisheries: Ecological, Economic and Social Evaluation* (eds Pitcher, T.J. & Hollingworth, C.E.), Chapter 10, this volume. Blackwell Science, Oxford, UK.

Storaas, T. & Punsvik, T. (translated by Oystein Aas) (1998) Hunting and fishing for joy! *Jakt og Fiske*, **10**, 68–70.

Teisl, M., Boyle, K.J. & Roe, B. (1996) Conjoint analysis of angler evaluations of Atlantic salmon restoration on the Penobscot river, Maine. *North American Journal of Fisheries Management*, **16**, 861–871.

Toivonen, Anna-Liisa, Lic. Tech., Research Scientist, Finnish Game and Fisheries Research Institute. Personal E-mail.

Turunen, T. & Suuronen, P. (1996) Hooking mortality of small brown trout and grayling in Finnish rivers catch-and-release fisheries. *Boreal Environment Research,* **1**, 59–64.

Ulster Angling Federation (1998) Welcome to the Ulster Angling Federation home page. http://www.gamefishing.co.uk/UAF/welcome.htm (last cited on 10 May 2001).

Vacha, F. (1998) Information on Czech Republic fisheries. In: *Recreational Fisheries: Social, Economic and Management Aspects* (Proceedings from the 19th session of the European Inland Fisheries Advisory Commission) (eds Hickley, P. & Tompkins, H.), Chapter 9. Food and Agriculture Organization of the United Nations. Rome, Italy.

Vainio, J. (1998) Fishing in Finland. Fish and regulations. http://www.iuakk.fi/~outdoor/Finland/Finland3.htm (last cited on 2 May 2001).

Weber, R. (1990) *Basic content analysis*. 2nd edn, The Sage University Papers Series, Vol. 49. Sage Publications, Inc, California.

Chapter 8

Maintaining Quality in Recreational Fisheries: How Success Breeds Failure in Management of Open-Access Sport Fisheries

Sean Cox and Carl Walters

Abstract

Where recreational fisheries are open to public access, there is a basic pathology in which success breeds failure: development of a quality fishing situation leads to increased fishing effort until quality is reduced to be no better than other situations with comparable costs and difficulties of access. Impacts of increased fishing effort on angling quality can be greatly exaggerated by fish behaviours that limit the proportion of total stock vulnerable to capture at any moment. In open-access fisheries, managers mainly react to the quality deterioration problem by trying to produce more fish and by using simple regulations such as bag and size limits. These tactics have never worked and as a result, high quality fishing is found only where fishing effort is severely restricted. There are three situations where high quality/low fishing effort occurs: (1) high cost/time required to access the fishery such as very remote lakes and ocean coastal areas far from major tourist routes; (2) control of access by private or local interests, such as guiding camps, fishing clubs, and lakeshore owners; or (3) an equitable strategy of limited access via a lottery system, as has been used routinely in big game management for decades. It is hoped that we will see a spread of lottery access systems rather than privatisation of access rights in North America as a means of providing angling quality. A key scientific challenge will be to design access management experiments that reveal critical uncertainties and potential benefits of novel management approaches.

Key words: angling quality, effort modelling, fisheries management, open-access, recreational fisheries.

Introduction

Most sport fisheries in North America remain open to public access, without effort or licence limitation. In such situations, fishery managers need to contend with two potentially conflicting dynamics: response of fish populations to fishing and enhancement (production), and response of fishers to changes in the abundance of fish (consumption). Fisheries science has mainly concentrated on the 'production-side' of this dynamic relationship, with a tacit assumption that the 'consumption-side' will somehow take care of itself. Consumption-side dynamics are usually ignored because we assume that recreational fishers are not efficient (or persistent) enough to generate severe biological overfishing, at least in part because we expect them to give up before depleting most stocks to biologically dangerous levels. Management activities aimed at sustaining quality of recreational fishing are usually targeted on production-side measures such as habitat enhancement, stocking, and regulation of per-angler impacts via season, bag, and size limits. In management planning we rarely account for the potential influence of these measures on the dynamics of fishing effort (i.e. consumption-side) and the subsequent effects of effort responses on angling quality and harvest. In the fisheries literature, this myopia leads to some peculiar arguments. For example, Shaner *et al.* (1996) complained that it was impossible to define an optimum stocking rate for channel catfish (*Ictalurus punctatus*) because harvest increased linearly with stocking rate. Their results showed quite strongly that increasing stocking rates just led to more effort without an improvement in quality of fishing as measured by catch per effort.

The central argument of this chapter is that production-side management in open access settings will almost inevitably lead to increased fishing effort but not to increased quality of fishing as perceived by the 'average' angler. While angling 'quality' obviously has many dimensions, it is clear from sociological studies that catch rates are important determinants of angling quality (Holland and Ditton 1992). Further, diversity of angling opportunity (i.e. species composition, wilderness fishing, trophy fishing) is also a key quality measure. Regardless of how quality is measured, mean angler satisfaction derived from the fishing experience likely decreases with increasing effort (Malvestuto and Hudgins 1996). In the discussion below, it does not really matter whether quality is defined and recognised by anglers in terms of catch rate, fish size, or diversity: we contend that provided there is freedom of movement via open access, anglers will concentrate wherever they see higher quality, until they no longer see it [see Policansky 2002 for further comments on angler crowding – eds]. We suggest that effort dynamics are particularly important considering that, in most populations, relatively few fish are behaviourally reactive or accessible to fishing gear at any moment, so that anglers 'see' and compete for much smaller numbers of fish than we measure in biological sampling.

Abundance–effort relationships and depression of catch rates

Generally our empirical experience in recreational fisheries is that fishing effort is roughly proportional to abundance as measured by indices like stocking rates (Fig. 8.1, Moring 1993, Fraley 1996, Shaner *et al.* 1996). Generally we do not find clear relationships between abundance and catch rate except across regional gradients of accessibility (lower effort, higher catch rates, and sometimes higher abundance in less accessible places). In this section, we argue that quality of fishing is more sensitive to effort than often supposed due to (1) competition among anglers for a limited stock of vulnerable fish; (2) production limitation mechanisms like density-dependence in growth/survival rates that limit numbers of quality fish; and (3) cumulative impacts of fishing pressure that also limit numbers of quality size fish.

At any particular time in a recreational fishery (or any other for that matter), not all fish are expected to be available to the fishing gear for various reasons. Some fish will be within the effective depth range of the gear and will react to it (a 'vulnerable' state) while others are simply not behaviourally reactive or else remain in sites/habitats where angling gear cannot

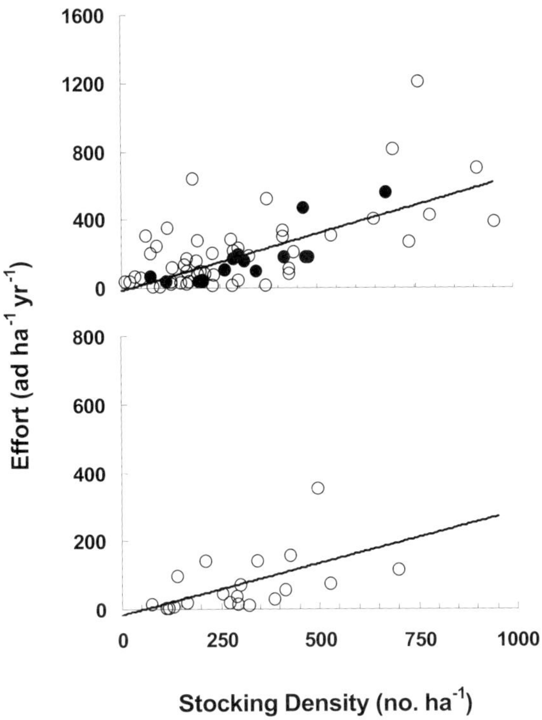

Fig. 8.1 Fishing effort on British Columbia lakes as a function of rainbow trout stocking rate. Predicted relationship (solid line) is based on least-squares regression between Equation 8.3 and observed data (solid circles). Regression relationships for regions 3 (open circles; top panel) and 8 (closed circles; top panel) are not significantly different from one another, but both are significantly different from region 5 (bottom panel). ad = angler days.

reach (an 'invulnerable' state). Exchange between these two states likely occurs on a variety of time scales due to a variety of processes, ranging from hours–days for diurnal movements and feeding rhythms to weeks–months for recovery from previous hooking (i.e. catch-and-release). Fish exchanging between vulnerable and invulnerable states results in a vulnerable 'stock' seen by anglers that is generally much smaller than the stock seen by biological sampling (Fig. 8.2).

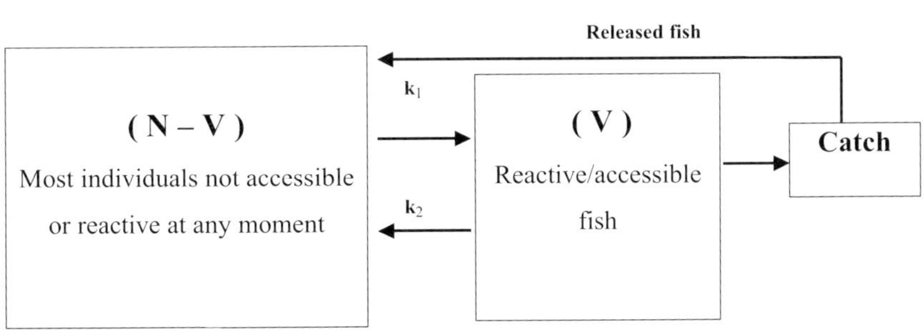

Fig. 8.2 Schematic representation of recreational fishery dynamics. The total stock, N, is made up of an invulnerable component (N – V) and a vulnerable component V. Exchange rates between the two states are given by k_1 and k_2.

Exchange of individuals between vulnerability states, and removals by fishing (including fish released but not immediately vulnerable) imply a strong inverse relationship between density of vulnerable fish and fishing effort. If exchange between behavioural states is rapid compared to removal and addition of fish by population dynamics processes (recruitment, mortality, growth), a simple 'variable speed splitting' (Walters and Bonfil 1999, Walters and Korman 1999) model can be used to predict the form of the relationship between effort and abundance of available fish. Suppose the total number of fish potentially available is N, and that V of these are vulnerable (accessible and reactive to gear) at any moment. Suppose further that fish enter the vulnerable pool at rate $k_1(N - V)$, and leave this pool at rates k_2V (movement into invulnerable state) and qEV (catch, where q = catchability and E = fishing effort). If we then write a rate equation for V [$dV/dt = k_1(N - V - k_2V - qEV)$] and set the rate equal to zero to represent rapid equilibration of the 'instantaneous' number of fish actually available to anglers, it is easy to see that 'fast' exchange compared to depletion (high k_1, k_2 exchange coefficients compared to mortality rate for N as a whole) will result in V behaving as

$$V = k_1N/(k_1 + k_2 + qE) \quad [8.1]$$

This model implies that catch-per-effort (qV) will decline sharply with increasing effort E even if the fishery has no impact on N (Fig. 8.3).

Equation 8.1 has an important implication for predicting how fishing effort will be distributed over a collection of fishing sites (e.g. lakes, stream reaches) that have similar access characteristics (travel time from population centres, etc.). If site i has catch rate q_iV_i, we expect anglers to 'detect' (by direct experience and word of mouth) differences among sites and move about so that every site ends up displaying the same average fishing quality or

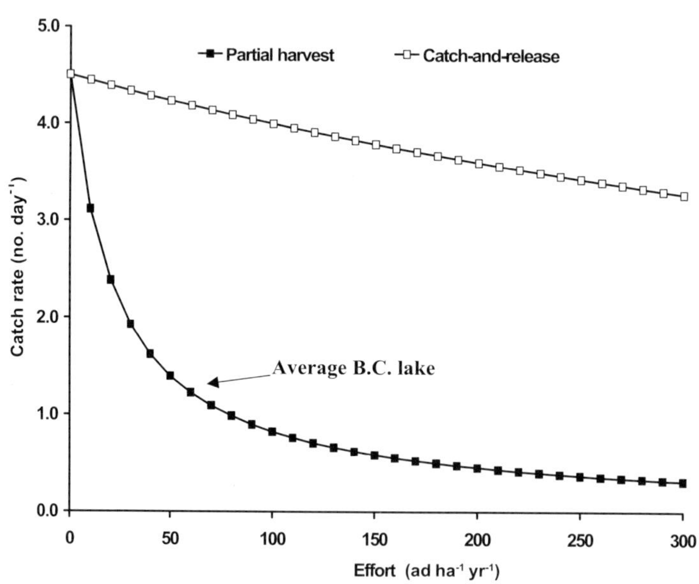

Fig. 8.3 Predicted relationship between fishing effort and catch rate for fixed fish abundance. Curves are shown for partial harvest fishery with a 50% retention rate (solid squares) and a catch-and-release fishery with only 5% hooking mortality (open squares).

catch rate, c_o. That is, we expect $q_i V_i = c_o$ for every site on average (an 'ideal free distribution' prediction). Substituting this expectation into Equation 8.1 and solving for instantaneous effort e_i on site i, we predict that effort will vary over sites as

$$e_i = k_1 N_i / c_o - (k_1 + k_2)/q_i \quad [8.2]$$

That is, we expect efforts e_i to vary linearly among sites with variation in total abundance N_i, provided catchabilities (q_i) are not too variable among sites. We expect lower effort e_i in any site when the regional average catch rate c_o is high (or conversely, that high regional average catch rate will occur only when efforts e_i are low in general, due to factors like high access cost).

Equations 8.1 and 8.2 represent 'instantaneous' predictions of vulnerable abundance, catch rate, and effort density, which can be integrated over time so as to predict annual total catches, efforts, and impact of fishing on particular sites. The equation for total catch is similar to standard fisheries catch equations, as used for example by Engstrom-Heg (1986), while the equation for total effort E is

$$E_i = (N_i - N_\infty)(1 - \exp[-k_1 T])/c_o \quad [8.3]$$

where $N_\infty = c_o(k_1 + k_2/q)$ is the regional minimum fish abundance below which fishing effort will no longer be attracted (i.e. x-axis intercept in plot of effort vs. abundance). The data and model fit in Fig. 8.1 show that this model is consistent with effort–abundance data on British Columbia (BC) lakes.

An alternative test of the basic assumption that effort tends to move about so as to level catch rates among sites (at rate c_o in Equation 8.3) is to plot total catch and total effort against an independent index of total abundance (N) or stocking rate, for a set of sites with strongly contrasting abundance. If both catch and effort are on average proportional to abundance (straight line pattern with near zero intercept and slope greater than zero), then catch per effort must on average be independent of abundance (if catch = aN and effort = bN, catch/effort = a/b) and we can conclude that fishers have 'succeeded' at detecting and eliminating any really outstanding fishing opportunities. In such plots we expect to see considerable unexplained variation, due to site factors such as access (higher catch per efforts for relatively inaccessible sites), catchability, and behaviour/site morphometry differences leading to variation among sites in k_1 and k_2. But if proportionality is observed as an average response despite variability among sites, then the central conclusion would be that increasing abundance is not a way to ensure increasing quality of fishing, at least as measured by catch per effort.

One might argue that Equation 8.3 ignores potential limits on regional fishing effort due to the size of the angler population available to generate effort and to limits on how much time each angler can potentially spend fishing. This is certainly a valid argument for 'regions' that are very inaccessible or costly to access (e.g. fly-in lakes, remote northern areas), but it is deeply incorrect for most angling situations, which are now within a few hours' drive from at least one major urban centre. In these situations, if we keep producing more fish, we will keep attracting more effort. For example, hundreds of BC trout lakes within a few hours' drive of Vancouver, BC are mainly sustained through annual stocking of fingerling/yearling trout or natural reproduction; in these lakes we would consider a 'high' effort level to be 80–100 angler days/ha/year. In the Sierra Nevada mountains of California where there are also many lakes, at considerably greater travel distances from centres like Los Angeles and San Francisco, some lakes are stocked weekly or even biweekly with catchable trout. During the early 1960s, when C.W. worked as a creel census clerk in California, such lakes commonly received as much effort *per day* as we currently see *per year* in BC lakes. These effort differences are orders of magnitude larger than we would expect on the basis of differences in angler population size or number of lakes available for fishing.

Catch-and-release fishing does not prevent an inverse relationship between vulnerable fish density and fishing effort as predicted by Equation 8.1 even where total abundance remains high. Released fish undoubtedly suffer a certain amount of 'handling trauma' after being caught and released by anglers. There is also evidence to suggest that for some length of time after being released, they may behave differently from other fish that have either never been caught, or have been caught and released in the distant past (Lewynsky and Bjornn 1987). Depending upon the length of time to full recovery from capture (up to 3–4 weeks in some studies), Equation 8.1 will still predict an inverse relationship between effort and catch per effort because released fish do not remain in the vulnerable state. Such direct effects of angling pressure on catch rates were particularly evident following the opening of a catch-and-release fishery for largemouth bass in a Florida lake (Champeau and Denson 1987) where test fishery catch rates declined by half immediately following the opening. Such declines in catch rate could not be explained solely on the basis of hooking mortality losses (Champeau and Denson 1987).

Effects of fishing effort on catch rates can be particularly strong in situations where the vulnerable fish 'pool' V is defined only in terms of larger, older 'quality' fish. In such cases, there are severe limits on the potential total number (N) of quality fish that can be produced by aquatic environments, for at least two reasons. First, the rearing environment must support and grow many small fish for every quality one that is ultimately produced. In places lacking natural reproduction, low stocking rates may be necessary to ensure good growth and survival to large sizes, and reduction in stocking rates causes a direct reduction in V. Second, in most sport fisheries there is considerable 'incidental' and cumulative mortality on small fish caused by anglers attempting to catch the larger ones. Even a low annual mortality rate on smaller fish can have a very large cumulative impact on the number of fish reaching quality sizes, especially if these smaller fish are subject to the annual rate for several years before reaching quality size.

The bottom line of this section is quite simple: success at producing more vulnerable fish (increasing V in Equation 8.1, by producing higher N) is no guarantee of producing higher quality of fishing as measured by catch rate or fish size. Rather, in open-access situations we expect anglers to detect unusually good fishing opportunities and to respond by adding and redistributing effort so as to 'flatten' the quality of fishing across sites or opportunities. Where these effort responses are particularly strong, ultimate levels of fish abundance, angling quality, or angler satisfaction are largely outside of management control. This flattening can only be prevented by somehow preventing the 'ideal free distribution' process of fishing effort redistribution (i.e. by directly limiting fishing effort for at least some angling sites). This admonition is particularly important for situations where large fish size is considered a key element of quality, since there are severe ecological limits on the total number of large fish that can be produced.

Revealing hidden costs and benefits of open-access

While it is not our intention to determine what the objectives of recreational fisheries management should be, it is clear that most decisions involve a compromise between the objectives of management agencies (maximising effort, provide quality) and those of individual anglers (maximise catch rate). In the preceding sections, we have argued that (1) increasing fishing effort suppresses catch rates in the short term by removing vulnerable/reactive fish and in the long term by increases in fishing mortality and (2) fishing effort increases with natural or artificial increases in fish abundance. If these arguments hold, then recreational fishery managers should not expect to achieve both effort maximising and catch rate maximising objectives at the same time. Instead, a trade-off relationship should exist between effort and catch rate such that the maximum socioeconomic benefits or returns occur at some intermediate level of the two. In this section we develop a mathematical framework for quantitative evaluation of such trade-offs and apply this to our BC rainbow trout lakes example.

For the purpose of the following analysis we must assume that anglers are willing to pay (*WTP*) for different levels of angling quality either through direct travel expenditures, i.e. travel to more remote regions, or by purchasing time at fishing camps and resorts [see Rudd *et al.* 2002 for a basic description of WTP models – eds]. Over the range of travel costs to fishing sites and remote lodges in British Columbia's southern interior, the relationship between fishing quality (observed *CPUE*) and the associated travel costs (*WTP*; dollars) is given approximately by

$$CPUE = \alpha + \beta \cdot WTP \qquad [8.4]$$

Substituting this relationship into the effort Equation 8.3, and ignoring lake-specific subscripts i, we can solve for the value (*WTP*) of limiting effort to some arbitrary total E^*

$$WTP_{E^*} = \frac{N_o C - \alpha\left[E^* + \frac{u_{\max} C}{q}\right]}{\beta\left[E^* + \frac{u_{\max} C}{q}\right]} \qquad [8.5]$$

where N_0 is the stocking rate, $C = (k_1 + k_2)/k_1$ is a constant function of the exchange rate parameters and $u_{\max} = 1 - \exp[-k_1 T]$ is the maximum exploitation rate in the presence of infinite fishing effort. Total fishery value is then taken as the product of *WTP* times E^*. To be useful for policy analysis we proceed one more step by solving (8.5) for the optimum effort level E_{opt} for a given initial stock size, N_0, which is

$$E_{opt} = \frac{-\alpha u_{\max} + 2\sqrt{\alpha N_0 q u_{\max}}}{\alpha q} \qquad [8.6]$$

For stocking rate $N_{0,i}$, on lake i, we then compute (1) the value $V_{\text{opt},i}$ of limiting effort to the optimum (Equation 8.5 multiplied by Equation 8.6 where $E^* = E_{\text{opt}}$) and (2) an estimate of the observed value under open access $V_{\text{obs},i}$ using the observed effort on lake i (E_{obs} multiplied by

Equation 8.5 where $E^* = E_{obs}$). The percentage difference between these values thus provides an estimate of the relative loss (RL) due to open access, which is computed as

$$RL = 100\ (V_{opt,I} - V_{obs,i})/V_{opt,i.} \qquad [8.7]$$

Figure 8.4 shows the optimum effort curve and open-access effort relationship for BC region 3 lakes. Interestingly, observed fishing effort on most lakes was quite near the estimated optimum effort. In general, relative losses decreased as stocking rates increased (Fig. 8.5) and losses greater than 20% occurred on only 30% of all lakes. Most losses occurred due to excess effort (65%) while others (35%) had effort levels below optimum presumably due to low quality access or lake conditions that result in particularly poor rainbow trout survival (i.e., our stocking rate index grossly over-estimates relative abundance of fish and thus optimum effort). If access conditions could be upgraded, these lakes would be prime targets for management strategies that attempt to increase fishing effort up to, but not exceeding, the optimal level.

It is encouraging to see that access/effort management on only a small proportion of fisheries could potentially lead to substantial gains in fishery value relative to expected optimal value. However, as we have pointed out in this chapter and discuss further below, management strategies that attempt to manipulate fishing effort levels or fish abundance must carefully consider the consequences of effort redistribution and subsequent quality impacts on unmanaged waters (Walters and Bonfil 1999).

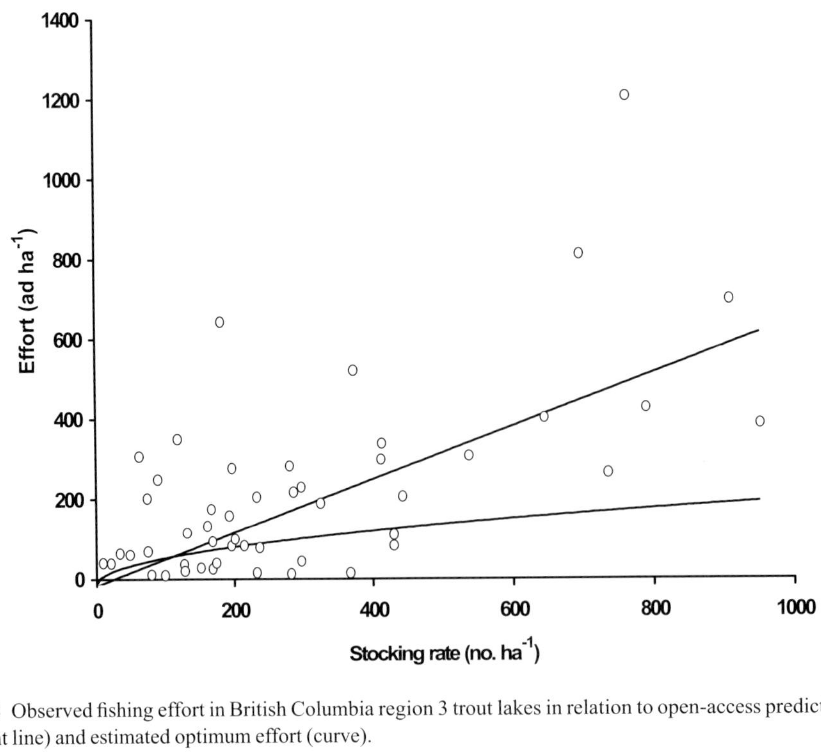

Fig. 8.4 Observed fishing effort in British Columbia region 3 trout lakes in relation to open-access predicted effort (straight line) and estimated optimum effort (curve).

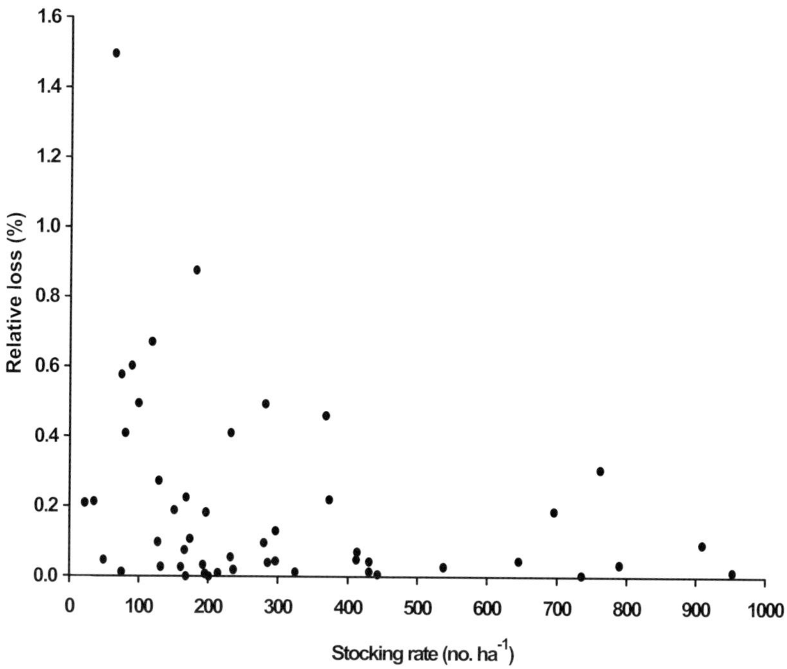

Fig. 8.5 Estimated loss percentages as a function of rainbow trout stocking rate in BC lakes. Losses are calculated using differences between value calculations for observed effort ($V_{obs,i}$) and optimum effort predictions ($V_{opt,i}$) for each stocking rate data point in Fig. 8·4.

Conflicting stakeholder interests and management paralysis

Most management jurisdictions can reasonably claim to provide a diversity of opportunities in terms of fishing quality. However, in most cases, the quality opportunities are created by high access costs and/or direct access limitation by private landowners, so that high quality opportunities are accessed mainly by fishers willing to spend more time and money. A key option for management is to create more opportunities for quality fishing at relatively low access cost (for less wealthy anglers), by directly limiting fishing effort on some sites using a lottery process that gives every angler a 'fair' chance at these opportunities. However, in our experience, managers have been loath to adopt this approach because of the bitter opposition they see whenever it is suggested. Here we point out that this opposition reflects real and fundamental conflicts of interest among angling stakeholders, and these conflicts almost inevitably result in paralysis of management and inaction.

In our experience conducting public meetings and talks presented to angling groups, one of the surest ways to elicit screams of outrage is simply to mention the idea of 'limited entry' fishing (most North American anglers recognise this terminology because of its widespread use in big-game harvest management). Arguments that typically follow limited entry proposals are:

(1) Fish aren't like large mammals – there are millions of fish
(2) I have a fundamental right to go fishing wherever I want in public waters [see Policansky 2002 on this issue – eds]

(3) If there aren't enough fish, it is your job to produce more
(4) Anglers aren't responsible for low catch rates, it is the (substitute your favourite from the following list: commercial fishers, habitat damage by loggers, natives, poachers, biologists with nets, loons, seals, pollutants, El Niño)
(5) The good old days are gone forever and there are just too many people now

Probably the most pervasive and difficult for fisheries managers to deal with is argument (2); there really is no answer to an angler who either denies that there is a quality problem in the first place or else denies that public agencies have any right to deal with it even if it does exist. Argument (3) is one that biologists have obviously responded to far too often in the past. Arguments of type (4) serve mainly to waste time and deflect attention from the real quality management issues, and it is our impression that most of the people who bring them forth are very well aware of this. Fatalistic arguments (5) come mainly from the very best anglers, who in fact are feeling relatively little personal impact from competition.

Experience with limited entry hunting, and a very few tests of limited entry fishing (e.g. Atlantic salmon (*Salmo salar,* Salmonidae) on the St. John River, New Brunswick), indicate that opposition by anglers dies away very quickly if quality benefits become evident. But that is little comfort to the fishery manager who must deal with the initial opposition, particularly if it is in a small community setting where the most outraged anglers are liable to be the manager's neighbours.

There are also conflicting viewpoints within and among angling groups and the economic stakeholders who depend on recreational fishing (tackle shops, lodges, guides, etc.). Opposition to such programs from within the angling community arises mainly from local anglers who have easy access to all fishing opportunities while more distant anglers (such as Vancouver angling clubs travelling to BC interior) view limited entry programs with interest because they now spend large amounts of time and money travelling to increasingly distant waters to access good fishing. On the business side, tackle shops, boat sales, and service industries are a few interests best served by policies that maximise total fishing effort without regard for whether anglers are satisfied by the quality of fishing. Still other businesses (remote lodges, guides) are best served by ensuring that regional fishing quality is low enough to force some anglers into paying for the special opportunities that these businesses provide. Still others would be very well served by local effort limitation; for example, some lodges in British Columbia are located on trout lakes with public access, and their effective operating seasons can be severely curtailed by early season impacts of nonlodge anglers on abundance.

The easiest and most common way for managers to react to these conflicts is to adopt a 'minimum whinge' approach (Pope 1984), which basically involves taking no more action than is needed to quiet the most vocal stakeholders. This leads to 'fire-fighting' (take decisive action only where/when complaints are loudest), to 'shifting baselines' over time (Pauly 1995) where progressively poorer overall quality of fishing is accepted provided stakeholders do not notice it, and to emphasis on less controversial measures like production-side enhancement and stocking programs.

Alternative paths to effort limitation

If we are correct in asserting that management aimed at improving angling quality must take as careful account of consumption-side (fishing effort) dynamics as production-side dynamics, then recreational fisheries managers will have to start thinking much harder about how to directly, fairly, and effectively limit fishing effort for those places where quality of fishing is a high priority. Access and effort limitation is in fact happening across North America, but not as a planned and deliberate management strategy. While short-sighted anglers and economic stakeholders compete for too few fish while bickering about their 'rights' and demanding more fish production, the smart money is finding various ways to acquire access rights and prevent general public access to good fishing sites. These ways range from purchase of stream bank properties to acquisition of fishing rights on large private properties to development of cheap and efficient schemes for accessing remote regions. If this trend continues, it will not be long before some parts of North America look much like Austria, where most of the good fishing waters are tightly controlled (and generally very well managed) by relatively exclusive fishing clubs and private interests. The trend may be limited in western North America by the presence of large public land areas (e.g. National Forests and parks), but the best of these areas are extremely inaccessible and there are considerable areas of private riparian land embedded in the public holdings as a legacy from the early days of settlement (river bottom and lakeside properties were prized by ranchers and other early settlers). Further, as in Austria, we will likely see growth in angling clubs as 'corporate buyers' of riparian land, to make access acquisition and limitation more affordable for groups of anglers with moderate incomes.

While it is not difficult to argue that reducing effort should improve angling in general, there remains much uncertainty about lake-specific effort levels and regional level impacts of effort limitation programs. Should the idea begin to gain popularity of providing at least some fishing locations where effort is severely limited and high catch rates maintained, a critical policy issue will be how many of such locations to provide. Limiting access to all locations could severely impact businesses that depend on maintaining high total angling effort (tackle shops, boat sales, and service sectors), even if the majority of anglers come to prefer far fewer but much better fishing days each. For typical BC trout lakes, we estimate that effort reduction of roughly 60–70% would be needed to substantially improve catch rates and value on 'overfished' lakes (see Fig. 8.3). Such a drastic reduction in fishing opportunity would not likely be acceptable (or economically wise) for more than a small percentage of lakes. A further complication in BC is whether to deliberately target lakes with both fishing lodges and public access as sites for effort limitation, with the dual objective of improving angling quality and protecting the economic interests of lodge owners.

Further, we do not understand the dynamics of angling effort well enough to confidently predict whether or not limiting effort for a large number of locations would cause anglers to redistribute their activity onto unmanaged sites, intensifying quality impacts in those locations. That is, would anglers still spend as much time fishing if they were even more crowded into fewer open-access locations? Would they redistribute their activity differentially into less accessible locations where fishing quality is presently higher, so as to differentially impact those locations?

So there is great uncertainty about both the optimum socioeconomic mix of open access versus limited entry locations, and about possible large-scale side effects of effort limitation on remaining open-access areas. Considering these uncertainties, the best management approach may be to gradually increase the number of limited-access locations while monitoring regional impact on angling quality and quantity. Such an adaptive management approach could be viewed as a 'titration experiment' (J. Kitchell, U. Wisconsin, personal communication), where limited entry locations are added over time until some desirable balance or endpoint is detected. Along the way, we would learn a great deal about the dynamics of fish vulnerability and about how anglers respond to changes in quality fishing opportunities.

Acknowledgements

We particularly thank Eric Parkinson, Jim Kitchell, and Steve Carpenter for stimulating discussions about the behaviour of fish and fishers. We also thank an anonymous reviewer for critical comments on the manuscript. Funding for S.C. was provided by the Natural Sciences and Engineering Research Council of Canada and the British Columbia Ministry of Fisheries.

References

Barnhart, R.A. & Roelofs, T.D. (eds) (1987) *Catch-and-Release Fishing: a Decade of Experience*. Humboldt State University, Arcata, CA.

Champeau, T.R. & Denson, K.W. (1987) Effectiveness of a catch-and-release regulation for largemouth bass in a Florida lake. In: *Catch-and-Release Fishing: a Decade of Experience* (eds Barnhart, R.A. & Roelefs, T.D.), pp. 241–252. Humboldt State University, Arcata, CA.

Engstrom-Heg, R. (1986) Prediction of wild brown trout catch rates from estimated yearling population density and fishing intensity. *North American Journal of Fisheries Management*, **6**, 410–417.

Fraley, J. (1996) Cooperation and controversy in wilderness fisheries management. *Fisheries*, **21**(5), 16–21.

Holland, S.M. & Ditton, R.B. (1992) Fishing trip satisfaction: a typology of anglers. *North American Journal of Fisheries Management*, **12**, 28–33.

Lewynsky, V.A. & Bjornn, T.C. (1987) Response of cutthroat and rainbow trout to experimental catch-and-release fishing. In: *Catch-and-Release Fishing: a Decade of Experience* (eds Barnhart, R.A. & Roelefs, T.D.). Humboldt State University, Arcata, CA.

Malvestuto, S.P. & Hudgins, M.D. (1996) Optimum yield for recreational fisheries management. *Fisheries*, **6**, 6–17.

Moring, J.R. (1993) Effect of angling effort on catch rate of wild salmonids in streams stocked with catchable trout. *North American Journal of Fisheries Management*, **13**, 234–237.

Pauly, D. (1995) Anecdotes and the shifting baseline syndrome of fisheries. *Trends in Ecology and Evolution*, **10**(10), 430.

Policansky, D. (2002) Catch-and-release recreational fishing: a historical perspective. In: *Recreational Fisheries: Ecological, Economic and Social Evaluation* (eds Pitcher, T.J. & Hollingworth, C.E.), Chapter 6, this volume. Blackwell Science, Oxford, UK.

Pope, J.G. (1984) Fisheries resource management theory and practice. In: *New Zealand Finfish Fisheries: the resources and their management* (eds Taylor, J.L. & Baird, G.G.), pp. 56–62.

Rudd, M.A., Folmer H. & van Kooten G.C. (2002) Economic Evaluation of Recreational Fishery Policies. In: *Recreational Fisheries: Ecological, Economic and Social Evaluation* (eds Pitcher, T.J. & Hollingworth, C.E.), Chapter 3, this volume. Blackwell Science, Oxford, UK.

Shaner, B.L., Maceina, M.J., McHugh, J.J. & Cook, S.F. (1996) Assessment of catfish stocking in public fishing lakes in Alabama. *North American Journal of Fisheries Management*, **16**, 880–887.

Walters, C.J. & Bonfil, R. (1999) Multispecies spatial assessment models for the B.C. groundfish trawl fishery. *Canadian Journal of Fisheries and Aquatic Scienc*e, **56**, 1–29.

Walters, C. & Korman, J. (1999) Cross-scale modeling of riparian ecosystem responses to hydrologic management. *Ecosystems*, **2**(5), 411–421.

Chapter 9

The Importance of Angler Motivations in Sport Fishery Management

Barbara Calvert

Abstract

Increasing pressures in many sport fisheries have resulted in the need for management strategies that encourage conservation, accommodate angler desires, and support the sport fishing industry. Strategies that promote a low-consumptive style of sport fishing may be effective in meeting these challenges. This paper describes angler motivations and suggests how they can be used to develop effective low-consumptive management strategies.

Key words: angler behaviour, catch-and-release, low-consumptive, management, motivations, sport fishing.

Introduction

In many sport fisheries, the growing demand by anglers for further sport fishing opportunities and the concurrent decline in fish stocks have resulted in the need for management strategies which encourage conservation, whilst simultaneously accommodating angler desires. Strategies that promote a low-consumptive style of sport fishing may be effective in meeting these challenges. Low-consumptive sport fishing is a form of sport fishing in which the angler practises voluntary catch-and-release of nonrestricted fish, and is attracted to the noncatch aspects of the experience. Because it is unlikely that anglers who practise voluntary catch-and-release fishing will release *all* nonrestricted fish, some consumption of fish will occur. In addition, some risk of mortality exists for released fish; hence the term 'low-consumptive' (Bielak 1988). In order to develop effective low-consumptive management strategies, it is necessary to understand the role of angler motivations in the overall sport fishing experience.

The sport fishing experience

As shown in Fig. 9.1, the sport fishing experience consists of at least three dimensions. These are motivations, behaviour and satisfaction. Motivations are the incentives that inspire a person to go fishing, and that bring about specific angler behaviour (motivations are discussed in more detail below). Satisfaction with the completed experience results from both motiva-

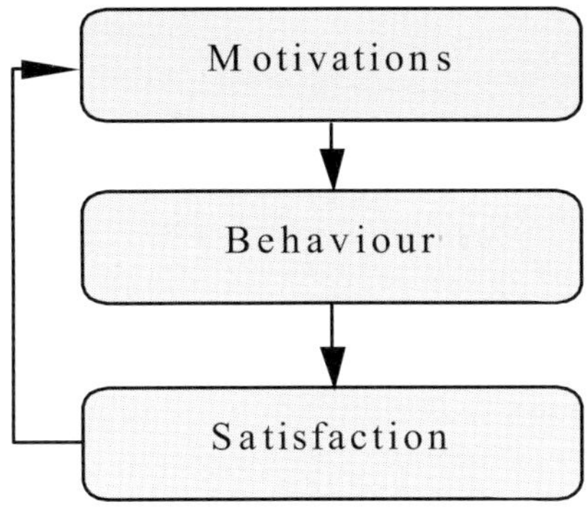

Fig. 9.1 Dimensions of the sport fishing experience.

tions and behaviour (Naito 1992), and influences future motivations to go fishing (Dawson and Wilkins 1980).

Angler behaviour is comprised of catch, social and economic components. The catch component encompasses the angler harvest rate and the fishing effort (angler days) spent by anglers. The harvest rate is influenced by management catch restrictions, voluntary release by anglers of nonrestricted fish, and lack of compliance with regulations by anglers. The fishing effort may be determined by angler motivations, management regulations and fish availability (Naito 1992). The social component of angler behaviour may include behaviour derived from motivations, and from other activities linked with a fishing trip. The economic component of angler behaviour consists of financial investments related to sport fishing (Fisheries and Oceans Canada 1994).

Satisfaction with a sport fishing experience is dependent upon the fulfilment of motivations (Holland and Ditton 1992, Naito 1992). Satisfaction 'is determined by the difference between the outcomes one desires or thinks should be received (motivations) and the perceived fulfilment of those outcomes' (Holland and Ditton 1992: 28). Satisfaction may result in personal benefits to anglers, such as improved mental and physical health (Hendee and Bryan 1978). Determinants of fishing trip satisfaction that are common to many angler populations may be ranked differently in importance according to individual angler subpopulations. Radomski *et al.* (2001) suggest that understanding angler behaviour in order to optimise angler satisfaction will be increasingly important in future management of sport fisheries.

Motivations

Motivations for sport fishing comprise catch and noncatch categories, as depicted in Fig. 9.2. Catch motivations include retention and nonretention factors. Anglers may be inspired to go fishing by the desire to retain caught fish for consumption, or to keep as a trophy. Nonretention incentives which may also entice anglers to go on a fishing trip include the desire to fish for sport, to improve fishing skills, to test fishing gear, and to experience a challenge and thrill; management regulations and fish availability may also influence angler decision making.

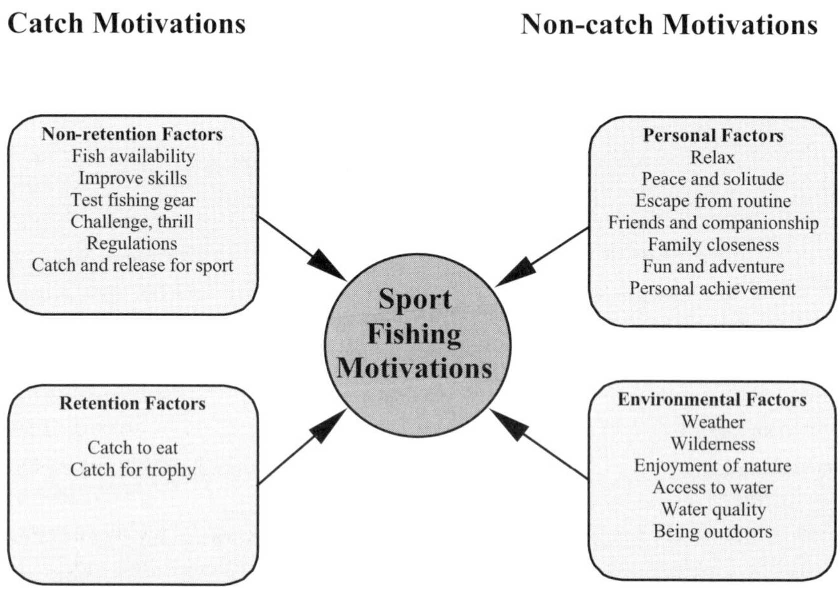

Fig. 9.2 Motivations for sport fishing (Ditton *et al.* 1978, Dawson and Wilkins 1981, Fedler 1984, Loomis and Ditton 1987, Falk *et al.* 1989, Naito 1992, Fedler and Ditton 1994, Walrond 1997).

Noncatch motivations, such as personal and environmental factors, may also play a role in tempting anglers to go on a fishing trip. Personal factors may include the desire for family closeness, for fun and adventure, for personal achievement, for peace and solitude, and to enjoy the company of friends. Anglers may also want to relax or to escape from the daily routine. Environmental factors that prompt anglers to go sport fishing may include the wilderness setting, the anticipated weather, and the water quality and accessibility. The chance to spend time outdoors and enjoy nature may also induce anglers to go fishing.

Angler sectors

The importance of specific motivations to anglers has been debated in the sport fishing literature for over two decades. Some authors suggest that angler populations place a higher priority on noncatch motivations relative to catch motivations (Moeller and Engelken 1972, Bryan 1974, Fedler 1984). Other authors argue that the motivation to catch a fish may be more important or just as important to angler populations as noncatch motivations (Hicks *et al.* 1983, Matlock *et al.* 1988).

Most authors however, concede that an angler population is not a homogenous group characterised by a defining set of motivations. Rather, an angler population is made up of subpopulations or sectors, with each sector distinguished by a characteristic combination of catch and noncatch motivations. The importance of catch motivations relative to noncatch motivations may vary with each angler sector. Individual angler sectors can be targeted by fishery managers in order to accommodate specific group desires (Bryan 1977, Loomis and Ditton 1987, Chipman and Helfrich 1988, Quinn 1992, Gigliotti and Peyton 1993, Fedler and Ditton 1994). Managers can also direct management strategies towards specific angler sectors in order to achieve conservation objectives.

Anglers may be categorised according to target species preferred, fishing method, fishing locale, and fishery management regime (Fedler and Ditton 1994). Club membership (Gigliotti and Peyton 1993, Fisher 1997), financial investments in angling activities (Chipman and Helfrich 1988), gender and age may also classify angler groups.

Angler motivations may also vary according to level of fishing experience. Bryan (1977) split anglers along a specialisation spectrum, from occasional or novice anglers at one end, to technique-setting specialists or experienced anglers at the other end. Occasional anglers were generally unconcerned with environmental setting and had unspecialised fishing skills. The primary goal of this group was to catch fish. At the other end of the spectrum were technique-setting specialists or experienced anglers. These anglers were very specific in their choice of environmental setting and had very specialised fishing skills. This group was interested in catching specific sizes of fish. However, the noncatch aspects of the fishing experience were of primary importance to these anglers.

In between these two angler sectors were the generalists and the technique specialists. The generalists had some preferences for environmental setting and some specialised fishing skills. This angler group was primarily interested in catching a certain number of fish. Technique specialists had specific preferences for environmental setting and had specialised fishing skills. Like the technique-setting specialists, these anglers were interested in catching certain sizes of fish. The noncatch aspects of the fishing experience were also important to these anglers.

Anglers may also be grouped according to the degree of consumptive or catch orientation. Fedler and Ditton (1986) split anglers along a scale of low-, mid- and high-consumptive orientation. Low-consumptive anglers were primarily interested in the noncatch aspects of the experience, but still viewed catching a fish to be somewhat important. Mid- and high-consumptive anglers placed a higher priority on catching fish relative to the noncatch variables. The low-consumptive anglers were more satisfied overall with their fishing trips in comparison to the satisfaction levels of the mid- and high-consumptive anglers. This was due to the higher priority that low-consumptive anglers placed on the noncatch components of the experience. However, Fedler and Ditton (1986: 225) suggested that, although the low-consumptive anglers were primarily interested in the noncatch variables, 'the presence of fish and the possibility of catching one or more is a necessary component of any fishing trip'. A recent study by Ross and Loomis (2001) found that anglers rated relaxation, being outdoors, and experiencing natural surroundings as more important than catching fish.

Catch-and-release

For over two decades in both saltwater and freshwater sport fisheries throughout North America, many managers have used catch-and-release management regimes to meet specific conservation objectives. For example, release of fish by anglers may be required in order to comply with regulations restricting the size or quota of fish caught. Mandatory release of all fish may also be required to satisfy no-kill regulations in areas of serious stock depletion (Barnhart *et al.* 1987, Barnhart 1989). However, this approach to conservation may not be sufficient to address the challenges of increasing pressures on sport fisheries. Cook *et al.* (2001) have shown that mandatory creel limits in Minnesota are not effective in managing the fishery since they often cause anglers to have unrealistic expectations of their potential harvest. What is needed is the formal incorporation of voluntary catch-and-release into management frameworks.

Unfortunately, voluntary catch-and-release practices have not usually been formally promoted by fishery managers, but have traditionally been promoted informally by angling organisations, the sport fishing industry, governments (Bielak 1987) and sport fishing tournaments (Barnhart 1989). Many angling magazines have also championed voluntary catch-and-release practices to anglers in order to encourage angler ethics and to maintain local fishing opportunities (Washburn 1989, Hiltner 1991, Wagner 1992, Fullum 1994, Thomas 1995). In addition, voluntary catch-and-release sport fishing has grown into a personal philosophy for many environmentally conscious anglers.

However, a case study by Schmied (1993) provides an example of the use of voluntary catch-and-release as a management tool. In that case, rising demand for increased sport fishing opportunities in the face of depleting fish stocks presented a challenge for fishery managers in tidal sport fisheries in the south-eastern United States. The myriad of regulations in place was not adequately managing these increasing pressures. Consequently, a promotional campaign was established by fishery managers to encourage anglers to limit their nonrestricted catches voluntarily through catch-and-release practices, in addition to complying with the catch regulations. This option had the advantage of less government interference, and less cost to taxpayers through the reduced need for enforcement programs. The campaign was favourably received by anglers. Schmied (1993) concluded this case study with the suggestion that the promotion of responsible angling behaviour, in conjunction with the use of catch restrictions, has the capability to substantially improve the effectiveness of sport fishery conservation programs.

There is a dearth of literature on the effects of voluntary catch-and-release sport fishing within catch restricted fisheries. However, one study by Clark (1983) on the effects of the voluntary release of nonrestricted fish in four catch restricted Michigan freshwater sport fisheries determined that total mortality rates of fish populations were reduced with voluntary catch-and-release practices. In this study, increasing voluntary release rates were associated with increasing total catch rates of all fish, including trophy fish. Clark (1983) concluded from this study that voluntary release of nonrestricted fish within a catch restricted sport fishery should be promoted by managers.

Low-consumptive sport fishing

In British Columbia, the growing demand by nonresident anglers for further tidal sport fishing opportunities, and the concurrent decline in fish stocks, have resulted in the need for management strategies which encourage conservation, whilst simultaneously accommodating angler desires. A recent case study by Calvert and Williams (1999) identified the principal characteristics of low-consumptive tidal anglers in British Columbia, and suggested ways of promoting low-consumptive sport fishing. In this study, data from a survey on nonresident tidal anglers of British Columbia were interpreted to determine the differences in behav-

Table 9.1 Examples of results from the 1990 nonresident tidal sport fishery survey for British Columbia showing significant differences found for consumptive and low-consumptive anglers. From Calvert (1998).

Survey Criteria	Consumptive	Low-consumptive
Average age	49.2	46.0
Experience (median response)	Average	Experienced
Average no. of fishing days in BC (1990)	3.9	6.8
Average no. of years fishing in BC	4.2	4.6
Average outdoor trips (nonfishing) taken	1.4	1.9
Coho salmon kept per day	1.42	0.95
Importance of consuming fish (0 = low, 5 = high)	3.48	3.34
Importance of enjoying nature (0 = low, 5 = high)	3.89	4.21
Support for reduced bag limits (0 = low, 5 = high)	1.05	1.29
Support for minimum size limits (0 = low, 5 = high)	2.08	2.15
Average amount spent for fly-in lodge packages	$2061	$2909
Satisfaction (median response)	Good	Excellent

ioural, social and economic characteristics of low-consumptive and consumptive anglers. A selection of the key differences found in this study are presented in Table 9.1.

Not surprisingly, low-consumptive anglers were more interested in fishing for sport than for consumption. Compared to consumptive anglers, low-consumptive anglers contributed more significantly to the conservation of fish stocks by practising voluntary catch-and-release of nonrestricted fish. The low-consumptive anglers kept lower average numbers of fish per day than did their consumptive counterparts. As well, the low-consumptive group showed greater support for more catch-and-release regulations and reduced bag limits.

The low-consumptive anglers also valued the nonconsumptive aspects of the sport fishing experience more highly than did their consumptive counterparts. In particular, they were most different from their consumptive counterparts with respect to the importance they placed on such fishing experience attributes as enjoyment of nature, family togetherness, and challenge and excitement.

However, what was surprising in this study was that the low-consumptive anglers contributed significantly more to the tourism economy of British Columbia than did their consumptive counterparts. In particular, they fished more often, stayed longer, visited more regions of the province, and spent more on goods and services in British Columbia, than did the consumptive anglers.

Calvert and Williams (1999) proposed strategies to encourage further low-consumptive sport fishing in British Columbia's tidal waters (Table 9.2) which included angler education, improved regulations, and the marketing of sport fishing packages which incorporate a range of recreational experiences. While these strategies are relevant to the British Columbia tidal sport fishery, they are also applicable to other fishery jurisdictions.

Support among researchers in the sport fishing literature for a sector approach to sport fisheries management lends weight to the argument that managers in pressured sport fisheries should direct specific conservation strategies towards low-consumptive anglers. In addition though, conservation strategies that are more appropriately directed towards the entire angling population may still be necessary. This paper has shown that in order to develop low-consumptive management strategies, it is necessary to understand the importance of angler motivations. Management strategies that encourage low-consumptive sport fishing

Table 9.2 Management strategies promoting low-consumptive sport fishing in British Columbia's tidal waters. From Calvert and Williams (1999).

Goals	Strategies	Measures
Increase angler participation in voluntary catch-and-release of nonrestricted fish.	Institute promotional campaigns to educate anglers on the need for conservation. Institute promotional campaigns to teach anglers the correct release techniques. Establish and promote a code of ethics for anglers.	Public/private partnerships for education between governments, angling organisations and the sport fishing industry. Use of educational tools such as brochures, posters, leaflets, media advertisements, videos, public information meetings/presentations and sporting/outdoor magazines.
Lessen angler emphasis on preferred species. Prevent depletion of preferred species. Increase catch rates and sizes for all species.	Include catch-and-release of nonrestricted fish in daily bag/retention limits. Reduce bag limits for all species. Severely curtail bag limits for depleted species.	Promotion and implementation of these stricter regulations through the federal government's Department of Fisheries and Oceans. Promotion of these stricter regulations by angling organisations.
Attract more low-consumptive anglers who are primarily interested in the nonconsumptive benefits of sport fishing. Accommodate the desires of low-consumptive anglers. Further angler emphasis towards the nonconsumptive aspects of sport fishing.	Promote enjoyment of nature, family togetherness and challenge and excitement as the main components of a sport fishing experience. Promote sport fishing holiday packages offering a variety of recreational activities in conjunction with fishing.	Promotions by the sport fishing industry, angling organisations and the provincial government.
Attract more low-consumptive anglers who are primarily interested in the sporting aspect of fishing. Attract nonanglers who are interested in adventure tourism.	Promote the British Columbia tidal sport fishery as a catch-and-release trophy fishing destination.	Promotion by all sport fishing sectors. Use of promotional tools such as awards, trophy alternatives and tag-and-release fishing.

can facilitate the sustainability of the sport fishing industry not only through social and environmental benefits, but also through economic benefits.

Acknowledgements

Special thanks to Dr Peter Williams (School of Resource and Environmental Management, Simon Fraser University) and Dr David Brunckhorst (School of Rural Science and Natural Resources, University of New England) for providing supervision for this project. Thanks also to two anonymous referees for helpful and constructive comments on an earlier version.

Barbara Calvert carried out this work as part of her Master's thesis for the School of Rural Science and Natural Resources at the University of New England, New South Wales. She is currently working as an analyst at the Department of Fisheries and Oceans, Ottawa, Ontario. The views expressed in this paper are the author's and do not necessarily reflect her employer's views.

References

Barnhart, R.A. (1989) Symposium review: Catch-and-release fishing, a decade of experience. *North American Journal of Fisheries Management*, **9**, 74–80.

Barnhart, R.A., Higgins, P.T., May, R.H. & Roelofs, T.D. (1987) Foreword. In: *Catch-and-Release Fishing A Decade of Experience: A National Sport Fishing Symposium* (eds Barnhart, R.A. & Roelofs, T.D.). Humboldt State University, Arcata.

Bielak, A.T. (1987) Promoting catch-and-release of Atlantic salmon. In: *Catch-and-Release Fishing A Decade of Experience: A National Sport Fishing Symposium* (eds Barnhart, R.A. & Roelofs, T.D.), pp. 126–142. Humboldt State University, Arcata.

Bielak, A.T. (1988) Promoting catch-and-release: The ultimate low consumption fishing technique. In: *Recreational Fisheries Conference Proceedings 1986*. pp. 57–69. Information and Publication Branch, Fisheries and Oceans Canada, Ottawa.

Bryan, H. (1977) Leisure value systems and recreational specialization: The case of trout fishers. *Journal of Leisure Research*, **9**, 174–187.

Bryan, R.C. (1974) *The Dimensions of a Salt-Water Sport Fishing Trip*. Southern Operations Branch, Environment Canada, Vancouver.

Calvert, B. (1998) Management Strategies to Promote Low-consumptive Sport Fishing in British Columbia's Tidal Waters. Unpublished Masters Thesis. University of New England, Armidale, New South Wales.

Calvert, B. & Williams, P. (1999) Low-consumptive angler behaviour and preferred management strategies: the case of sport fishing in British Columbia's tidal waters. In: *Evaluating the Benefits of Recreational Fishing* (ed Pitcher, T. J.). Fisheries Centre Research Reports. **7**(2), 58–63.

Chipman, B.D. & Helfrich, L.A. (1988) Recreational specializations and motivations of Virginia river anglers. *North American Journal of Fisheries Management*, **8**, 390–398.

Clark, R.D., Jr. (1983) Potential effects of voluntary catch-and-release of fish on recreational fisheries. *North American Journal of Fisheries Management*, **3**, 306–314.

Cook, M.F., Goeman, T.J., Radomski, P.J., Younk, J.A. & Jacobson, P.C. (2001) Creel limits in Minnesota: a proposal for change. *Fisheries*, May 2001: 19–26.

Dawson, C.P. & Wilkins, B.T. (1980) Social considerations associated with marine recreational fishing under FCMA. *Marine Fisheries Review*, **42**(12), 12–17.

Dawson, C.P. & Wilkins, B.T. (1981) Motivations of New York and Virginia marine boat anglers and their preferences for potential fishing constraints. *North American Journal of Fisheries Management*, **1**, 151–158.

Ditton, R.B., Mertens, T.J. & Schwartz, M.P. (1978) Characteristics, participation, and motivations of Texas charter boat fishers. *Marine Fisheries Review*, **40**(8), 8–13.

Falk, J.M., Graefe, A.G. & Ditton, R.B. (1989) Patterns of participation and motivation among saltwater tournament anglers. *Fisheries*, **14**(4), 10–17.

Fedler, A.J. (1984) Elements of motivation and satisfaction in the marine recreational fishing experience. In: *Marine Recreational Fisheries*, 9 (ed Stroud, R.H.). pp. 75–83. Sport Fishing Institute, Washington, DC.

Fedler, A.J. & Ditton, R.B. (1986) A framework for understanding the consumptive orientation of recreational fishers. *Environmental Management*, **10**(2), 221–227.

Fedler, A.J. & Ditton, R.B. (1994) Understanding angler motivations in fisheries management. *Fisheries*, **19**(4), 6–13.

Fisheries and Oceans Canada (1994) *1990 Survey of Recreational Fishing in Canada: Economic and Commercial Analysis Report No. 148.* Communications Directorate, Fisheries and Oceans Canada, Ottawa.

Fullum, J. (1994) If you're not going to eat the fish put it back. *The Conservationist*, **48**(4), 18–21.

Gigliotti, L.M. & Peyton, R.B. (1993) Values and behaviors of trout anglers, and their attitudes toward fishery management, relative to membership in fishing organizations: A Michigan case study. *North American Journal of Fisheries Management*, **13**, 492–501.

Haworth, J.T. (1983) Satisfaction statements and the study of angling in the United Kingdom. *Leisure Sciences*, **5**(3), 181–196.

Hendee, J.C. & Bryan, H. (1978) Social benefits of fish and wildlife conservation. In: *Proceedings of the Western Association of Fish and Wildlife Agencies*, **58,** 234–254.

Hicks, C.E., Belusz, L.C., Witter D.J. & Haverland, P.S. (1983) Application of angler attitudes and motives to management strategies at Missouri's trout parks. *Fisheries*, **8**(5), 2–7.

Hiltner, R. (1991) Size restrictions and catch-and-release: Tools for maintaining a healthy fishery. *North Dakota Outdoors*, **53**(9), 10–12.

Holland, S.M. & Ditton, R.B. (1992) Fishing trip satisfaction: A typology of anglers. *North American Journal of Fisheries Management*, **12**, 28–33.

Loomis, D.K. & Ditton, R.B. (1987) Analysis of motive and participation differences between saltwater sport and tournament fishers. *North American Journal of Fisheries Management*, **7**, 482–487.

Matlock, G.C., Saul, G.E. & Bryan, C.E. (1988) Importance of fish consumption to sport fishers. *Fisheries*, **13**(1), 25–26.

Moeller, G.H. & Engelken, J.H. (1972) What fishers look for in a fishing experience. *Journal of Wildlife Management*, **36**(4), 1253–1257.

Naito, G. (1992) *Modelling the Dynamics of Fish and Anglers for Management of Rainbow Trout Trophy Lakes in British Columbia.* Unpublished Masters Thesis. Simon Fraser University, Vancouver.

Quinn, S.P. (1992) Angler perspectives on walleye management. *North American Journal of Fisheries Management*, **12**, 367–378.

Radomski, P.J., Grant, G.C., Jacobson, P.C. & Cook, M.F. (2001) Visions for recreational fishing regulations. *Fisheries*, May 2001, 7–18.

Ross, M.R. & Loomis, D.K. (2001) Put and take fisheries: investigating catch and retention assumptions. *Fisheries*, February 2001, 13–18.

Schmied, R.L. (1993) The use of angler ethics education to support marine fisheries management. *Trends*, **30**(2), 42–47.

Thomas, G. (1995) Catch-and-release: It works! *Louisiana Conservationist*, **47**(5), 10–13.

Wagner, S. (1992) Trophy alternatives: Anglers can have their catch-and-release it, too. *Outdoor Oklahoma*, **48**(3), 31–35.

Walrond, C. (1997) Caples River survey. *Trails in the Third Millennium*. Cawthron Institute. December 2–5, 369–377.

Washburn, L. (1989) Catch and release fishing: a concept worth using. *Iowa Conservationist*, **48**(5), 20–21.

Chapter 10
Evaluating Recreational Fishing in Germany

Werner Steffens and Michael Winkel

Abstract

The number of anglers in Germany may be estimated to be 1.4–2 million, corresponding to 1.7–2.4% of the total population. Prerequisites for recreational fishing in Germany are a governmental fishing licence and an angling permit from the owner of the fishing rights. The total annual catch of anglers in Germany is estimated at 35 000–50 000 tonnes, which is much more than the average yield of commercial fishers in freshwater (4000–5000 tonnes). The most important motivation to go fishing in Germany is enjoyment of nature (77%), followed by relaxation (69%), catching fish for consumption (65%), family outing (60%), meeting friends (50%) and enjoyment of the catch (50%). About 92% of recreational fishers consume the fish they have caught. The annual turnover of recreational fisheries in Germany is at least €1.2bn.

Key words: angling in Germany, catch by anglers, legislation for angling, motivations for angling, possibilities for angling, recreational fishing.

Geographic and demographic situation

The Federal Republic of Germany is situated in the temperate zone between 47 and 55°N in Central Europe and covers an area of 358 000 km^2. The precipitation in the northern lowlands is about 500–700 mm; in the mountains (Alps) in the south of the country it comes to more than 2000 mm. The average temperature during January, the coldest month of the year, in the lowlands ranges from +1.5–-0.5°C. In July the average temperature in northern Germany is +17–+18°C. The mean temperature during the year is about +9°C.

Inland surface waters cover about 800 000 ha, that is 2.2% of the total area. The length of flowing waters is about 7500 km. Lakes are concentrated mainly in the northern, eastern and southern parts of the country.

The country comprises 16 states (Länder) of very different size. The population totals 82 million and the density of population is high, at 230 inhabitants per km^2.

Fish fauna

Today about 70 fish species live in German inland waters. Nearly half of them are cyprinids. Percids and salmonids are also species-rich families. Important species for fishery, commercial as well as recreational, are eel (*Anguilla anguilla*), pike (*Esox lucius*) and pikeperch (*Stizostedion lucioperca*).

The most popular fish species for angling in coastal and marine waters is cod (*Gadus morhua*), but herring (*Clupea harengus*), mackerel (*Scomber scombrus*), garfish (*Belone belone*) and flat-fishes are also popular.

Number of anglers and angler associations

Although exact figures are unfortunately not available, the number of anglers in Germany may be estimated to be about 1.4–2 million (Hilge 1998, Steffens and Winkel 1999). This corresponds to 1.7–2.4% of the total population of the country and is less than the average across the whole of Europe, which amounts to 4.7% (Pintér and Wołos 1998).

About 900 000 of the German recreational fishers are members of local clubs joined in two federal anglers' associations. These are the Association of German Sport Fishers (VDSF) and the German Anglers' Association (DAV). There is a slow but steady increase in the number of anglers in most of the states of the Federal Republic of Germany. Recreational fishing is carried out in inland waters as well as in brackish and marine waters.

Legislation

A prerequisite for recreational fishing in Germany is a governmental fishing licence (Fischerei-schein). This can be acquired after having passed an examination and requires attending a training programme for a period of about 30 hours. In most cases these preparatory courses are organised by the angling associations and the examinations are carried out by or under the supervision of the government authorities. This mode of proceeding ensures that anglers are well qualified for responsible fishing. Since fishery legislation is determined individually by the 16 states there are some differences with respect to the training courses and the examinations. However, generally the main fields of education are water ecology, ichthyology, legal regulations (fishery act, nature conservation, and animal protection), angling equipment and angling practice (see Lukowicz 1998). The minimum age for fishing varies between 7 and 12 years, depending on the legislation of the respective state, and a special children's licence is necessary.

Additionally, an angling permit (Angelerlaubnis) from the respective owner of the fishing right is necessary for fishing. Often, local fishing clubs or regional associations are stakeholders or leaseholders of waters, which are then managed and used by the club members. Regional clubs and state associations linked together in the German Anglers' Association agreed that their waters form a connected water pool (gemeinsamer Gewässerfonds) that is open to all of their members.

Recreational fishing is only possible by rod. Catch-and-release is not accepted. Approved honorary fishery inspectors, usually members of fishing clubs, are engaged in controlling the anglers with regard to responsible fishing (e.g. species, number and size of fish, fishing time).

Fish catches

Catches by individual anglers in Germany may vary considerably from 1 to 200 kg per annum. According to the data of Hilge (1998) the total catch of anglers from inland waters in Germany is estimated to be about 19 000 tonnes per year. Supposing a number of 1.4 million anglers the average individual yield is 13.6 kg per annum. From other estimates the average catch is, however, calculated to be 25 kg per annum. Then the total annual catch for 1.4–2 million anglers would be 35 000–50 000 tonnes. Assuming a €5 (*c*. DM10) per kg basis, this represents about €175–250 m (at present about US$160–225 million).

Figure 10.1 shows the results of surveys in three German states concerning the fishing yield of recreational fishers. An inquiry in Berlin revealed an average catch of 25.8 kg fish per annum for organised fishers (Grosch *et al.* 1977). In Bavaria, the average individual catch of anglers was 16.5 kg per annum (Lederer 1997) and an investigation in Saxony-Anhalt resulted in an average fishing yield of 21.2 kg per annum (Wedekind 2000).

Compared to the catch of recreational fishing, the catch of commercial fishers in natural waters (mainly lakes and rivers) is only 4000–5000 tonnes (Steffens and Winkel 1999, Wedekind *et al.* 2001). Whereas in Germany in the early 1900s, commercial fishery in lakes and rivers was predominant, the situation has changed 100 years later and angling is increasing in significance. Although in the past there was competition at times, the future development of inland fishery in Germany will be characterised by a close co-operation between professional fishery and recreational fishers. Well-educated commercial fishers specialising in the

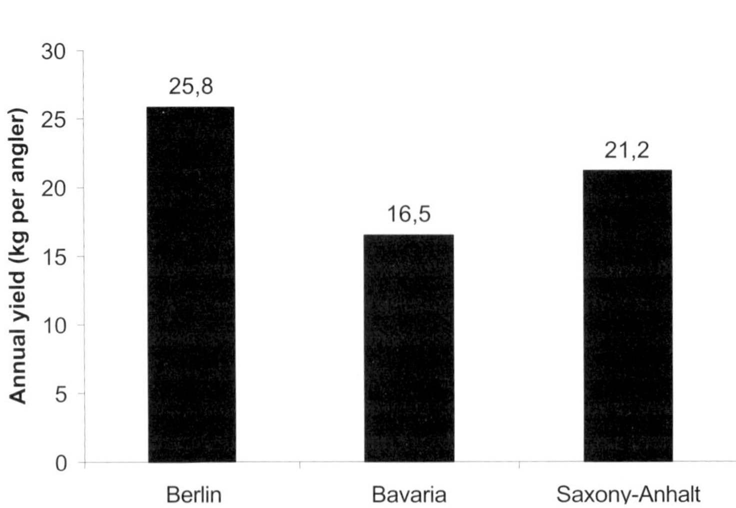

Fig. 10.1 Estimated annual yield of German anglers according to surveys in different German states.

Table 10.1 The ten top species preferred for fishing and for consumption by German anglers (data from the German Anglers Association 1999).

Species preferred for fishing	%	Species preferred for consumption	%
Esox lucius	21	*Stizostedion lucioperca*	18
Cyprinus carpio	20	*Salmo trutta f. fario**	17
Stizostedion lucioperca	17	*Cyprinus carpio*	15
Anguilla anguilla	9	*Anguilla anguilla*	10
*Salmo trutta f. fario**	9	*Esox lucius*	8
Gadus morhua	6	*Gadus morhua*	6
Perca fluviatilis	6	*Perca fluviatilis*	5
Tinca tinca	2	*Salmo salar*	2
Rutilus rutilus	2	*Clupea harengus*	2
Silurus glanis	2	*Rutilus rutilus*	2

* Including *Oncorhynchus mykiss*

management of fishery resources can provide optimal conditions for a large number of recreational fishers.

In an inquiry more than 700 anglers in Germany were asked to nominate the top target species that they would like to fish (Steffens and Winkel 1999). Simultaneously they were requested to name the species they liked best for eating. The results concerning the top ten species are listed in Table 10.1. According to these data, the preferred species for fishing in Germany are pike, common carp and pikeperch. Also of interest to anglers are eel and trout, cod (in marine waters) and perch (*Perca fluviatilis*). For consumption, anglers like pikeperch, trout and common carp best, followed by eel, pike, cod and perch.

Of course, there are local differences with regard to the preferred fish species of recreational fishers. A survey in Saxony-Anhalt revealed that in this state the common carp was the most favoured fish species for anglers followed by pike, eel, pikeperch and tench (*Tinca tinca*).

Motivations for angling

The most important motivation to go fishing in Germany (77% of replies) is enjoyment of nature (Fig. 10.2). Relaxation is also a very essential reason (69%). These results from a federal inquiry of the German Anglers' Association (Steffens and Winkel 1999) are in accordance with the findings of Lederer (1997) in Bavaria, where relaxation (89%) and enjoying nature (87%) were the main motives for angling too. Wedekind (2000) also ascertained in an investigation in Saxony-Anhalt that the most mentioned motive for angling was the desire to be in a natural environment (83%). The following significant reasons in this survey were the engagement in the hobby itself including the desire to catch fish (67%) and the need for relaxation (62%).

Ninety-two per cent of the recreational fishers consume the fish they have caught. This underlines the findings considering the motivations for fishing given in Fig. 10.2. German anglers enjoy nature and use their hobby for relaxation. However, another significant reason for fishing is to capture fish for consumption.

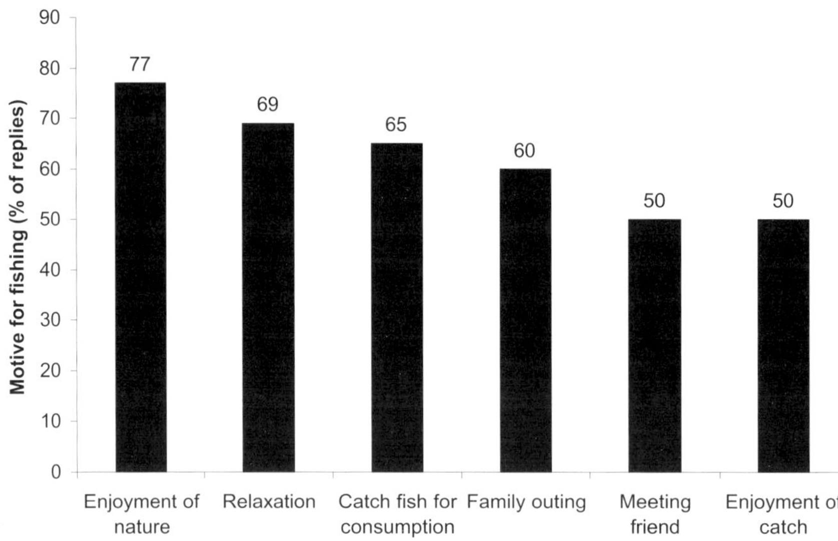

Fig. 10.2 Motives for fishing as shown by the percentage number of replies in which the particular reason was given (data from the German Anglers Association 1998).

According to the recent investigation in Saxony-Anhalt the average age of anglers in this German state is 45.3 years ranging from 13 to 84. Forty-eight per cent of recreational fishers in Saxony-Anhalt are employed in business, trade and industry, 25% are pensioners and 10% are unemployed.

Thirty-eight per cent of anglers go fishing between 10 and 30 days per year, 26% between 30 and 50 days and 20% more than 50 days. The distance to be covered to and from the fishing sites is 10–50 km for more than 56% of anglers, 1–10 km for 32% and more than 50 km for about 10% of anglers.

Economic aspects

There is only little information available concerning the economic aspects of recreational fishing in Germany. Hilge (1998) has given some data obtained from an angling organisation representative (Table 10.2). It has to be assumed that the total amount of DM1753 m (*c.* €900m) per year represents the lower limit. According to Hilge it can be supposed that more than 20 000 people are employed in jobs having direct links to recreational fishing. On the basis of a (very low) annual gross income of at least DM30 000 this would yield DM600 m (*c.* €307 m) which has to be added to the figures in Table 10.2. Thus it may be calculated that the annual turnover of recreational fisheries in the Federal Republic of Germany is at least €1.2bn. However, the chances are that this is a substantial underestimate. The value of club properties (buildings, waters, boats) was estimated to be DM380 m (9500 clubs – each on average DM40 000).

Calculations of average annual expenses of anglers in Saxony-Anhalt resulted in an amount of DM1310. This is based on 30 fishing days, expenses of DM27/angling day and annual costs for equipment of DM500 (Wedekind 2000). Compared to the estimated value

Table 10.2 Economic aspects of recreational fisheries in Germany (Hilge 1998).

Item	Million DM (values per annum)
Value of total annual catch of anglers (1.4 million anglers × 25 kg fish/angler × 10 DM/kg)	350.0
Sales of rod and line, accessories, clothing, etc. (in about 3000 specialised shops)	583.0
Fish for stocking (anglers clubs' part only)	116.7
Labour supply of anglers (cleaning the environment, etc.) (850·000 anglers × 8 hours per annum × 15 DM)	102.0
Honorary functions in clubs and associations at various levels	41.8
Sales of fishing permits by owners of waters	140.0
Fishing tax (as part of fishing licence) (of angling club members only)	6.6
Fees for clubs and associations at district, state and federal level	92.0
Sales of journals and books (350 000 journals per month at 7 DM; 8 DM per book and angler)	40.6
Angling tourism (at home and abroad)	280.0
Total	1 752.7

of the anglers' catch of DM210 per year (annual catch of 21 kg, DM10/kg fish) the expenses are by far higher. In a similar investigation the annual expenses of Bavarian anglers were calculated to be DM1104 (Lederer 1997).

Constraints to recreational fishing

Water resources in Germany are very limited and are, therefore, generally hard pressed at existing levels of demand. Strong measures are undertaken to protect the environment and to maintain or improve the quality of lakes, dams and rivers. This can lead to excessive restrictions, to the prejudice of anglers and other groups of water users. Green movements and animal welfare concerns affect recreational and commercial fisheries as well as agricultural practices and hunting to an increasing extent.

A very serious problem in Germany and other European countries are legally protected fish-eating birds, especially cormorants (*Phalacrocorax carbo*). These birds have multiplied considerably and expanded their distribution during the last two decades all over Europe. They cause high losses in fish populations and reduce the possibilities for fishing in many waters (Steffens 2001).

The lack of viable data collecting systems and the failure of science to deliver practical management actions is disadvantageous for the evaluation and developing of recreational fishing in Germany. Research with respect to biological and socioeconomic problems of angling is urgently needed. Responsible fishing has to be considered a legitimate use of nature and waters, and the image of fisheries must be improved in the public eye.

References

Grosch, U.A., Buchin, H.-E. & Brandt, G. (1977) Zusammensetzung, Fangaufwand, -ziel und -ertrag der Berliner Sportfischerei. *Arbeiten des Deutschen Fischerei-Verbandes*, No. 22, pp. 129–145.

Hilge, V. (1998) Data on recreational fisheries in the Federal Republic of Germany. In: *Recreational Fisheries: Social, Economic, and Management Aspects* (eds Hickley, P. & Tompkins, H.), EIFAC Symposium Dublin, Ireland, 11–14 June 1997, Fishing News Books, Oxford, pp. 10–14.

Lederer, M. (1997) *Die sozio-ökonomische Bedeutung der Angelfischerei in Bayern.* Dipl.-Arbeit Technische Universität Weihenstephan.

Lukowicz, M. v. (1998) Education and training in recreational fishery in Germany. In: *Recreational Fisheries: Social, Economic, and Management Aspects* (eds Hickley, P. & Tompkins, H.), EIFAC Symposium Dublin, Ireland, 11–14 June 1997, Fishing News Books, Oxford, pp. 287–293.

Pintér, K. & Wołos, A. (1998) Summary report of the symposium topic session on the current status and trends in recreational fisheries. In: *Recreational Fisheries: Social, Economic, and Management Aspects* (eds Hickley, P. & Tompkins, H.), EIFAC Symposium Dublin, Ireland, 11–14 June 1997, Fishing News Books, Oxford, pp. 1–4.

Steffens, W. (2001) Beratung der Arbeitsgruppe 'Fischfressende Vögel' auf der 21. EIFAC-Tagung in Budapest. *Fischer & Teichwirt* 52, pp. 164–167.

Steffens, W. & Winkel, M. (1999) Current status and socio-economic aspects of recreational fisheries in Germany. In: *Evaluating the Benefits of Recreational Fishing* (ed Pitcher, T. J.) pp. 130–133. UBC Conference Vancouver, Canada, Fisheries Centre Res. Rep. Vol. 7 (2).

Wedekind, H. (2000) Investigations on recreational fisheries in Saxony-Anhalt, Germany. Paper EIFAC/XXI/2000/Symp. E 31, 8 pp. *EIFAC Symposium on Fishery and Society: Social, Economic and Cultural Perspectives of Inland Fisheries.* Budapest, Hungary, 1–3 June 2000.

Wedekind, H., Hilge, V. & Steffens, W. (2001) Present status, and social and economic significance of inland fisheries in Germany. *Fisheries Management and Ecology* **8** (in press).

Chapter 11

A Survey of the Economic Value of Nordic Recreational Fisheries

Anna-Liisa Toivonen

Abstract

This chapter focuses on experiences in organising a multinational fisheries survey. Concurrence and uniformity was prerequisite for comparisons of the results between countries. In this chapter special attention is on how to acquire commensurate samples in different countries, on the mailing procedure and on the price of the survey. A speculative test was run with estimates of three strategic variables of the survey to see if the sample size could have been smaller. The quality of the obtained data is reviewed using confidence limits as the measure.

Key words: contingent valuation, data quality, multinational survey, sample size.

Recreational fishing in Nordic countries

There is a strong tradition of recreational fisheries in the Nordic countries. Fishing used to be an important part of private livelihood and it has gradually developed into a popular form of spending spare time in the natural world. Vast number of lakes, interesting rivers with migrating salmonid fish populations and long coastlines provide good possibilities for the pursuit. Access to the fishing sites is easy in general.

In spite of the common cultural background there are certain national differences concerning recreational fisheries to be considered in connection with evaluation and surveying. Participation in recreational fisheries varies in the Nordic countries. According to national statistics the participation percentages are as follows: Denmark 12.5%, Finland 40%, Iceland 31.5%, Norway 50% and Sweden 35% (Toivonen *et al.* 2000). Fishing rights in inland waters belong to the landowner in each of the countries. There are often complicated national rules on how the landowner is compensated for leasing the fishing rights to individual fishers. In addition to the fee to the landowner, there is another fishing fee to be paid to the state in Denmark, Finland and Norway but not in Iceland or Sweden. Moreover, common right is applied to recreational fishing: in Finland angling and ice fishing are free for everybody whereas fishing with rod and line is free in coastal areas and in the big lakes in Sweden. In addition, Denmark, Iceland and Norway apply common right in the sea but with restrictions.

Determining economic information

The need for information on economic issues of recreational fishery had been established in at least three preceding conferences or workshops (Petersson 1994, EIFAC 1998, Toivonen and Tuunainen 1997). To set up an international survey on economic valuation of recreational fisheries a working group was formed. Naturally, every country was represented in the group. We headed for governmental organisations and expertise on several subjects like contingent valuation, recreational fisheries management, fisheries economy and statistical surveys. Financing the survey required substantial lobbying and a lot of time and effort, because recreational fisheries in the Nordic countries are managed by different administrative sectors. During the two years of intensive fund-raising the contents of the survey and the practical processes were designed in detail (Toivonen *et al.* 1999) and the survey was implemented from October 1999 to January 2000.

The subject was economic valuation and the method was contingent valuation (Rudd *et al.* 2002), therefore population registers were used as sampling frames to include both fishers and nonfishers in the sample. Not all fishers are members of fishing clubs, for instance, and not all fishing requires a licence, so fishing licence frames result in under-coverage. Nonmembers and nonlicence holders, even nonfishers, can put a value on recreational fishery. Systematic samples were drawn from geographically sorted population registers. Every individual in the age group 18–69 had an equal chance of being chosen in the sample. The research unit was the individual person instead of the whole household. If the whole household had been the research unit, it would have been necessary to ask about the structure of every participating household in the questionnaire. The definition of household varies by country, too.

In practice, however, serious difficulties were encountered due to legislation on person registers in Denmark and Norway. Both countries forbid the export of population register samples for mailing or any other purposes. The problem was settled by conducting the mailings nationally in Denmark and by using telephone catalogue as the sampling frame instead of the population register in Norway. One of the legislative consequences in Denmark was that no background information about the nonrespondents was available and the background information about the respondents originated from themselves, so cross-checking was therefore not possible. In Norway the sample was drawn from a telephone catalogue frame that claims to cover 99.5% of households. No legislative restrictions existed for exporting those addresses. When the representativeness of the response was checked (Roth *et al.* 2001), it was found, however, that young persons, men in particular, were under-represented in the Norwegian response in relation to the whole population. This resulted in higher weights (maximum 9153) in the under-represented groups than in general (mean 1327) in Norway. High weights may be destructive to confidence limits in unfavourable cases. Stratification by demographic variables could have increased the precision of the estimates. We did not have the necessary information on either the share of fishers or the response activity in different age and sex groups in every country until afterwards.

Method and results

The mail survey was executed centrally from Finland in three contacts. The subcontracting printing house produced the translated questionnaires and cover letters on demand, sent out the envelopes and received the reply mail. The addresses were printed on the questionnaires from the address files and windowed envelopes were used. The address files were updated after the first and second wave. The second and the third contact were sent to those who had not replied by the deadline. The data was stored optically by the printing house. This procedure ensured similarity and simultaneity. In Denmark, the survey was very strictly controlled by the Danish working group member to synchronise with the other countries. The functions were also co-ordinated by both Finland Post Ltd and Post Denmark Ltd because the national Danish return mail was routed unopened to Finland for return mail control.

The original sample size was 25 192 Nordic citizens and the final number of respondents was 11 404 (Fig. 11.1). After deletion of the nonreachable, the response rate of the mail survey was 45.8%, Iceland having the lowest percentage of 34.2 and Finland the highest of 51.3. The price of the survey, including only variable costs like addresses, materials, printing, mailing

Fig. 11.1 All 11 404 questionnaires sorted by country and record number (photo Anna-Liisa Toivonen).

Table 11.1 Costs of the survey. Unit costs are calculated for sampled items and for response items. Countries are listed in alphabetical order by their English names.

	Total cost in US$ (1)	Sample size (2)	Unit cost in US$ (3) = (1)/(2)	Response (4)	Unit cost in US$ (5) = (1)/(4)	Response rate as a percentage (6) = 100 × (4)/(2)
Denmark	20 900	5192	4.0	2376	8.8	45.9
Finland	18 700	5000	3.7	2550	7.3	51.3
Iceland	11 800	2500	4.7	840	14.0	34.2
Norway	22 700	5000	4.5	2182	10.4	44.6
Sweden	34 000	7500	4.5	3456	9.8	46.7
Total	108 100	25 192	4.3	11 404	9.5	45.8

services, data storage and some incidental items, was US$108 000 (Table 11.1). The OECD purchasing power parities of 1999 have been used for currency conversions.

The unit cost for the whole sample (column 3) is lowest in Finland and Denmark where the mailings were conducted nationally. The difference between the lowest and highest unit cost is US$1. The unit cost for the response (column 5), i.e. the final unit cost, is high for Iceland in particular, because the response rate was low there and the sample size was small compared to the other countries. The difference between the lowest and highest unit price for the response is US$6.7.

The results of the survey were computed by weighting each value with a weight that was specific to the respondent. Each respondent represents several persons depending on the country, sex and age group. The basic weight is the ratio of population/respondents in respective age and sex groups. Additionally, the weights were calibrated to return the true participation percentage in recreational fisheries in each country. This was necessary because fishers are more likely to respond than nonfishers due to the salience of the topic (Dalecki *et al.* 1993).

The economic value of a nonmarket commodity like recreational fisheries comprises use value and nonuse value. To avoid overestimating the economic value of recreational fisheries, it was assumed that fishers represent the use value and nonfishers the nonuse value respectively. In cases where contingent valuation results are applied to cost–benefit analyses, the willingness-to-pay over and above what has actually been paid is the correct measure for the benefit. We first asked the fishers to count their annual fishing expenses and next we asked how much more they would have been willing to pay for the same fishing experience until it would have been too expensive and they would have stopped fishing. A scenario was depicted on present and future threats to fish stocks and fishing possibilities. Both fishers and nonfishers were asked their willingness to pay for conserving the current state of fish stocks and current quality of recreational fisheries.

Some of the main findings of the survey (Toivonen *et al.* 2000) suggest that in Iceland both annual fishing expenditures and additional willingness-to-pay for the same fishing experience are far higher on average than in the other Nordic countries (Table 11.2). In addition, the nonfishers' mean willingness-to-pay for current state of fish stocks and current quality of recreational fisheries is higher in Iceland than elsewhere. The nonfishers' willingness-to-pay for current state of fish stocks and current quality of recreational fisheries leads to a high estimated total in Denmark where the share of nonfishers is high in the population. The results

Table 11.2 Annual expenditures of fishers and both fishers' and nonfishers' willingness-to-pay for recreational fisheries.

	Fishers' actual expenditures		Fishers' additional willingness-to-pay		Nonfishers' willingness-to-pay	
	Estimated total million US$	Mean/person US$	Estimated total million US$	Mean/person US$	Estimated total million US$	Mean/person US$
Denmark	60	135.3	29	71.3	191	77.2
Finland	201	153.4	82	73.2	81	47.1
Iceland	23	418.5	7	139.8	14	133.1
Norway	193	139.8	106	82.4	79	62.3
Sweden	281	151.9	106	56.5	145	46.1

imply that in Iceland recreational fishery is a business. In the other countries, the cost of the hobby is marginal and fishers could easily spend relatively much more on it. Reasons behind the nonfishers' willingness-to-pay are more complicated. In Iceland the nonfishers may feel that recreational fisheries are of indirect use value to them through tourism, and that could partly be the case in Norway, too. One possible motive for the relatively high willingness-to-pay of Danish nonfishers is that they can already see some of the threats coming true in eutrophic lakes. The lower figures in Finland and Sweden probably reflect pure existence value. The aggregates tell that there is high potential willingness to pay in general, and if that potential can be utilised within the recreational fisheries sector, all will benefit from the increased activity.

When counting the means, standard errors of means were multiplied by 1.96 to obtain the 95% confidence limits. An experiment was conducted to see if it would have been possible to retrieve reliable results with a smaller sample size and thereby obtain savings in the survey costs. Random samples from the data set of 11 404 replies were drawn using the jack-knife technique, 50 replications by country and variable. New weights, means and confidence limits were calculated for each replication. The results (Table 11.3) show that for these particular variables, a pain barrier of ± 30% in the confidence limit is exceeded in Iceland already with the full sample and in Denmark with the random sample size of 75% of the response. In Finland, Norway and Sweden, 50% of the response would still have ensured a decent confidence limit of under ± 30%. This result is due to the low participation percentage in Denmark and low response rate in Iceland, moderate response rate and participation percent in Finland, large sample in Sweden and even share of recreational fishers and nonfishers in Norway, respectively.

In conclusion, for conducting international or national surveys, population register information should be available for scientific purposes. The scientific community should take initiatives to reform and equalise national legislation to the direction of free transfer of information in accordance with the EU Directive 95/46/EC. In implementing a survey, every possible action should be taken to increase the response rate. Where there is a doubt of low participation per cent both a large sample and a reasonable response rate are needed for reliable results. Apart from plain money and work, a multinational survey takes more time than you think.

Table 11.3 Mean 95% confidence limits (±%) of means for random replications of the response. Confidence limits exceeding ± 30% in bold.

	100% of the sample			75% of the sample			50% of the sample		
	Actual expenditure	Additional willingness-to-pay	Nonfishers' willingness-to-pay	Actual expenditure	Additional willingness-to-pay	Nonfishers' willingness-to-pay	Actual expenditure	Additional willingness-to-pay	Nonfishers' willingness-to-pay
Denmark	26	27	12	**31**	**32**	14	**34**	**34**	17
Finland	11	11	14	12	13	16	15	16	19
Iceland	18	23	**39**	21	25	**41**	26	**31**	**43**
Norway	9	13	13	11	15	14	13	17	17
Sweden	13	21	10	15	24	12	18	28	14

Acknowledgements

The Nordic recreational fisheries survey is an outcome of several years' intensive co-operation of the working group members Håkan Appelblad, Bo Bengtsson, Peter Geertz-Hansen, Gudni Gudbergsson, Dadi Kristofersson, Hilde Kyrkjebø, Ståle Navrud, Eva Roth, Pekka Tuunainen and Gösta Weissglas. The survey was made possible by the financial support of the Nordic Council of Ministers, Ministry of Agriculture and Forestry (Finland), The Agricultural Productivity Fund (Iceland) and the participating institutions, University of Southern Denmark, Danish Institute for Fisheries Research, Finnish Game and Fisheries Research Institute, Institute of Freshwater Fisheries (Iceland), Agricultural University of Norway, Directorate for Nature Management (Norway), National Board of Fisheries (Sweden) and Umeå University (Sweden). Rauno Yrjölä from Ympäristötutkimus Yrjölä Oy has helped out with the data processing.

References

Dalecki, M.C., Whitehead, J.C. & Blomquist, G.C. (1993) Sample nonresponse bias and aggregate benefits in Contingent Valuation: an examination of early, late and nonrespondents. *Journal of Environmental Management*, **38**, 133–143.

Directive 95/46/EC of the European Parliament and of the Council of 24 October 1995 on the protection of individuals with regard to the processing of personal data and on the free movement of such data. http://europa.eu.int/eur-lex/en/lif/dat/1995/en_395L0046.html. 21 pp.

EIFAC (1998) Summary report of the symposium topic session on the current status and trends in recreational fisheries. p. 1–4. In: *Recreational Fisheries. Social, economic and management aspects.* (eds Hickley, P. & Tompkins, H.) FAO, Bodmin.

Petersson, Å. (ed) (1994) Fritids- och turistfiske – rapport från Nordisk konferens om Fritids- och Turistfiske Resursförvaltning och Samhällsekonomi, Östersund 11–12 April 1994. TemaNord 1994: 651.

Roth, E., Toivonen, A.-L., Navrud, S., *et al.* (2001) Methodological, conceptual and sampling practices in surveying recreational fishery in the Nordic countries – experiences of a valuation survey. *Fisheries Management and Ecology*, **8**, 355–67.

Rudd, M.A., Folmer H. & van Kooten, G.C. (2002) Economic evaluation of recreational fishery policies. In: *Recreational Fisheries: Ecological, Economic and Social Evaluation* (eds Pitcher, T.J. & Hollingworth, C.E.), Chapter 3, this volume. Blackwell Science, Oxford, UK.

Toivonen, A.-L. & Tuunainen, P. (eds) (1997) Socio-economics of recreational fishery. Workshop 12–14 May 1997, Vaasa. TemaNord 1997: 604.

Toivonen, A.-L., Tuunainen, P., Navrud, S., Roth, E., Bengtsson, B. & Gudbergsson, G. (1999) Measuring the total economic value of recreational fisheries in Scandinavia. In: *Evaluating the Benefits of Recreational Fishing.* Papers, Discussion and Isssues: Fisheries Centre Research Reports, Volume 7 Number 2 (ed. Pitcher, T.J.), pp. 150–153.

Toivonen, A.-L., Appelblad, H., Bengtsson, B., *et al.* (2000) Economic value of recreational fisheries in the Nordic countries. Nordic Council of Ministers, TemaNord 2000: 604. http://www.norden.org/fisk/sk/publikationer.asp

Chapter 12
An Evaluation of Recreational Fishing in England and Wales

Jim Lyons, Phil Hickley and Stephen Gledhill

Abstract

The size, economic value, administration and use of the freshwater recreational fishery in England and Wales is described in the context of status and trends. About 26 000 km of river and 30 000 lakes are fished by 3.5% of the population. A case history of the River Trent shows that overall improvement in river quality and fish catch rates has conflicted with anglers' perceptions and appears not to have provided anticipated socioeconomic benefits. Studies on the economic and social values of fisheries in general showed that throughout England and Wales the annual expenditure by anglers is £2.4bn ($3.41bn) and that the existence values, benefits to rural economies and social benefits are significant. The future challenge for the Environment Agency is to meet its statutory obligations as the Government's environmental regulator whilst improving and developing recreational fisheries.

Key words: angling, economic management, fisheries management, social value.

Introduction

Within Europe freshwater angling is one of the most popular of sports with at least 20 million participants (Wortley 1995). This pattern holds good for England and Wales where almost 2.9 million people, 3.5% of the population, go fishing (National Rivers Authority 1995). [According to Steffens and Winkel 2002 the European average is 4.7% of the population – eds.] With approximately 26 000 km of river and 30 000 lakes being fished, the fishery resource is rich and varied and ranges from highland streams to lowland rivers, natural lakes to water supply reservoirs, and canals to old gravel pits.

All inland freshwater fisheries in England and Wales are in private ownership as, historically, all fishing rights were associated with the possession of the adjacent land. Fisheries regulation therefore falls into both public and private sectors according to the nature and level of control required (Hickley *et al.* 1995). The responsibility for implementation and enforcement of legislation rests with the Environment Agency (Environment Act, 1995) whereas site-specific rules can be laid down by the owners of fisheries. Anglers wishing to fish require both a rod licence from the Environment Agency and consent from the owner of the fishing rights, usually provided as a day permit or angling club membership ticket.

The European Inland Fisheries Advisory Commission (EIFAC) recognised an increasing perception of the wide extent and high value of recreational fisheries and went on to recommend that authorities responsible for the control of recreational fisheries take into account the full economic and social value of the aquatic ecosystem and, where necessary, should review policies (Hickley and Tompkins 1998). In this context the UK Government recently commissioned a review of policy and legislation relating to the salmon and freshwater fisheries of England and Wales (MAFF 2000). The principal conclusion from the review, which provided 195 detailed recommendations, was that the conservation of freshwater fish and the management of fisheries should aim to:

- Ensure the conservation and diversity of freshwater and migratory fish and conserve their aquatic environment
- Enhance the contribution freshwater fisheries make to the economy, particularly in remote rural areas and in areas with low levels of income
- Enhance the social value of fishing as a widely available and healthy form of recreation

Since the mid-seventeenth century when Izaac Walton (1593–1683) commended angling as a recreational pursuit (Walton 1653) fishing practices have evolved and developed. The traditional picture of the affluent game (salmonid) angler and the working class coarse (nonsalmonid) angler has changed with the evolution of a whole variety of different angling practices, particularly in the context of coarse fishing. Recreational fishing in certain sectors of the sport is now big business and can be important both in contributing to the local, rural economy and in providing social benefits in urban areas. It is also increasingly recognised that recreational fishing fulfils a valuable role in raising environmental awareness of wildlife and the environment. Comparison of results from two nation-wide angling surveys (NOP Market Research Ltd 1971, National Rivers Authority 1995) confirm that anglers' habits and preferences change with time. Whilst a complementary shift in attitudes and perceptions might be predicted, this can give rise to unreasonable expectations on the part of the angler [see also Cox and Walters 2002 – eds]. A major problem at present is the way that river fisheries are often reported in the angling press as not performing as well as they used to. The following case study for the River Trent illustrates the situation well.

Case study: the Lower Trent fishery

Recreational fishing and its management in England and Wales today is a reflection of historical changes to legislation, altered environmental conditions and the varied perception of anglers as to what constitutes a good day's fishing. This case study reflects both specific local issues and the wider changes seen in recreational fishing across England and Wales in recent years.

The River Trent (Fig. 12.1) is one of the UK's largest rivers and represents an important recreational fishery at both a local and national level. The Trent catchment has a population of six million people, it is 286 km long and drains an area 10 435 km^2. This study relates to fishing on a 169 km length of the lower river and includes the major East Midlands conurbations of Derby, Nottingham and Leicester.

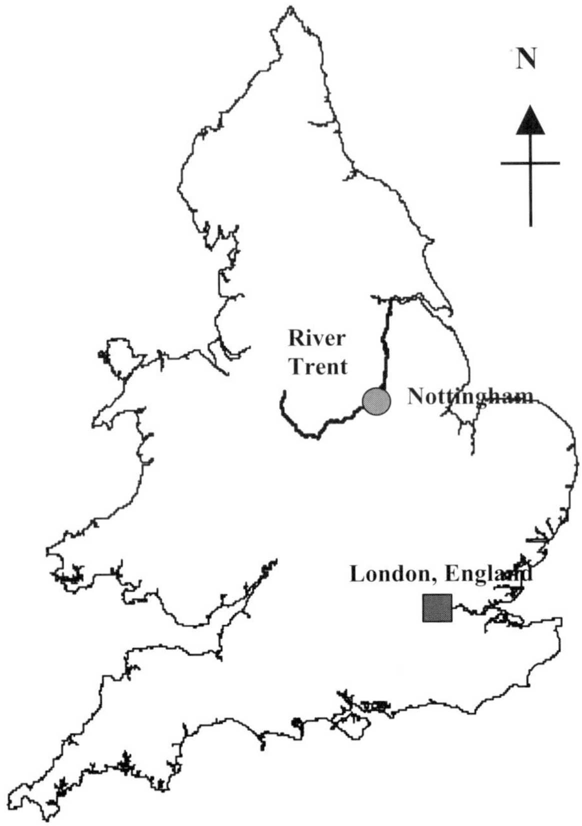

Fig. 12.1 Map showing the geographic position of the River Trent.

Historical perspective

Historically the sport fishery was a nationally popular venue for competition anglers. Traditionally the major target species were roach *Rutilis rutilis* (L.), bream *Abramis brama* (L.), chub *Leuciscus cephalus* (L.) and gudgeon *Gobio gobio* (L.), a reflection of both their numerical dominance within the fish community and the high regard in which they were held by Trent anglers. Between the 1960s and the early 1980s the river reached its peak in popularity with many National and All England Championships being hosted. Another measure of popularity is the number of rod licences sold during this period. Records for the river (Fig. 12.2) show that during this period over 300 000 licences were purchased annually. Since the mid-1980s there has been a steady decline in the river's popularity, culminating with a decision by the National Federation of Anglers to suspend holding National Championships on the river for at least the next five years.

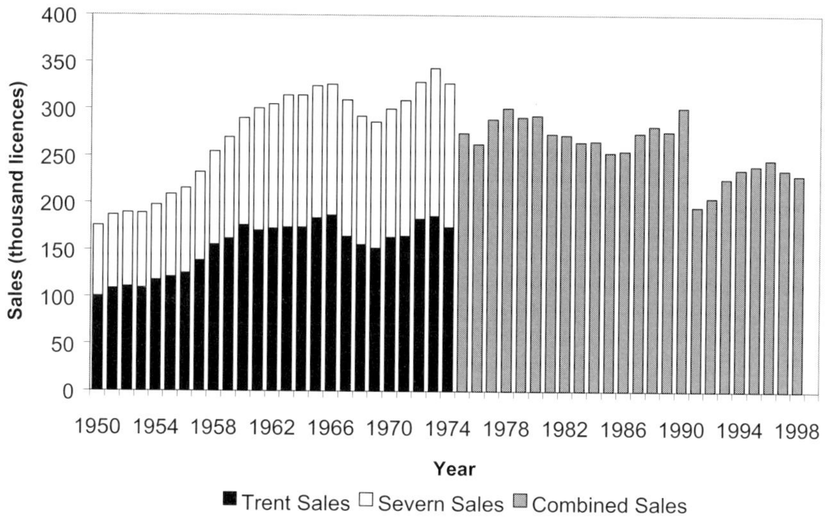

Fig. 12.2 Sales of rod licences for the Severn and Trent catchments 1951-1999.

The issues

Environmental conditions

Prior to the early 1980s a high organic load was found along most of the river (Fig. 12.3). In addition, water used at power stations in the cooling process was discharged to the river causing elevated river temperatures. Improvements in sewage treatment works and industrial effluents have led to both a significant reduction in organic loading and improved water clarity. Water temperatures have also returned to a more natural cycle following the closure of a number of power stations along the river and now reflect ambient air temperatures more closely. Hydroacoustic and angling census studies show that the fishery has responded to this environmental change. Prior to water quality improvements large numbers of roach and gudgeon, whose distribution was highly regular along the whole river, dominated the fish community. Today the fish population is more diverse in nature, distribution along the river is more sporadic and fish numbers are relatively lower. Reduced organic input, lower water temperatures, slower growth rates, lack of cover, and predation are all factors that may be contributing to this population shift.

Contest fishing usually requires a draw to be made at the start of a match to determine the angler's fishing position. The changes seen in the Trent fishery in recent years have meant a significant shift in the importance of this draw with anglers' skill and experience being in many cases replaced by luck.

The situation found today by contest anglers demonstrates that the presence of good fish populations does not necessarily lead to good fishing for all. Whilst there have been improvements in water quality and catch rate, the response of the fish community has led to a fishery that no longer provides the ideal competition fishing conditions that once existed. The result is that anglers who enjoy this type of fishing have left the river.

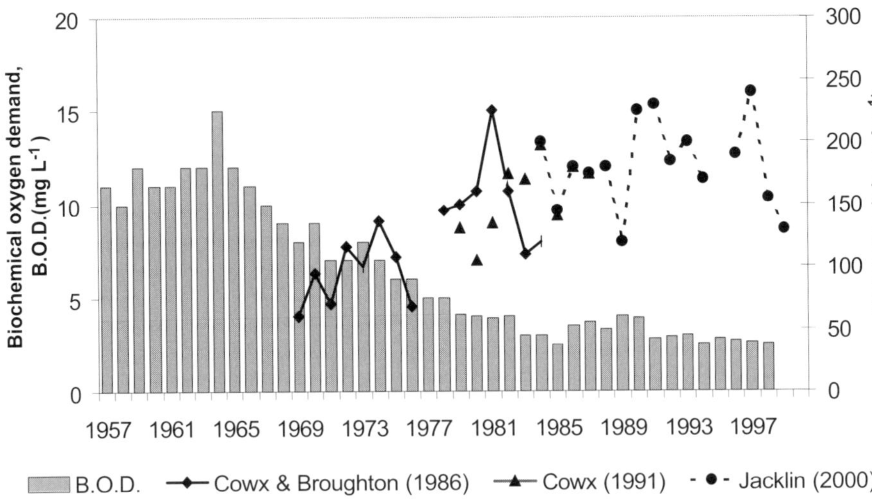

Fig. 12.3 Catch rate (g/man hr) and B.O.D. (mg/l) for the River Trent between 1957 and 1999.

Availability and access

The Lower Trent has traditionally attracted anglers from a wide catchment of neighbouring counties, especially South Yorkshire. The attraction was due to two main factors: the good reputation that the river commanded within the competition fishing fraternity and the absence of good angling closer to home due to the highly polluted, and often fishless, nature of local rivers. Recent improvements in the availability, access and water quality of local rivers together with the creation of new stillwater fisheries catering for the contest angler have combined with the loss of the River Trent's reputation as 'the UK's premier match fishery'.

Economic decline

Community

The dominant industry in many areas within the Lower Trent catchment was coal mining. For many towns and villages employment both direct and indirect relied on the prosperity of these mines. Since the early 1980s the majority of mines that worked the coalfields of North Nottinghamshire and South Yorkshire have closed, leading to high unemployment and economic decline in many of the affected areas. Within these communities fishing was the largest single participant sport with many fishing the Trent on a regular basis. A typical day's match fishing on the Trent would include transport, bait, entrance fees, food and drink, with little change being seen from £50 (approx. $70). With lower prosperity the amount spent on fishing became a primary concern with many anglers. One way to reduce this expense is to fish closer to home. With the increase in availability of local fishing, both riverine and stillwater, these are now realistic alternatives to the River Trent. The result is a significant decline in the river's lucrative angling trade.

Angling clubs

The demographic shift seen in recent years in the Trent catchment whether voluntary or enforced has had a dramatic and, in some cases, a terminal effect on clubs that own or rent fisheries on the river. The majority of these angling clubs has in the past relied on a constant stream of match booking to cover their costs. Increasingly match bookings have become sporadic and in many cases no longer provide sufficient income for these clubs. A primary example is the loss of annual National Championships, each match often attracting over 2000 anglers, and estimated revenue of £100 000. In addition, current rents largely reflect this 'boom period' of the 1970s and 1980s during which matches were held on every available reach. The current situation is one in which many kilometres of riverbank are devoid of anglers. For those clubs that own the riparian land and its fishing rights a financial buffer exists. However most clubs rent their respective fisheries and hence this buffer does not exist leaving many with a financial imbalance. At worst bankruptcy occurs and some clubs have already suffered this fate.

Social change

As an integrated member of the public the angler reflects many of the aspirations and changes in perspective seen throughout society. The increased expectation among the general public for quick success and instant entertainment now exists within the fishing community. In addition, competition for an individual's leisure time has also increased dramatically in recent years. Computer games and access to a wider range of sports and hobbies allied to improved transport facilities mean that local angling no longer enjoys the dominant position it previously held. A lack of new juvenile recruits to angling is not only confined to the Lower Trent area. Nationally the present age profile compared to 1980 indicates fewer anglers under the age of 16, and more over the age of 55 (National Rivers Authority 1995).

The requirements for remaining anglers have been met in many instances by the current explosion of small, highly stocked stillwater fisheries. These waters tend to contain a monospecific fish community of carp, *Cyprinus carpio* L., most fish weighing between 0.5 kg and 3 kg. In addition to fishing, these new fisheries also provide their customers with eating facilities, rest-rooms, bait and tackle supplies. Many also cater for the nonangler by providing entertainment for the whole family, creating a 'supermarket scenario' with its one-stop supply ethos.

Over one hundred of these fisheries have been constructed in the North Nottinghamshire and South Yorkshire area providing local, sought-after fishing conditions. These purpose-built fisheries provide a fishing experience far removed from that which a large, natural river such as the Trent has to offer. This trend towards fishing intensively stocked stillwaters is also seen nationally with over half of all coarse anglers preferring to fish stillwaters instead of rivers, with carp the most popular quarry (National Rivers Authority 1995). Whether this is good or bad for fishing is a hotly debated topic, its merits depending upon the individual's perspective of what constitutes a good day's angling. It is clear, however, that in the Lower Trent area this move towards stillwater fishing has contributed to the decline in the number of anglers fishing the River Trent.

Anglers who fish natural lakes and rivers soon learn about the different fish species, their behaviour and the wildlife that surrounds them. The experiences gained from fishing small, purpose-built, newly constructed stillwaters does not provide the same benefits, as many of these fisheries contain only one target species of fish. While this does provide relatively easy catches for the new angler it takes away much of the 'whole angling experience' and this may cause recreational fishing to shift from a life-long pursuit to a short-term activity.

Scientific information

Data have been collected from Trent fishing competitions since the early 1970s (Cowx and Broughton 1986). The collection of this information has provided valuable cost-effective data on fish populations. The approach has also meant that the Environment Agency has been in direct contact with the anglers on a regular basis, which has been of great value for promoting good public relations. However, quantitative scientific information on the Trent's fish stocks is still relatively sparse although the increased use of modern technologies such as hydroacoustics and acoustic telemetry are beginning to provide this information.

Conclusion

The study of the River Trent demonstrates that recreational fishing in England and Wales has some clear economic, social and environmental benefits. However, the recent trend for anglers to move away from the river emphasises a growing tension between natural self-sustaining fisheries and artificial stocked fisheries. With a move to purpose-built stillwaters, single species fisheries, a desire for larger fish and exotic species, many of the environmental benefits of recreational fishing are being lost. This highlights a situation of countering pressures that are facing the Environment Agency in selecting the right balance between providing 'what the angler wants' and managing fish and their environment (Environment Agency 1999). On the River Trent, where significant environmental improvements have been achieved, the fishery has in fact declined, creating new management issues that were unforeseen.

The England and Wales perspective

Amount of fishing

There are an estimated 2.3 million coarse anglers and 0.8 million game anglers in England and Wales. Coarse fishers make an average of 43 trips per year whereas trout and salmon fishers make 16 and 7.5 trips per year respectively (National Rivers Authority 1995). Approximately 15 000 salmon, *Salmo salar* L. (0.5 fish/angler/year) and 20 000 sea trout, *Salmo trutta* L. (0.8 fish/angler/year) are taken by rod and line and a typical angler's catch for a lowland coarse fish river would be a nominal 100 g/angler/hour (Hickley 1996). Game fishers tend to keep their catch, requiring stock management of some kind, whereas coarse fishers usually return their fish to the water after capture such that there is minimal impact on the fish resource (Wortley 1995).

Angler preferences

Preferences amongst coarse anglers for target species and type of fishery vary (National Rivers Authority 1995) with one quarter of anglers not minding which species was caught. Of those with a preference, 36% of coarse anglers expressed a preference for carp, 28% for roach, and 21% for bream. Regarding location, 52% of coarse anglers fished stillwaters most often, 35% rivers and 14% canals. Recent changes in angler preference have led to the intensive stocking of some waters together with the introduction of nonnative fish in others. Salmon anglers obviously prefer to catch salmon but more often catch sea trout and, similarly, trout anglers prefer to catch brown trout, *Salmo trutta* L. The opening up of a large number of still-water trout fisheries that principally stock rainbow trout, *Oncorhyncus mykiss* (Walbaum), has, however, made the sport much more accessible.

Economic and social value of fisheries

Recreational angling in England and Wales is an important business with an estimated total annual expenditure of £2.4bn ($3.41bn) on coarse angling (£1070/angler/year) ($1519/angler/year) and £0.92bn ($1.30bn) on game angling (£1093/angler/year) ($1552/angler/year). These figures reflect direct expenditure on fishing trips and include items such as travel, food and drink, bait, tackle and permits (National Rivers Authority 1995) but do not take account of expenditure on accommodation which is a particularly important area of expenditure that benefits rural economies (MacAlister, Elliott & Partners 1999).

Environmental economics is an important tool for the strategic management of the aquatic environment (Postle 1993), and in recent years attempts have been made to quantify the economic value of recreational fisheries (Postle and Moore 1998). In its review of fisheries legislation (MAFF 2000) the Freshwater Fisheries Review Group estimated that at least 12 000 full-time equivalent jobs are directly dependent on the sale of fishing tackle alone. Furthermore, the estimated total value of major items of fishing tackle, bait and magazines is £380m ($540m).

Radford *et al.* (1999) evaluated the salmon fisheries of the UK and, with capital asset values, assessed each salmon to be worth £3600 ($5112) in economic terms. In promoting the importance of wild trout as a resource, Elliot (1989) considered the value of each fish to be worth £500 ($710). Only limited studies have been carried out on UK coarse fisheries but work by Green and Turnbull (1992) estimated the consumer surplus value of a coarse angler day to be £6.67 ($9.47). Middlesex University (1994) showed that willingness to pay for a day's coarse fishing ranged from £5 to £8.80 ($7.10 to $12.50), according to river.

Peirson *et al.* (2001) describe the value of fisheries in terms of existence value, rural economics and the social benefit of urban fisheries using case studies. The existence value, that which is derived by an individual from knowing the resource exists, might be considered as a public interest in a resource regardless of whether or not it is exploited. Financial figures have been attributed to existence values; however in some circumstances it could be argued that it is their political rather than financial value that is of greater importance.

Benefits to the economy and social fabric of rural areas is apparent when rivers such as the Teifi are considered. This mixed salmon and sea trout fishery in rural Wales has not only injected money into the local economy but has also contributed to other substantial benefits including employment. It was estimated that as many as 15 jobs could be dependent on angling on the River Teifi. Scaling this figure up to cover all rural fisheries in England and Wales would, even with the most conservative of estimates, show a significant positive impact on rural employment and prosperity (Peirson *et al.* 2001).

An example of the wider benefits that angling brings to urban areas is seen in Leeds, a large city (population *c.* 725 000) in the north of England, where a recent survey (MacAlister, Elliott & Partners 1999) has found that an important reason for people going fishing is 'being with friends'. Many of the angling clubs are based at social clubs and places of work and the survey highlights how fishing plays an important social, communication and relaxation role in the lives of the survey respondents, who were principally adult men. In the inner city, recreational fishing is also important in raising social and environmental awareness amongst young people, giving them an alternative to crime and drugs. An annually organised fishing competition with the title 'Get hooked on fishing, not drugs and crime' has been driven by the Yorkshire Police and the Environment Agency, and is being replicated in other parts of the country. In Southern England the Environment Agency and the Hampshire Police Force have directly supported a number of inner city fisheries projects that have provided both angling opportunities for the young and access for the disabled.

Conservation issues

In England and Wales there are 30 riverine Special Sites of Scientific Interest (SSSIs) designated under the Wildlife and Countryside Act 1981. Also nine species of fish gain protection under the European Habitats Directive 1992 and six rivers have been put forward as Candidate Special Areas of Conservation for, amongst other ecological features, their fish populations.

The numbers of Atlantic salmon returning to UK waters have declined significantly over the last ten years, most notably for spring salmon. In response, new restrictive national legislation was introduced in April 1999 to meet international demands for action (NASCO 1998). Also conservation measures such as catch-and-release and the management of fisheries habitat to maximise in-river production are central tenets of Salmon Action Plans produced by the Environment Agency for each of the major salmon rivers in England and Wales (National Rivers Authority 1996).

For many people recreational fishing provides them with their first introduction to learning about wildlife and the environment. By raising people's environmental awareness wildlife and their habitats have a better chance of being protected and enhanced for the benefit of present and future generations. Anglers not only perform an important role for the nation by concerning themselves with issues such as diffuse pollution and over-abstraction but they also act as environmental watchdogs, often being the first to report a pollution incident.

The Environment Agency

As part of the sustainable and integrated management of air, land and water the Environment Agency has specific responsibilities for water resources, pollution prevention and control, flood defence, fisheries, conservation, recreation and navigation (Environment Agency 1997). In particular the Agency has a statutory duty to 'maintain, improve and develop the salmon, trout, freshwater and eel fisheries' of England and Wales.

Direct spend on the Agency's fisheries service is over £20m ($28.4m) of which approximately £15m ($21.3m) is income from fishing licence sales, with the remainder being provided by the Government as a grant from public funds. The Agency's total staff complement is 10 000 of which 480 are fisheries employees. Activities include monitoring fish stocks, enforcing fisheries legislation, carrying out habitat enhancement work and providing advice to fishery owners and occupiers. In addition, other Agency work (£590m annual expenditure) takes fisheries into account, e.g. minimum acceptable flows (water resources), setting environmental quality objectives (environmental protection), design of major schemes (flood defence), and protection of habitat (conservation). The concept of partnership with stakeholders is recognised and owners, users and general fisheries interests are consulted either formally through statutory committees or informally through local angling consultative organisations and *ad hoc* liaison groups.

Future challenges

As the statutory authority for environmental protection in England and Wales, the Environment Agency has a pivotal role to play in the future strategic planning of recreational fishing in England and Wales. Universal concerns are water resource management, land use practices, diffuse and point pollution, urban development, fish disease, non-native species and public attitudes toward angling (Hickley and Tompkins 1998). However, the only statutory objective for the management of freshwater fisheries is the Environment Agency duty to 'maintain, improve and develop' and this duty can be difficult to interpret satisfactorily and also lacks any sense of overall purpose. To address this difficulty the Agency has adopted a corporate goal that 'All waters in England and Wales will be capable of sustaining healthy and thriving fish populations and everyone will have an opportunity to experience a diverse range of good quality fishing' (Environment Agency 1999). In addition, the Agency has consulted on a vision for its contribution to sustainable development (Environment Agency 2000). Within this vision there are two important components that recognise the human and fish dimensions of recreational fisheries, namely improving the quality of life and enhancing wildlife. Thus to make any progress the regulation and management of recreational fisheries must address overall fishery performance, i.e. the total package of fish stocks, fish habitat, fish catches and the anglers' environment, and at the same time integrate with all the other environmental functions of the Agency (Environment Agency 1998a,b).

The challenge for the Environment Agency is to meet its statutory obligations as an environmental regulator whilst facilitating the improvement and development of recreational fisheries. The likely key to success will be to strive for a framework of integrated catchment management, sustainable resource development and appropriate levels of funding, and to work in partnership with the anglers and the wider community.

Acknowledgements

The authors wish to acknowledge the assistance of their colleagues at the Environment Agency in the compilation of this chapter. Thanks also go to the *Angling Times* for providing many of the slides used in the presentation that accompanied this paper and to Lawrence Talks who co-presented the paper. The views expressed are those of the authors and not necessarily those of their parent organisation.

References

Cowx, I.G. (1991) The use of angler catch data to examine potential fishery management problems in the lower reaches of the River Trent, England. In: *Catch Effort Sampling Strategies* (ed I.G. Cowx), pp. 154–165. Fishing News Books, Blackwell Scientific Publications, Oxford.

Cowx, I.G. & Broughton, N.M. (1986) Changes in the species composition of anglers' catches in the River Trent (England) between 1969 and 1984. *Journal of Fisheries Biology,* **28**, 625–636.

Cox, S. & Walters, C. (2002) Maintaining quality in recreational fisheries: how success breeds failure in management of open-access sport fisheries. In: *Recreational Fisheries: Ecological, Economic and Social Evaluation* (eds Pitcher, T.J. & Hollingworth, C.E.), Chapter 8, this volume. Blackwell Science, Oxford, UK.

Elliott, J.M. (1989) Wild Brown Trout *Salmo trutta;* an important national and international resource. *Freshwater Biology,* **21**, 1–5.

Environment Agency (1997) *An Environmental Strategy for the Millennium and Beyond.* Environment Agency, Bristol.

Environment Agency (1998a) *An Action Plan for Fisheries.* Environment Agency, Bristol.

Environment Agency (1998b) *Freshwater Fisheries and Wildlife Conservation – a good practice guide.* Environment Agency, Bristol.

Environment Agency (1999) *Coarse Fisheries Strategy.* Environment Agency, Bristol.

Environment Agency (2000) *An Environmental Vision.* Environment Agency, Bristol.

Green, C.H. & Tunstall, S.M. (1992) *Valuing Water Quality Improvements.* Report to Department of the Environment.

Hickley, P. (1996) Recreational fishing in England and Wales. European Inland Fisheries Advisory Commission Report of the Workshop on Recreational Fishery Planning and Management in Central and Eastern Europe, Zilina, Slovakia, 22–25 August 1995. *EIFAC Occasional Paper* **32**, 79–85.

Hickley, P., Marsh, C. & North, E. (1995) Ecological management of angling. In: *The Ecological Basis for River Management* (eds Harper, D.M. & Ferguson, A.J.D.), pp. 415–425. John Wiley & Sons Ltd, Chichester.

Hickley, P. & Tompkins, H. (eds) (1998) *Recreational Fisheries: Social, Economic and Management Aspects*: Fishing News Books, Blackwell Scientific Publications, Oxford.

Jacklin, T. (2000) Aspects of Fisheries Management on the River Trent – A large lowland river. PhD thesis. University of Nottingham.

MacAlister, Elliott & Partners Ltd (1999) *Economic Evaluation of Inland Fishing in England & Wales Case Study Reports: Thames, Teifi and Leeds.* Case Study Reports (W2–039).

MAFF (2000) *Salmon and Freshwater Fisheries Review.* London: Ministry of Agriculture, Fisheries and Food, PB 4602.

Middlesex University (1994) The Evaluation of the Recreational Benefits and Other Uses Values from Alleviating Low Flows. *NRA R&D Note* 258. Bristol: National Rivers Authority.

NASCO (1998) *North Atlantic Salmon Conservation Organisation Agreement on a Precautionary Approach*, CNL (98)46.

National Rivers Authority (1995) National Angling Survey 1994. *NRA Fisheries Technical Report* 5.

National Rivers Authority (1996) *A strategy for the management of salmon in England & Wales.* National Rivers Authority, Bristol.

NOP Market Research Ltd (1971) *National angling Survey 1969–70.* Natural Environment Research Council, London.

Peirson, G., Tingley, D., Spurgeon, J. & Radford, A. (2001) Economic evaluation of inland fisheries in England and Wales. *Fisheries Management and Ecology*, **8**, 415–24.

Postle, M. (1993) Development of environmental economics for the NRA. *National Rivers Authority R&D Report* 6.

Postle, M. & Moore, L. (1998) Economic valuation of recreational fisheries in the UK. *Recreational Fisheries: Social, Economic and Management Aspects.* (eds Hickley, P. & Tompkins, H.), pp. 299–304. Fishing News Books, Blackwell Scientific Publications, Oxford.

Radford, A.F., Hatcher, A.C. & Whitmarsh, D.J. (1999) An economic evaluation of salmon fisheries in Great Britain. *Report prepared for the Ministry of Agriculture, Fisheries and Food.* Centre for Marine Resource Economics, Portsmouth Polytechnic.

Rudd, M.A., Folmer, H. & van Kooten, G.C. (2002) Economic Evaluation of Recreational Fishery Policies. In: *Recreational Fisheries: Ecological, Economic and Social Evaluation* (eds Pitcher, T.J. & Hollingworth, C.E.), Chapter 3, this volume. Blackwell Science, Oxford, UK.

Steffens, W. & Winkel, M. (2002) Evaluating recreational fishing in Germany. In: *Recreational Fisheries: Ecological, Economic and Social Evaluation* (eds Pitcher, T.J. & Hollingworth, C.E.), Chapter 10, this volume. Blackwell Science, Oxford, UK.

Walton, I. (1653) *The Compleat Angler* (reprinted edition 1939), Penguin Books, Harmondsworth.

Wortley, J. (1995) Recreational Fisheries. In: Review of inland fisheries and aquaculture in the EIFAC area by sub-region and sub-sector (ed O'Grady, K.). *FAO Fisheries Report* 509, Suppl. 1, 60–73.

Chapter 13
Valuation and Policy in Alaskan Sport Fisheries

J. W. Duffield, M.F. Merritt and C. J. Neher

Abstract

Alaska is an important North American sport fishing destination that has had little previous economic study. A multiyear (1995–1998) survey-based economic and social analysis was conducted to estimate net economic values for a variety of enhanced and wild stock sport fisheries across northern, western and interior Alaska (Region III), and to undertake a benefit–cost evaluation of fisheries for program planning. A second goal of the research was to estimate the influence of proposed regulation changes on angler trip frequency. Methods applied were dichotomous choice contingent valuation (CVM) and contingent behaviour. Response rates were relatively low (37–50%) but were in the same range as other state and national agency mail surveys for this population. Estimated CVM models were consistent with economic theory (price, income and quantity responses statistically significant and of the correct sign), fit the data well, and met the precision objectives (standard errors averaging 8.2% of the mean NEV). Consistent with the existing economics literature, estimated NEVs ranged from US$817 per nonresident trip targeting salmon in remote sites to $34 per resident trip for road-accessible stocked waters. Overall, 1996 sport fishing for all species in Region III is estimated to have a total NEV of approximately $28.8m; the greatest benefits accrue from salmon sport fisheries. The mean benefit/cost ratio for stocked waters combined is 7.5; ratios for wild stock fisheries are greater. Proposed regulation changes would lead to small to moderate changes in angler trip frequency, however these changes would come from a minority of anglers.

Key words: Alaska, angler preference, contingent behaviour, contingent valuation, cost–benefit, net economic value, nonmarket valuation, sport fishery management.

Introduction

Sport fisheries managers need a clear picture of fishery participants, their experiences and preferences, the values they place on fishing and what motivates change in their participation in order to shape management policy that is responsive to those who benefit from the resource. This chapter presents the methods and findings from an economic valuation and policy study of Alaskan sport fisheries. As a context for the Alaska study, this chapter also provides a brief overview of economic methods. Where relevant, the findings from related research on North American fisheries are compared to provide a broader perspective.

The study region is the Alaska Department of Fish and Game's (ADF&G) Region III, which includes northern, western and interior Alaska (Fig. 13.1). This 788 410 km^2 region comprises about 85% of the Alaska land mass but has a total resident population just under 150 000 (about one-quarter of the total Alaska population). The population concentration is around Fairbanks (Fig. 13.2), and it is only this part of the region that is road-accessible from Anchorage, Canada and the lower 48 states. The astounding variety of fishing destinations in the region include the Kanektok and Goodnews rivers that flow west, the long, enchanting Arctic rivers both north and south of the Brooks Range such as the Noatak, Kobuk, and Coleville, the Copper River drainage, and the huge Yukon and Kuskokwim drainages and all their tributaries. Nonetheless, the paradox of Alaska fishing is that most use is concentrated in the road-accessible sites: for Region III this is the Fairbanks area fisheries including the Tanana drainage and Chena River. Of a total of about 163 000 angler fishing trips in the region in 1996, 64% were in the Tanana drainage (Howe *et al.* in press a). This pattern of use is driven by travel costs and the fact that there are only about 3200 km of road in the entire region. Most of the region can only be accessed by small commercial planes to outlying communities and from here access to the actual fishing sites is by air taxi or motor boat. To reach the headwaters

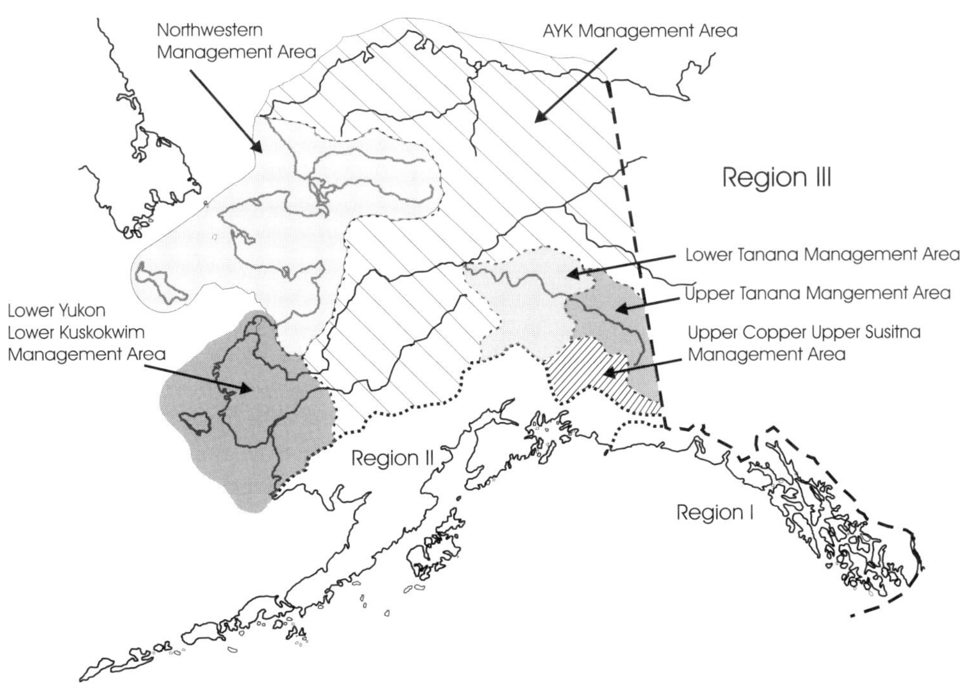

Fig. 13.1 Map of the three sport fish regions in Alaska and the six Region III management areas.

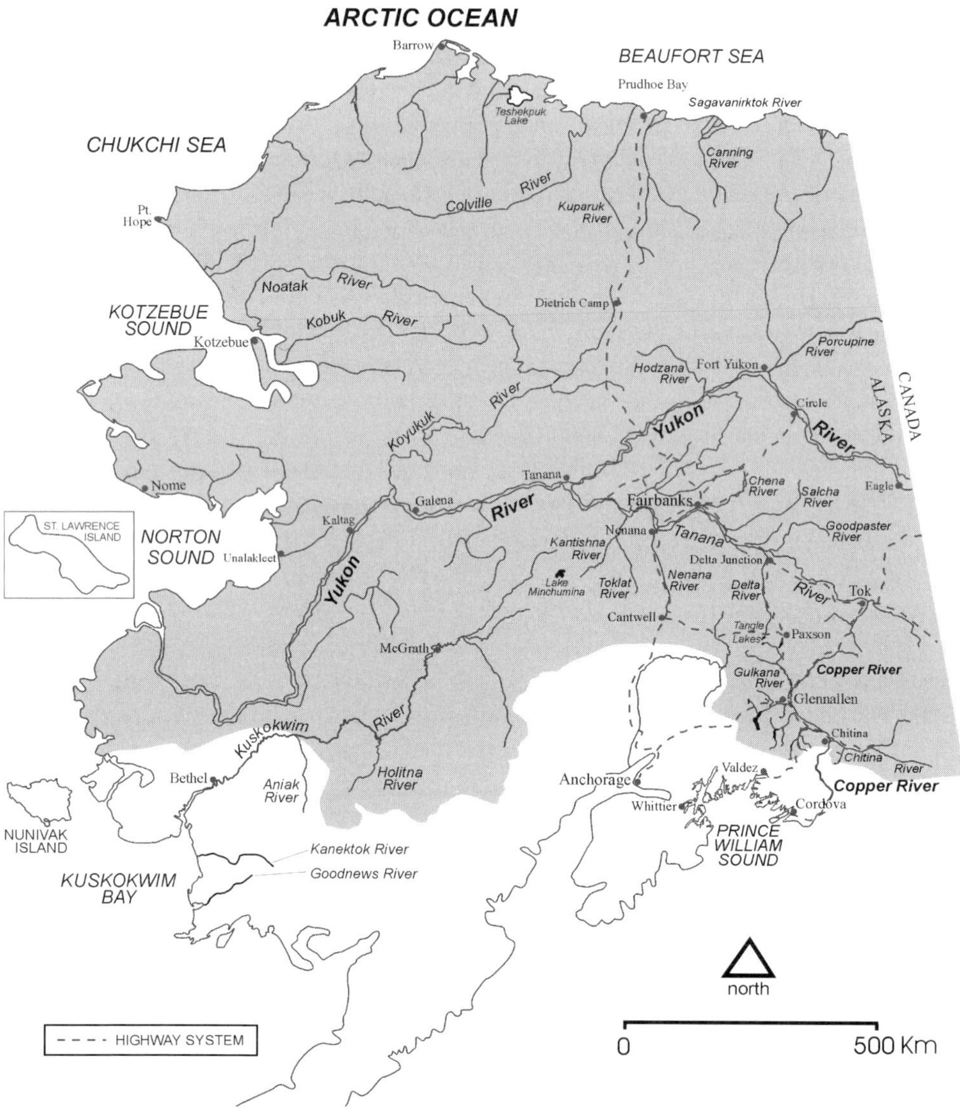

Fig. 13.2 Map of major communities and waters in Region III, Alaska.

of, for example, the Goodnews River, one would have air travel costs from Anchorage or Fairbanks of around $1000–1500 per person, depending on the party size. To stay at a remote lodge in western Alaska can cost up to $3500 per week. Given these kinds of costs, and despite the great abundance of sport fishing opportunities, Region III accounts for only about 10% of the total for Alaska.

Relative to most other fisheries in the United States, Alaskan fishing is characterised by very high quality resources, a relatively small resident angler population, and high trip costs. This combination leads nonresident anglers to comprise the majority of users [as in Northern Territories, Australia, see Lyle *et al.* 2002 – eds]. Based on the U.S. Fish and Wildlife Service's National Fishing and Hunting Survey (NFHS) (which is conducted at 5-year intervals),

62% of those who fished Alaska in 1996 were from outside the state. Only a few other regions in the USA are similar in being destination fisheries – for example, Wyoming at 73% nonresident use and Montana at 53% (U.S. Fish and Wildlife Service 1997). Also, relative to other US fisheries, Alaskan fishing is primarily for wild stocks. A special aspect of Alaskan fisheries is substantial subsistence use by Alaska natives and rural residents – often on waters where sport fishing is also common. As in most North American fisheries, there is free public access to most waters – unlike the private access and fee fishing common in Western Europe and the British Isles. Accordingly, when allocation issues arise that would trade-off fishing quality with other commercial uses – such as conflicts with commercial fishing or hydroelectric development – the value of any given sport fishery is not obvious. Previous research on net economic values (NEV) for sport fisheries (the amount of money a person would be willing to pay, WTP, to fish over and above their actual costs [see Rudd *et al.* 2002 – eds]) in Alaska is fairly limited. The NFHS provides NEVs by state for the entire USA, but the samples are too small for disaggregation beyond the state average level – for example, the sample in Alaska in the 1996 study was limited to about 100 anglers. An excellent study was conducted on south-central Alaska sport fishing in the late 1980s (Carson *et al.* 1987). A finding from that study was that sport fishing trips to Alaska were considerably more valuable than the average NEVs reported for most US sites – on the order of $200 per day.

Beginning in 1995, the ADF&G agency chose to initiate a multiyear survey-based economic and social analysis of current and alternative conditions for Region III sport fishing, an excellent study region since it represents the full diversity of Alaskan fishing from remote fly-out wilderness trips to road-accessible fishing in stocked waters. A major goal of the research was to estimate public benefits in dollar terms for a range of sites and species targeted. A specific application for the estimated NEVs was to undertake a benefit–cost evaluation of fisheries for program planning [see Rudd *et al.* for a full description of the economic basis of this method – eds]. The objectives of fisheries management in the Sport Fish Division state that, in addition to managing for sustainable harvests and maintaining access, public benefits will outweigh management costs. A second research goal was to estimate changes in the frequency of angler trips that might result from the hypothetical implementation of preferred management options. Sport fisheries managers have few options available to increase angler satisfaction, these being diversification of fisheries through stocking, improvement of site facilities and access, habitat enhancement and regulations. The Sport Fish Division's mission is to conserve wild stocks, provide for diverse sport fishing opportunities and to optimise social and economic benefits from sport fisheries. Hence, the question prompting this component of the research was: can one identify optimal policy through its influence on public benefits? When values per trip are roughly constant, trip frequency is one indicator of public welfare. In addition to these two primary goals, information was collected on angler preferences for alternative fishing experiences and angler and trip characteristics.

The first study year (1995) focused on waters in the Fairbanks area where rainbow trout, *Oncorhynchus mykiss*, Arctic char, *Salvelinus alpinus*, and landlocked salmon, *O. kisutch,* are stocked. In 1996 all sport fisheries were evaluated region-wide, with a particular reference to Arctic grayling *Thymallus arcticus* fisheries. In subsequent years the focus was on region-wide sport fishing by species type: salmon, *Onchorhychus* spp. fisheries (1997) and fisheries targeting northern pike, *Esox lucius*, burbot, *Lota lota* and lake trout, *S. namaycush* (1998). Most recently a policy study was undertaken of a proposed change in subsistence fishing regulations (1999). To obtain detailed study results, refer to Duffield *et al.* (2000 and 2001a,b,c,d).

Theory and methods

This section describes the study methods, including the approach chosen for estimating NEVs, angler response to management changes and survey procedures [see Rudd *et al.* 2002 for a detailed account of the methods of economic evaluation of recreational fisheries using examples from Puerto Rico, Costa Rica and elsewhere – eds].

Economic methods for valuing sport fisheries

Economic analysis of sport fisheries can generally be divided into two broad types: (1) regional economic analysis of the economic impact of recreational activity on jobs and income in a given local or regional economy and (2) benefit–cost analysis of the fishery including benefits from the standpoint of the angler. These are two very different accounting frameworks and answer very different types of questions. The first type is essentially an accounting enterprise that tracks the market transactions in a given region that flow from an influx of expenditures from outside the region – e.g. nonresident anglers buying anything from lemon pie to fishing tackle to gasoline and its implications for jobs and incomes. By contrast, the second type of analysis is applied microeconomics – essentially identifying the economic demand for fishing activity. The underlying demand (price–quantity) relationship is the basis for computing NEV that can be used in a benefit–cost framework for comparison to the values of marketed goods and services. This latter type of study is the focus here.

There are a wide variety of problems where economic demand analysis of recreational activity may have application, one being to forecast use or to estimate the influence of pricing policies. For example, the authors have applied nonmarket models to predict appropriate prices for a new recreational use permit to allow fishing and hunting on trust lands in Montana (Neher and Duffield 1993). The most common application for NEV estimates is to inform allocation decisions. Rather than making decisions based on intuition or tradition it may often be possible to apply an objective economic criteria of maximising net economic benefits to society as a whole. Applications include studies of dam construction impacts on recreation and fisheries (Krutilla *et al.* 1972, Duffield 1980, Duffield 1984), as well as evaluation of dam removal (Loomis 1996). In the Western US competing uses of water necessitate computing fishery-related NEVs for instream flows (e.g. Duffield *et al.* 1992) and water rights allocations between anglers and irrigators (Duffield *et al.* 1990). Other applications include reservoir management – both for downstream flow levels (as on the Grand Canyon, Boyle *et al.* 1988) and for reservoir-based recreation (Ward *et al.* 1996). Net economic values are required in cases of natural resource damages due to oil spills or hazardous substance releases, as in the *Exxon Valdez* oil spill litigation for damages to subsistence (Duffield 1997) and sport fisheries (see Chapter 22 in Ward and Duffield, 1992). Last, but not least, NEVs are necessary for evaluating public investment in fishery management. The current study compares total angler benefits to the costs of fishery research and management.

Given the motivation for being able to value sport fisheries in economic terms, what are the available methods? Economists have developed two distinct approaches to this problem. One general approach is termed revealed preference; essentially one relies on observed behaviour and data from related markets to infer a value for recreational activity. The most widely applied revealed preference model for recreation demand is the travel cost model. Anglers who live at a greater distance from a given water face higher costs of travel and are less likely to visit (on a per capita basis) than those who live nearby. Hotelling (1947) was the first to recognise that travel costs are like a spatially varying price that, along with the number of observed visits from an origin, could be used to infer the economic demand for a site. Over the last five decades, Hotelling's idea has spawned a very large economics literature including hundreds of empirical applications to a variety of recreational activities, including fishing and hunting (Clawson and Knetsch 1966, Herriges and Kling 1999, Ward and Beal 2000). The strength of the travel cost model is that it relies on observed behaviour regarding angler trips (the quantity variable from the standpoint of economic demand modelling); a weakness is that the actual price experienced by recreationists must be inferred – do true travel costs include just the variable costs of travel? Depreciation of vehicles? The value of travel time?

The other general type of models for valuing recreational use are stated preference models, in which respondents to a survey are essentially asked to state the value they place on the service at issue – such as fishing. The most common approach is called contingent valuation (CVM). In these types of models one poses a hypothetical market-like setting to a respondent, and, contingent on their acceptance of the situation, the respondent provides a valuation response. For example, an angler might be asked to suppose that their travel costs to the site might increase over time, and 'what is the most you would pay in increased travel costs before you would choose to not fish here?' This specific question format is called open-ended; the respondent states the maximum they would pay. This is a difficult kind of question to answer. An alternative format is to ask 'suppose the costs of fishing at this site increased by $10 over what you actually paid, under these circumstances would you still have chosen to make this trip?' This question format is called dichotomous choice; the respondent is faced with a given dollar amount (in this example, $10) and gives a 'yes' or a 'no'. This is a widely used specific format in the economics literature since it is most like a real market in terms of the response being 'buy' or 'not buy' (Bishop and Heberlein 1979, Cameron 1988). By asking different subsamples of an angler population different bid amounts (for example some are asked $10, some $20, some perhaps $50) one can identify a price response function and an underlying economic demand relationship for the service at issue. The first well-known application of contingent valuation was by Davis (1963) in a Ph.D. dissertation investigating the value of recreational visits to the Maine woods. There have since been well over a thousand applications reported in the literature, many of these to recreation. An excellent reference is Mitchell and Carson (1989), which was the first major text on contingent valuation [see also Rudd *et al.* 2002 for methods and application to sport fisheries – eds].

Both travel cost and CVM are among the very short list of legally approved methods for economic valuation (U.S. Department of Interior 1986, 1991). In some ways the methods are mirror images of each other. As noted, travel cost relies on observed quantity demanded, but price must be inferred – not an easy matter. Conversely, dichotomous choice provides a known price (the stated bid amount), the problem is inferring the quantity demanded from the proportion of respondents who say 'yes'. The economics literature provides evidence that these very different methods can provide comparable estimates. Walsh (*et al.* 1992) reviewed 20 years worth of recreation NEV estimates. They identified 287 estimates including 156 travel cost model estimates and 129 CVM estimates. A general finding of their review was that the mean for the CVM estimates (particularly for dichotomous choice) was similar in magnitude to mean estimates based on travel cost. A recent review of fishing and hunting valuation studies has similar findings (Markowski *et al.* 2001). In one of the earlier state-wide fishery valuation studies (Duffield *et al.* 1987, Duffield and Allen 1988) both travel cost and CVM were used to value 17 wild-trout fishing streams in Montana. The substantial variation in values across sites (ranging from about \$55 per trip for local fisheries to in excess of \$200 per trip for the Madison River) was replicated by both models – using separate data sets collected in sequential years. Correlation coefficients of NEVs from the two methods showed a high correlation (0.80).

Parenthetically, it may be noted that CVM has been applied to a much wider range of services than travel cost. CVM is the only available method for valuing indirect use or passive use where individuals may place a value on knowing a resource exists in a viable state independent of their own direct use. For example, many individuals contribute money to support the preservation of wildlife – giant pandas in China or grey wolves in Yellowstone – independent of their ever directly viewing or using these resources (Boyle and Bishop 1987, Duffield and Neher 1996). Because CVM can be used to value such services as a clean Prince William Sound (for example, in the context of the *Exxon Valdez* oil spill) the approach has been challenged in court by industry and has been the subject of a special panel commissioned by NOAA (Arrow *et al.* 1993). In both instances CVM has been upheld as a valid method for estimation of passive use values. However, this issue is a point of contention among economists. The point here is that the application of CVM to measuring the value of direct recreational use is a less challenging undertaking and is widely accepted. As a recent example, the method selected by the U.S. Fish and Wildlife Service (1998) for application in the NHFS in both 1991 and 1996 is dichotomous choice CVM.

Application to Alaska Region III sport fishing

The choice of methods for the Alaska application were based on study goals, budget constraints and the availability of agency staff to participate in the study. The key tasks are: (1) survey design, (2) data collection and (3) analysis and report writing. The arrangement in this study was to share survey design responsibilities between the agency and contract economists, data collection was the agency responsibility, and analysis and report writing was done by the contract economists. The total cost of the four years of studies was roughly \$200 000; about half of this was agency staff and operations costs that were allocated to the project. The per study budget for contract economists was fairly limited given the ambitious scope of the projects – on the order of about \$25 000 per year. In-person or on-site sampling was prohibitively costly. This led to the choice of a mail survey, which is a common and satisfactory approach.

Contingent valuation methodology

Our research used the CVM as the survey technique to estimate the values that people would place on traditionally nonmarket goods and services, specifically sport fishing trips. The application of CVM involves three elements: (1) a description of the resource which is to be valued; (2) the payment vehicle, or method by which the respondent will pay for the resource; and (3) the question format or specific method by which the value of the resource will be elicited. The trip is the logical unit of fishing services to value since the angler's main economic decision in this context is whether or not to take a trip. The payment vehicle was an increase in travel costs to the site. In dichotomous choice an additional survey design task is to identify the appropriate bid range and to allocate the sample among bids. In anticipation of a log transformation of the bid variable, a general strategy is to use bids at approximately equal log intervals. The range of bid amounts is pretested to verify that it covers a good range of the distribution of WTP. A fairly small number of bids, perhaps 5–7, are used to allow tests of significance across models at a given bid level. As an example, for the grayling study year, the bids used for resident anglers were \$10, \$25, \$50, \$100, \$200, and \$500. The allocation of the total sample affects the precision of the estimates (Duffield and Patterson 1991, Cooper 1993).

Hanemann (1984) and Cameron (1988) have investigated the theoretical motivation for dichotomous choice models. The probability of accepting a specific bid offer amount, given the individual's true (unobserved) valuation WTP is:

$$\pi(t) = \Pr(WTP > t) = 1\text{-F}(t) \quad [13.1]$$

where F is a cumulative distribution function (c.d.f.) of the WTP values in the population. In the logit model F(.) is the cumulative density function of a logistic variable. The specific logistic model applied here is:

$$\pi(t, \overline{\chi}) = [\ 1 + \exp(-\alpha t - \tilde{\gamma}' \overline{\chi})\ ]^{-1} \quad [13.2]$$

where $\pi(t, \overline{\chi})$ is the probability that an individual with covariate vector $\overline{\chi}$ is willing to pay the bid amount t. The parameters to be estimated are α and $\tilde{\gamma}'$ (the constant term is included in $\tilde{\gamma}'$). The equation to be estimated can be derived as:

$$L = \ln[p/(1-p)] = \alpha t + \tilde{\gamma}' \overline{\chi} \quad [13.3]$$

where L is the logit or log of the odds of a 'yes' and p are observed response proportions. Alternatively a log-logistic model may be estimated, depending on goodness-of-fit (t is replaced with log t). Maximum likelihood estimates of the parameters in equation 13.3 can be obtained with a conventional logistic regression program (SAS Institute 1988). A variety of welfare measures for dichotomous choice models have been proposed in the literature including a truncated mean (Bishop and Heberlein 1992), the overall mean, and percentiles of the distribution, including the median (Hanemann 1984, 1989). We utilise the truncated mean and several different percentiles in this application. The truncated mean is defined by:

$$M_T = \int_0^T [1 - F(\chi)]\, dx \quad [13.4]$$

where F(χ) is the probability density function of the distribution. The truncated mean is more conservative than the overall mean. T is generally set equal to the highest bid offer; as a result the integrand in equation 13.4 is within the range of observed data. The pth quantile (100 pth percentile) of the distribution is given by F-1(p). For the log-logistic model, the pth quantile is given by:

$$\eta_p(\bar{\chi}) = \exp(-\tilde{\gamma}'\bar{\chi}/\alpha)[p/(1-p)]^{-1/\alpha} \qquad [13.5]$$

When $p = 0.50$, equation 13.5 provides an estimate of the median WTP. The WTP or NEV values reported from these models are measures of compensating variation (Just *et al.* 1982). Standard errors are estimated by bootstrapping (Park *et al.* 1989, Duffield and Patterson 1991).

Contingent behaviour methodology

Contingent behaviour questions ask respondents to predict how their behaviour would change given a hypothesised change in the attributes of a commodity or service (such as a fishing trip). There is a large scientific literature that fits within this general definition, including market research and polls to predict voting behaviour. Contingent behaviour data has been used in a variety of ways in the resource economics literature, usually in conjunction with travel cost or CVM models (Morton *et al.* 1995, Cameron *et al.* 1996, Cooper 1997). Some economic studies have used contingent behaviour questions to measure changes in visitation rates and to derive demand curve shifts. McConnell (1986) asked respondents how visits to local beaches would change if pollution of New Bedford Harbor, Massachusetts by polychlorinated byphenyls (PCBs) could be eliminated. Thayer (1981) asked recreationists how their choice of sites to visit would be altered by construction of a geothermal plant in the vicinity of the recreation sites. Narayanan (1986) and Duffield *et al.* (1990) used a conceptually similar approach to estimate values associated with instream flow. While the literature on using contingent behaviour models to measure valuation changes is fairly limited, there is a very large and varied literature on the basic problem of using surveys to predict future behaviour – including the issue of validation of survey predictions (as summarised in Table 13.1).

Table 13.1 Summary of literature on validation of contingent behaviour.

Type of study	Citation	Finding
1. Actual and predicted recreational visitation	McConnell (1986)	Survey over-predicted beach use by 30%*
2. Comparison of polls and voting behaviour	Mitofsky (1996)	Actual and predicted votes very similar
3. Voter turnout surveys and actual voter records	Traugot and Katosh (1979), Belli (1997)	Actual turnout overstated
4. CVM referendum study and actual referendum	Carson *et al.* (1986)	Actual vote accurately predicted
5. Market research and buying intentions	Theil and Kosobud (1968), Ferber and Piskie (1965)	Buying intentions overstate actual purchases – large error for consumer durables, small for vacations and education choices

* Studies done for purposes of litigation or rebuttal, if any, have not been evaluated for purposes of this summary.

Angler preference and motivation

Approaches to measuring angler preference, motivation and satisfaction is another important methodological issue for sport fisheries studies (Bryan 1979). An example of a fairly comprehensive angler preference study is the work by a social psychologist (Allen 1988) on the primary reasons for fishing Montana trout streams. Allen developed a list of 17 specific reasons and measured the importance of each possible motivation (for example, ranging from being with family and friends, to catching fish to eat, to testing fishing skills) with Lichert-scaled questions. The responses were used in a cluster analysis to identify four distinct types of anglers including an occasional angler group, several generalist groups and a specialist group. Specialists were twice as likely to be fly fishermen only compared to the population as a whole. When average net economic values per trip for the cluster-defined groups were estimated they ranged dramatically from $7.56 per trip for the occasional users to $91.03 and $117.07 for the generalists and $170.28 for the specialist group (Duffield and Allen 1988).

Given survey length constraints, only a few preference measures with management implications were adopted. Respondents were asked to rank alternative fishing experiences on a five-point scale from least preferred (1) to most preferred (5). Six generic experiences were ranked: fishing easily accessible sites near a road, fishing in a wilderness setting, harvesting fish, catching and releasing fish, having good trail access to fishing waters, and having more developed camping facilities. Anglers were also asked to rate their fishing experience on their most recent trip as poor, below average, average, above average, or excellent. These measures are of interest in themselves and can also prove to be important explanatory variables in the CVM model.

Survey and sample design

Designing surveys is a challenging undertaking. Questions must be clear, easily understood by respondents and have an unambiguous interpretation. An important constraint is having the instrument short enough that one does not exceed the attention span of the respondent. It is always necessary to pretest a survey to verify comprehension by the respondents. The general design of the survey instrument was similar for each year of study and consisted of four sections: (1) recreational fishing patterns and trips to Region III waters; (2) the respondent's most recent fishing trip, including CVM; (3) the respondent's preferences for management; and, (4) socioeconomic characteristics.

The logical sample frame is the list of anglers holding State of Alaska sport fishing licences. An important sampling problem was created by the fact that this was a regional study. A very large number of state-wide licence holders would have to be sampled to obtain even a small probability sample of anglers who actually fished the region. The approach taken was to conduct a stratified sample by residence location within Region III and to sample only the nonresident and Region I and II anglers who purchased licences within the region. Resident 1995 licence holders residing in the Tanana Valley comprised the population sampled for the stocked waters study. The three remaining surveys sought estimates for wild species of sport fish targeted in Region III, so nonresidents and resident populations were sampled, depending upon the distribution of species across geographic areas (Table 13.2). Sample size was established from the number of licence holders who had indicated on the Alaska state-wide

Table 13.2 Survey mailing statistics: number of surveys mailed by population and year.

Population sampled	Sample	1995	1996	1997	1998
Non-residents who bought licences in Region III	Random	— *	2000	1999	1271
Residents from Region I (south-east Alaska, including Juneau) and Region II (south-central Alaska, including Anchorage) who bought licences in Region III	Census	—	749	2305	1166
Seward Peninsula residents	Census	—	1105	1243	—
North-west residents	Census	—	258	—	—
Copper River area residents **	Census	—	—	2659	531
Remainder of Region III ***	Random	3497	4000	1999	2500
Total mailed		3497	8112	10 201	5468
Total delivered		2931	7709	9133	5019
Total returned		1441	2826	3546	1957

* Does not apply; survey not administered to this population.
** In 1997 Region III was expanded in size with the jurisdictional transfer of the upper Copper and Susitna River drainages, hereafter referred to as the Copper River area; the 1997 and 1998 surveys included this management area.
*** Composed primarily of Tanana River area residents near the urban centre of Fairbanks, and hereafter referred to as the Tanana population. In 1998 Region III was expanded in size with the jurisdictional transfer of the lower Yukon and Kuskokwim rivers; the 1998 survey included this management area.

harvest survey (Howe *et al.* 1996, and Howe *et al.* in press a,b,c) they had sport fished for the targeted species in the survey year, and by the desired level of precision. The objective criteria for precision were ±25% of the mean for a 95% confidence interval around the NEV based on the relationship of total sample and precision in previous dichotomous choice CVM applications to fisheries (e.g. Duffield and Patterson 1991). The questionnaires were pre-tested on a randomly drawn sample of 200 sport fish licence holders to determine the top bid level for the CVM question, and to test the effectiveness of the wording and question sequencing. A total of three mailings were conducted: an initial survey mailing, a postcard reminder mailed two weeks later, and a second survey to nonrespondents several weeks after the postcard mailing to increase the overall response rate (Dillman 1978).

Results

Response rates

Achieving adequate response rates is important for CVM model estimation and to reduce the likelihood of nonresponse bias. The overall response rates (completed surveys as a percentage of delivered surveys) were 49.2% for the stocked lake survey, 36.7% for the Arctic grayling survey, 38.8% for the salmon survey and 39.0% for the pike, burbot and lake trout survey. The Alaska state-wide harvest survey for 1996 (Howe *et al.* in press a) reports similarly low response rates of 49.2% for its 'standard' angler participation survey and 44.9% for its 'supplemental' survey, both using three mail contacts.

The U.S. Fish and Wildlife Services NFHS for 1996 was examined to see if Alaskan response rates were lower than the US average. The NFHS, implemented by the Bureau of

Census, is conducted at two levels. A screening phone call is made to identify anglers and hunters in the general population, then anglers and hunters are followed up with surveys. From an initial US sample of about 80 000, the screening call response rate was 71% and the response rate for the actual survey was 80% for an overall response rate of 57%. The Alaska screen response was 65.9% and the survey response was 77% for an overall response rate of 50%, which is below the US average. By contrast, the Montana screen response was 76% and the survey response was 89% for an overall response of 67%, well above the US average. It is interesting to note the NFHS response rate for California, where intensive telemarketing is reputed to have made phone surveying difficult. The California screen rate was only 57.5%, the survey rate was 77.5% for an overall 44.6% response. From these data there appear to be systematic differences in co-operation with agency surveys across states [see Lyle *et al.* 2002 for an Australian experience – eds].

We speculate that the Alaska response rates are in part a reflection of the independent spirit and attitudes toward government of those who live in the 'last frontier'. Within the survey subsamples, rural residents have among the lowest response. Regions I and II residents also have low response – these anglers may be less interested in the management of Region III fisheries.

Nonresponse bias was examined by comparing responses between waves of survey mailings and in the subsistence rod and reel study by sampling nonrespondents through a follow-up phone survey. Responses were generally consistent between the initial mailing and reminder survey, and between phone and mail contacts. In answer to the question of why nonrespondents did not return their subsistence rod and reel mail survey, 69.2% of the Anchorage sample did not remember receiving a survey, and 15.4% either didn't have time or didn't feel like responding (Duffield *et al.* 2000). While the results from our research should be interpreted in the context of the response rates, nonetheless based on examinations of nonresponse, we feel that responses adequately represent sampled populations.

To conclude on this point, the response rates obtained in the current study are similar to the rates obtained in other carefully administered agency surveys in Alaska and not much different from those obtained by the Bureau of Census in Alaska in a much more costly and intensive effort.

Net economic values and benefit/cost analysis

Nonmarket economics is used to determine the values (NEVs for sport fishing in this case) that people would place on traditionally nonmarket goods and services if markets did exist for these commodities. Net economic values are the appropriate values to use in benefit/cost analysis of a given policy (Herrick *et al.* 1994). Since enactment of the US Magnuson Act, which requires an analysis of proposed actions in economic and social terms, fishery managers are being asked more often to examine the efficiencies and impacts associated with management actions and program decisions. Our study is not intended to be a rigorous analysis of the benefit/cost problem, however it will provide managers with guidance in their policy-making.

A large number of specific estimates by population, species and site were developed for Region III during several years of study. We first present an example of a typical model including the actual raw data showing response by bid level and the fitted logistic model proportions. This material provides an illustration of how the actual NEV are computed. While bivariate models (just the bid as the single exogenous variable) were used for the welfare estimates reported here, we also present several multivariate models to show that the responses are consistent with the predictions of economic theory. Following this brief discussion, we turn to summary results from all four economic studies.

A typical model: Region III resident sport fishing trips for all species (1996)

Table 13.3 provides a summary of the actual response by bid level for the Region III resident subsample on their sport fishing trips in 1996. This is one of the larger subsamples in the grayling study with a total sample of 742. As shown in the table, 88% (or 125 of 142) respondents faced with a bid of $10 (in hypothetical increased travel costs for their most recent fishing trip) responded 'yes', they would still have chosen to make the trip. As one would predict from the law of demand, as the asking price or bid is increased, the proportion who would 'buy' declines. As the bid increases to $25, the percentage declines to 82%. At the highest asked bid of $500 only 11% indicate that they would still have chosen to make the trip. This data indicates there is a distribution of values held by the population sampled. At a bid of $100, about half the population says 'yes' (indicating their values per trip are greater than $100) and about half says 'no', indicating the converse. Thus, it is apparent that the median NEV for this subsample is about $100. Based on the chi-square statistic, the model provides a good fit to the data. Given the fitted distribution it is possible to compute a variety of measures of central tendency; for example, the truncated mean for this data set is $122, with a standard error of $7.22. This implies that the 95% confidence interval for our estimated NEV for resident anglers in 1996 is about $108 to $136.

Table 13.4 shows more complex (multivariate) models for the grayling data set: one each for nonresident and resident subsamples. A nonresident model was estimated with bid and income estimated coefficients both highly statistically significant and with the signs predicted by economic theory: bid has a negative coefficient indicating the expected price response (as price goes up, quantity demanded goes down); income has a positive coefficient as one expects for 'normal goods' (the higher the respondent's reported income, the more likely they are to purchase the good). The resident model includes variables for bid, the number of

Table 13.3 Bid distribution and response percentages, Region III resident anglers sport fishing for all species in 1996.

Bid amount ($)	Ln (bid)	Sample	Percentage 'yes'	Predicted p*
10	2.30	142	88.0	0.910
25	3.21	114	81.5	0.787
50	3.91	125	61.6	0.630
100	4.60	127	51.2	0.443
200	5.29	126	23.0	0.271
500	6.21	104	10.6	0.119

* Fitted logistic model: $\ln(p/1\text{-}p) = 4.852 - 1.1043 \ln(\text{bid})$, $n = 742$, chi-square goodness-of-fit 6.26, df 5, p-statistic 0.28.

Table 13.4 Multivariate models of net economic value for a grayling sport fishing trip in Region III, for nonresident and resident respondents, 1996.

Model variables	Coefficient	Standard error	p-value
(A) Nonresident model of fishing in Region III for any species			
Constant	4.9437	0.7336	< 0.001
Ln (BID)	-0.8185	0.1230	< 0.001
Ln (INCOME)	0.8414	0.3023	0.0054
Sample size		412	
% Correct predictions		76.7%	
% Correct W/only BID		68.6%	
(B) Region III Resident model of fishing for grayling in Region III			
Constant	4.6417	0.6278	< 0.001
Ln (BID)	-1.2903	0.1548	< 0.001
Ln (DAYS)	1.3767	0.3076	< 0.001
QUALITY	0.6107	0.3602	0.0900
Sample size		270	
% Correct predictions		84.5%	
% Corr. W/only BID		76.2%	

Note: Ln(BID) is the log of the bid amount; Ln(INCOME) is the log of the discrete income variable; Ln(DAYS) is the log of the number of days fished at site on the current trip; QUALITY is the dummy variable that takes the value of 1 if the respondent rated the current trip as either 'above average' or 'excellent', and a value of 0 if the trip was rated 'average' or below; % Corr. W/only BID is the percent correctly predicted with bid as the only variable for a bivariate specification.

days fished, and the self-reported trip quality (poor to excellent). Again all variables have the expected sign. Other things being equal, the higher the trip quality and the longer the trip, a higher value is expected. Table 13.4 also reports an alternative goodness-of-fit statistic for the estimated models which is the per cent of correct predictions (a given respondent answering 'yes' or 'no'). The bid variable alone is quite a powerful explanatory variable and results in a model that correctly classifies 69% of the nonresident responses and 76% of resident responses. The addition of the other variables increases the predictive power to 77% and 85% for the resident and nonresident models respectively. The ability of these models to correctly predict responses compares favourably with those reported in the literature. For example, the 1996 NFHS reports 22 separate state and regional sport fishing CVM models with the percent correct prediction ranging from 61% to 72%. These findings generally indicate that the CVM responses here are consistent with economic theory.

Summary NEV results

This section provides a summary of the major NEV findings from each study. For stocked waters, the NEV per trip estimates range from \$34 to \$69, which is considerably lower than the average values reported for region-wide trips to wild stock fisheries. Estimates of angler trips to the five stocked waters (Howe *et al.* 1996) were multiplied by their respective NEV to estimate total annual NEV of sport fishing, with bootstrapped standard errors in parentheses computed using a standard variance formula. The Quartz Lake fishery is significantly higher in total annual NEV at \$1.7m, followed by the Birch Lake fishery at about \$1.0m; the remaining three fisheries have similar mean values around \$400 000 (Table 13.5). Skaugstad *et al.* (1995) estimated that while all fishing occurring at Quartz, Birch and Chena lakes is for

Table 13.5 Estimates of adjusted mean NEV per fishing trip, angler trips, and total annual NEV of sport fishing at five major stocked waters of the Tanana Valley, 1995 (standard errors are in parentheses).

Stocked fishery	Adjusted mean NEV/trip in US$	1995 angler trips	Total annual NEV in US$
Quartz Lake	69 (5.3)	25 179 (1 721)	1 729 794 (178 524)
Birch Lake	59 (6.3)	16 970 (1 574)	997 524 (141 476)
Harding Lake	47 (7.3)	8 753 (876)	408 550 (76 425)
Chena Lake	36 (5.4)	11 034 (961)	397 658 (69 497)
Piledriver Slough	34 (6.5)	13 763 (840)	464 932 (93 299)
Total			3 998 457 (266 949)

stocked fish (native fish were eradicated from these lakes prior to stocking) about 50% of fishing at Harding Lake and Piledriver Slough is for nonnative fish. Harding Lake has indigenous populations of pike and burbot. Piledriver Slough has indigenous populations of grayling and pike. In order, therefore, to estimate the total annual NEV of fishing associated with stocked fish in these waters, the estimates for Harding Lake and Piledriver were reduced by 50%. This results in an estimated NEV associated with angling for stocked fish at the five enhanced waters of $3 561 765.

Turning to the 1996 study, of particular interest is whether fishing trips that specifically target grayling are valued differently than those targeting nongrayling species, comprised primarily of salmon. The answer is no, fishing trips are valued much the same, for both nonresidents and residents. For example, the adjusted mean NEV for a fishing trip specifically targeting grayling taken by Tanana residents is $123 (se = 15) and for a trip targeting nongrayling is $128 (se = 10). Nonresidents value their fishing trips for all species significantly more highly than do Region III residents (Table 13.6). This is consistent across surveys and with the findings of other recreational NEV studies. Overall, 1996 sport fishing for all species in Region III is estimated to have a total NEV of approximately $28.8m (Fig. 13.3). Of the total $28.8m, 33.9% is attributable to nonresident fishing trips. Of the 66.1% of estimated value attributable to Alaska residents, only 11.2% is from residents residing outside of the region. More than 50% of the total regional value of sport fishing for all species occurs in the Tanana area – the estimated total NEV of sport fishing for all species in the Tanana area is roughly $15.7m. Within the Tanana area, the Chena River's total NEV is estimated at $3.4m.

Table 13.6 Estimates of adjusted mean NEV for a sport fishing trip to any area in Region III, angler trips, and total annual NEV by population for all species, 1996 (standard errors are in parentheses).

Population	Adjusted mean NEV/trip in US$	1996 angler trips	Total annual NEV in US$
Nonresidents	591 (23.4)	16 531 (701)	9 767 176 (566 963)
Regions I and II	192 (30.0)	11 077 (639)	2 129 553 (354 496)
Tanana	122 (7.2)	126 310 (2678)	15 392 137 (968 783)
Seward Peninsula	150 (12.0)	8 618 (627)	1 290 028 (139 538)
North-west	275 (38.7)	841 (183)	231 090 (60 325)
Total		163 377	28 809 984 (1 186 914)

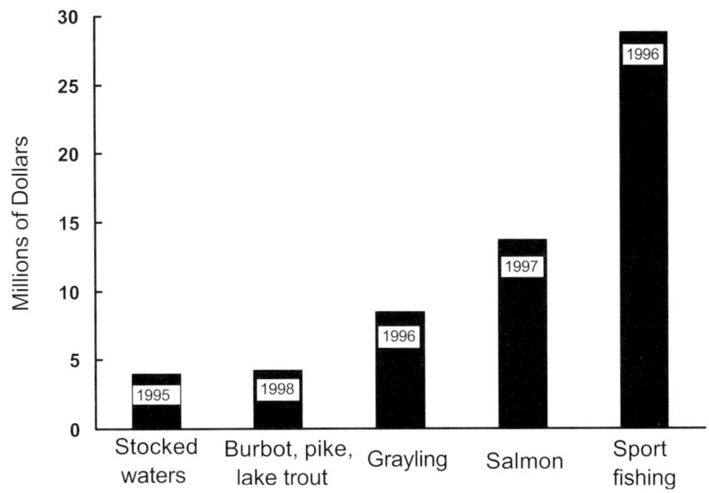

Fig. 13.3 Estimated total annual net economic value by survey targeting specific species and all sport fishing in Region III, Alaska.

Total NEV for grayling fishing within Region III in 1996 is estimated at a little over \$8.0m, or about 28% of the region's total sport fishing value. The Tanana population accounts for about \$5.8m of this total value, primarily because of the number of trips this population takes to fish for grayling (an estimated 47 240 trips in 1996). The estimated total NEV of sport fishing for grayling in the Tanana area is approximately \$3.5m. Fishing for grayling in the Chena River is estimated to have a total NEV of \$1.1m.

With regard to the salmon study findings, a question is whether fishing trips targeting chinook salmon vary in value from salmon species other than chinook. Although the adjusted mean NEVs per trip are higher for chinook salmon trips than for salmon other than chinook (primarily sockeye in the Copper River area and coho on the Seward Peninsula), there is no statistical difference. Adjusted mean NEVs for water-specific models of area residents varied from \$134 at the Gulkana River to \$102 at the Chena River for all salmon species, however the variation was not statistically different. As in the other studies, the NEV for nonresidents are substantially higher than resident values. Sport fishing for salmon in Region III is estimated

Table 13.7 Estimates of adjusted mean NEV for a salmon sport fishing trip to any area in Region III, angler trips, and total annual NEV by population, 1997 (standard errors are in parentheses).

Population	Adjusted mean NEV/trip in US\$	1997 salmon angler trips*	Total annual NEV in US\$
Non-residents	817 (47.7)	11 311	9 235 420
Regions I and II	192 (16.3)	3 567	683 330
Tanana	136 (11.5)	20 790	2 825 350
Seward Peninsula	137 (14.0)	4 423	604 000
Copper River	122 (10.0)	2 871	349 400
Total		42 962	13 697 500

* Standard errors for angler trips targeting a specific species are not available from the Alaska statewide harvest survey. Estimated percent salmon fishing trips was computed from Howe and Fleischmann unpublished data.

to have a total NEV of approximately $13.7m in 1997 (Table 13.7). Of this total, 67.4% is attributable to nonresident fishing trips. Of the 32.6% of estimated value attributable to Alaska residents, 15.3% is from residents residing outside of the region. About 31% of the total regional value of sport fishing for salmon occurs in the Copper River area – the estimated total NEV of sport fishing for salmon in the Copper River area is roughly $4.3m.

With regard to the pike, burbot and lake trout study, it is of interest to know whether fishing trips targeting either pike, burbot or lake trout vary in value to Tanana residents. While the adjusted mean NEV for lake trout ($211) is higher than for burbot ($139) or pike ($130), the differences are not statistically significant. Sport fishing for pike, burbot and lake trout combined in Region III is estimated to have a total NEV of approximately $4.3m in 1998 (Table 13.8). The majority (62.2%) of this value is contributed by Tanana area residents. As a whole, nonresidents are not particularly drawn to sport fishing for these species in Region III, contributing 18.2% to the total regional value.

The total annual NEVs in the stocked waters survey were used against hatchery production, stocking and evaluation costs (Table 13.9) to calculate the benefit/cost ratio (Fig. 13.4). The highest mean benefit/cost ratios are for the sport fisheries at Quartz Lake (20.6) and Birch Lake (8.0). The mean benefit/cost ratios for the remaining three stocked fisheries range from 2.3 to 4.2. The mean benefit/cost ratio for the five stocked fisheries combined is 7.5.

The problem of conducting benefit/cost analyses for wild stock fisheries is in identifying the actual marginal benefits that are a consequence of agency research and management efforts. For example, if all research was terminated, what is the real consequence for angler

Table 13.8 Estimates of adjusted mean NEV for a sport fishing trip to any area in Region III, angler trips, and total annual NEV by population for pike, burbot and lake trout, 1998 (standard errors are in parentheses).

Population	Adjusted mean NEV/trip in US$	1998 pike, burbot and lake trout angler trips*	Total annual NEV in US$
Non-residents	371 (105.5)	2090	779065
Regions I & II	238 (55.2)	2469	588165
Tanana	161 (23.1)	16559	2669973
Copper River	159 (47.5)	1604	254715
Total		22731	4291918

* Standard errors for angler trips targeting a specific species are not available from the Alaska statewide harvest survey. Estimated percent pike, burbot and lake trout fishing trips was computed from Howe and Fleischmann unpublished data.

Table 13.9 The cost basis used to evaluate the NEV/cost comparison in US$ for program planning in Region III, by survey, fiscal year (FY, July 1–June30) and budget component.

Estimated budget component	FY96 Stocked waters	FY97 Grayling	FY98 Salmon	FY99 Pike, burbot, lake trout
Research	N/A*	288000	178200	193100
Management**	N/A	78200	231900	12200
Total	475008	366200	410100	313300

* Not applicable
** Management costs were estimated from the proportion of time individual managers spent in managing the fisheries by geographic area. Not included are indirect costs associated with supervision and administration.

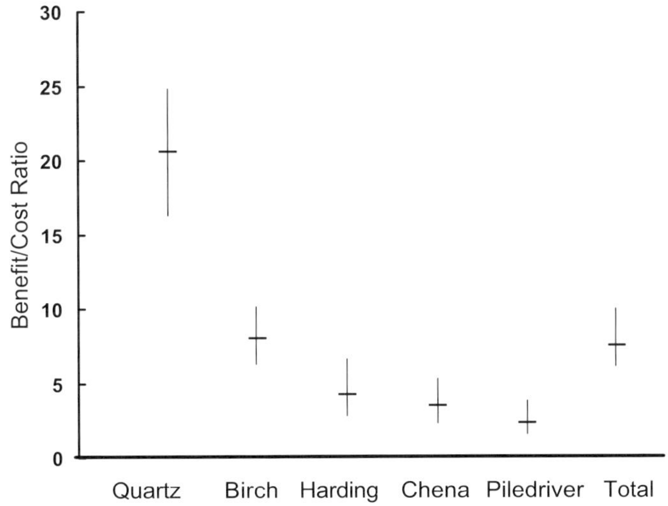

Fig. 13.4 Estimated mean benefit/cost ratios (horizontal bars) and 95% confidence intervals (vertical lines) for five stocked waters in the Tanana Valley, Alaska, 1995.

benefits in a wild stock fishery? Nonetheless, as a point of information, we report here the cost basis for the wild species research and management by species type (Table 13.9) and provide a comparison to total NEV by fishery. The majority of funds expended on grayling, pike, burbot and lake trout occur in the Tanana area, because this area supports the greatest angling pressure for these species. The Copper River area received the majority of funds spent on salmon research and management in 1998. The ratio of estimated angler NEV to research and management costs for salmon is 33.4, for grayling is about 21, and for pike, burbot and lake trout is 13.7.

Policy

Predicting angler responses to hypothetical management decisions *before* policy is made is a developing field of research. Angler trip frequency is one indicator of public welfare and can be directly tied to changes in total NEV of a fishery. Predicted changes in angler trips resulting from regulation changes can also be used in the context of a benefit/risk analysis. Benefits (angler trips) accrued or lost as a result of a management change can be weighed against the risks of over- or under-utilisation of the fishery resource. We applied contingent behaviour methods to predict changes in angler trips due to changes in species, number and size of fish at stocking, bag limit, catch and release management, length restrictions, legal fishing gear, and bait as a legal fishing means.

In the stocked waters survey, respondents were presented with the following statement, 'Fish and Game can produce limited numbers of fish for stocking. We would like to know your preferences for possible stocking options for Arctic char, rainbow trout, and salmon.'

Within each group of options relating to species, catch rates and size, respondents were asked to rank their most preferred. The results showed a clear preference for stocking char and rainbows, but not for stocking salmon.

Table 13.10 Top ranked combinations of stocking options.

Water ranked number 1 for stocking:				
Arctic char	Rainbow trout	Salmon	Number of anglers choosing this combination of preferred stocking options	Per cent of combinations*
Harding	Quartz	Quartz	147	16.0
Chena	Chena	Chena	137	14.9
Harding	Birch	Birch	134	14.6
Harding	Chena	Chena	61	6.7
Chena	Birch	Birch	58	6.3

* Column does not total 100% because other combinations were evaluated, however these are not shown due to low numbers of anglers selecting those stocking combinations.

Crosstabs were run on the stocking options to see if any combinations were consistently ranked as number one for all three species. The most popular combination was to stock Arctic char in Harding Lake, and stock rainbow trout and salmon in Quartz Lake (Table 13.10), and in the five top stocking options trips might increases 13–34%. Responses to the contingent behaviour questions indicate that anglers were receptive to proposed stocking changes and would likely fish more often if their preferred changes were made.

In one example from the grayling survey, respondents were presented with three management options for the Chena River, which is currently under catch-and-release management. Nearly half (48.1%) of the respondents support continued catch-and-release, but regulations allowing for harvest in a lower part of the river would lead to a 10.7% increase in trips. A take of one fish per day in the entire river would lead to a 16.3% increase in trips, while implementing the most liberal option, a two-fish daily bag with one fish over 38 cm (15 inches), would lead to a 28.7% increase in trips (Table 13.11). But this increase would come from a minority of anglers because, when the distribution of responses to the most liberal option is examined (Fig. 13.5), the vast majority (77.7%) indicated no change in their current average rate of annual fishing trips.

In the salmon survey, respondents were asked to predict how their visits to specific waters would be altered if bag limits and bait restrictions were to be changed. As in the grayling survey, a majority (over 80%) would not alter the number of trips regardless of regulation changes. But overall trips might decline from 2.5% to 14.3% with more restrictive regula-

Table 13.11 Responses to management options for grayling fishing in the Chena River from the Tanana population (standard errors are in parentheses).

Management options	Average reported increase in trip numbers per year under new regulations	Estimated per cent increase in trip numbers per year under new regulations	Sample size
Harvest 1 fish per day, no size limit	0.47 (0.09)	16.3	716
Harvest 2 fish per day, 1 over 38 cm (15 inches)	0.83 (0.17)	28.7	741
Catch-and-release with lower river open to harvest	0.31 (0.07)	10.7	686

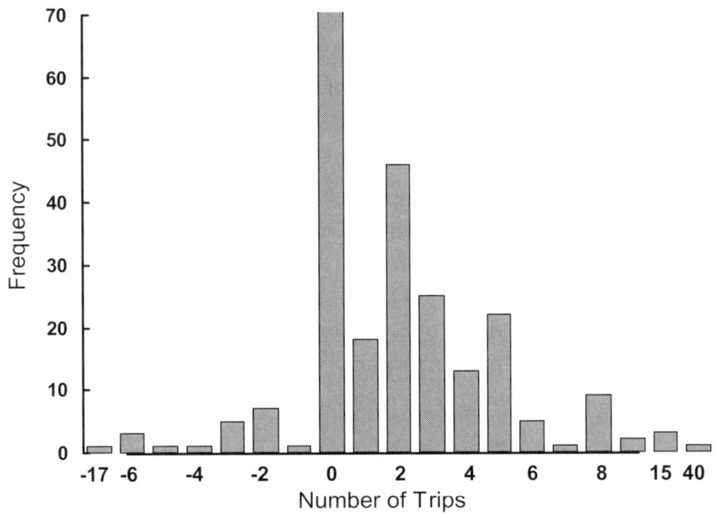

Fig. 13.5 Reported change in number of sport fishing trips to the Chena River under a two fish per day harvest limit (576; 77.7% respondents).

tions, and increase from 6.1% to 66.7% under liberal regulations. Support to decrease the bag limit from 5 to 3 chinook salmon in the Copper River drainage was strong, however reaction by the average Alaskan angler to the proposal to eliminate bait was mixed. Copper River and Tanana fishers opposed bait restriction, while nonresidents and Seward Peninsula residents supported it. In a similar pike, burbot and lake trout survey, more than 90% of respondents would not alter the number of fishing trips taken regardless of changes in regulations.

Angler and fishing trip characteristics

Angler characteristics and fishing trip interests are useful for understanding angler groups known as market segments. Market research provides information on resource users, and how to attract those segments with low rates of sport fishing participation. Additionally, angler characteristics and fishing trip interests can help to explain valuation and policy responses.

There were many similarities in angler characteristics between populations and years. However, on average, nonresidents had been fishing longer, were older, were comprised of a higher percentage of males, and fished more often than most residents (except Northwest and Seward Peninsula populations). Average years schooled was similar across populations. The percentage of respondents in households with an income before taxes over \$70 000 was highest for those residents of Northwest Alaska surveyed in 1996 (50.8%) followed by Regions I and II anglers and nonresidents (average of 40%). Copper River respondents reported the lowest percentage (20%) of households with an income before taxes over \$70 000.

Salmon was the wild species most often targeted by respondents of all populations on their recent fishing trip. More than 50% of nonresidents and up to 73% of Seward Peninsula residents targeted salmon. A much higher percentage of nonresidents, Regions I and II residents and Tanana residents specifically targeted grayling than did Seward Peninsula and Northwest

Table 13.12 Average annual fishing trips per respondent to Region III waters by population and survey*

Water	Respondent population					
	Nonresident	Regions I and II	Tanana	Copper	Seward Peninsula	North-west
Tanana drainage						
Chena R. grayling	0.60	0.83	**2.46**	—**	0.02	0.02
Other grayling	0.14	0.21	**0.71**	—	0.06	0.02
Chena R. salmon	0.12	0.11	**1.19**	0.08	0.01	—
Other salmon	0.04	0.05	**0.29**	0.15	0	—
Chena R. pike, burbot	0.70	0.24	**1.76**	0.10	—	—
Other pike, burbot	0.15	0.13	**0.52**	0.31	—	—
Copper drainage						
Gulkana R. salmon	0.45	0.15	0.61	**1.53**	0.02	—
Other salmon	0.14	0.10	0.09	**0.48**	0.03	—
Seward Peninsula						
Nome R. grayling	0.07	0.09	0.03	—	**6.90**	0.09
Other grayling	0.02	0.12	0.02	—	**2.58**	0.14

* Bold entries are those waters in the same area as the sampled population.
** Does not apply; survey not administered to this population.

residents. Copper River residents reported the highest percentage (51.6%) of those specifically targeting burbot, pike and lake trout, of the populations queried about fishing for these top level predators. Half (49.5%) of the Tanana Valley residents sampled in the stocked waters survey indicated they targeted primarily rainbow trout.

Not surprisingly, residents had much higher visitation rates to waters in their own areas than to waters in other areas (Table 13.12). In the Tanana drainage, the Chena River was most popular with nonresidents, Regions I and II and Tanana residents targeting grayling, salmon and pike or burbot. In the Copper River drainage, the Gulkana River was most popular with nearly all populations fishing for salmon. Of those waters listed on the Seward Peninsula, the Nome River was most popular with nonresidents and the local population for grayling fishing.

Because there is no conservation concern for stocked fish, bag limits are generous. Stocking of hatchery-reared fish occurs close to urban centres to facilitate angler access, thereby reducing fishing pressure on wild stocks. Conversely, releasing wild species is encouraged for conservation purposes through regulations (i.e. catch-and-release only areas or restricted bag limits), and public outreach programs. A large majority of grayling caught by all populations are released (Table 13.13). Seward Peninsula and Regions I and II populations reported keeping substantially more salmon than did respondents from other populations. All populations surveyed in 1998 released far greater numbers of pike and lake trout than were kept. Generally, fewer numbers of burbot and lake trout are caught per fishing trip across populations than other species.

Nonresidents who fished in Region III spent significantly more per trip than others (Table 13.14). Additionally, Regions I and II anglers spent more per fishing trip in Region III than those residing in Region III. Logically, anglers not residing in the region would be more likely to incur greater travel and lodging costs than residents of the region. Travel cost was generally the largest component of expenditures across populations and surveys.

Table 13.13 Average number of fish caught and per cent released per trip by population and survey.

	Respondent population					
Year/Species	Nonresident	Regions I and II	Tanana	Copper River	Seward Peninsula	North-west
1996–Grayling						
No. caught	8.7	5.1	4.4	—	2.7	4.2
% released	**93.1**	**92.2**	**86.4**	—	**88.9**	**85.7**
1997–Salmon						
No. caught	7.4	8.9	3.4	3.6	10.4	—
% released	**70.3**	**38.2**	**50.0**	**55.6**	**30.8**	—
1998–Pike						
No. caught	9.2	2.6	3.8	2.3	—	—
% released	**94.6**	**96.2**	**84.2**	**82.6**	—	—
1998–Burbot						
No. caught	0.1	0.2	0.5	1.2	—	—
% released	**0**	**100**	**40.0**	**33.3**	—	—
1998–Lake trout						
No. caught	2.3	2.7	1.4	1.8	—	—
% released	**87.0**	**74.1**	**71.4**	**61.1**	—	—

* Does not apply; survey not administered to this population.

Table 13.14 Average total expenditures in US$ per fishing trip for the respondent by population.

	Respondent population					
Year	Nonresident	Regions I and II	Tanana	Copper River	Seward Peninsula	North-west
1995	—*	—	55	—	—	—
1996	2152	429	182	—	167	269
1997	1893	199	163	164	93	—
1998	1198	263	151	84	—	—

* Does not apply; survey not administered to this population.

Angler preferences and motivation

Measuring angler preferences provides valuable input to the manager on whether current management is aligned with anglers' desires (Pollock *et al.* 1994). Angler preferences are also critical in estimating benefits of management changes. If managers are informed about preferred fishing experiences, and can plan accordingly, the more likely the angler is to give the fishery a higher satisfaction rating. The higher the angler satisfaction rating, the closer the fishery is to a social optimum.

Respondents were asked to rate their preferences on a scale of 1 to 5, with 1 being least preferred and 5 most preferred. Results presented in Figures 13.6 and 13.7 are the percentage of respondents in each population who rated an attribute as either 4 or 5. All respondents strongly preferred (60–80%) fishing in a wilderness setting to fishing an easily accessible site near a road. Having good trail access and more developed camping facilities was more preferred by nonresidents and Tanana residents than other populations. Releasing grayling was preferred (50–60%) to keeping grayling by all. Releasing salmon was less preferred

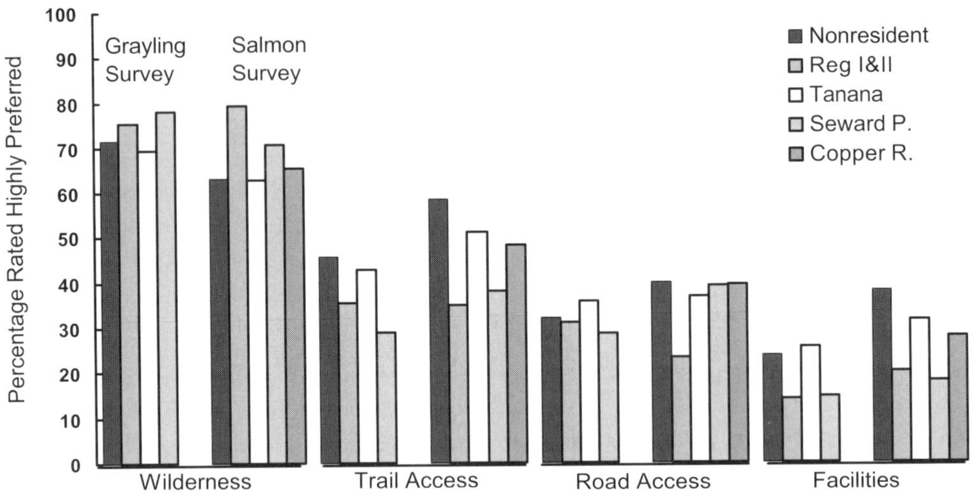

Fig. 13.6 Percentage of respondents in each population who rated their preference for fishing site attributes as either 4 or 5, with 5 being most preferred on a scale of 1-5.

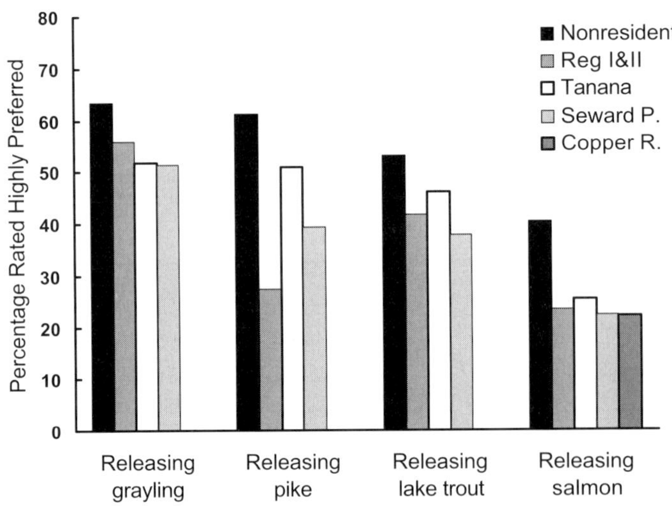

Fig. 13.7 Percentage of respondents in each population who rated their preference for catch-and-release as either 4 or 5, with 5 being most preferred on a scale of 1-5.

(20–25%) to keeping salmon by all except nonresidents. More nonresidents and Tanana residents preferred to release pike and lake trout.

There are conflicting preferences for specific management strategies, nearly evenly divided, for managing grayling in Tanana waters. Nearly half (48.7%) of Tanana residents preferred managing grayling for catch-and-release; on the other hand, 42.2% of the remaining residents preferred managing grayling for harvest. The Chena River was both the most frequently cited stream for catch-and-release (cited 124 times) and for harvest management (cited 98 times). Fewer residents on the Seward Peninsula (33.8%) favour catch-and-release management for grayling. The Nome River was the most frequently cited stream on the

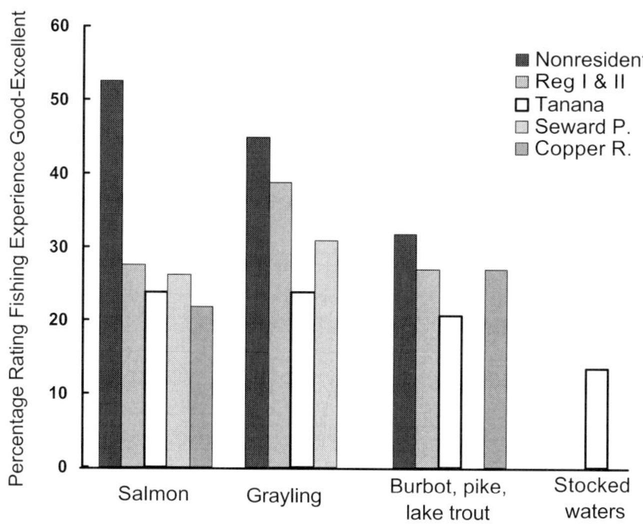

Fig. 13.8 Percentage of respondents in each population who rated their most recent sport fishing experience as good to excellent.

Seward Peninsula for both catch-and-release management (cited 30 times) and for harvest management (cited 29 times). Preference for status quo management of grayling is rated high by Seward Peninsula (60.9%) residents. Fewer nonresidents prefer managing for harvest than residents.

Satisfaction with the most recent fishing experience was rated by respondents on a scale of 'poor' to 'excellent'. Nonresidents fishing for salmon reported the highest percentage (52.5%) of fishing experiences rated above average or excellent (Fig. 13.8). North-west Alaskans and nonresidents reported the highest percentages (49.0 and 44.9%, respectively) of above-average and excellent experiences when fishing for grayling. Over 30% of nonresidents and 26.9% Copper River residents enjoyed their sport fishing experiences for burbot, pike or lake trout. As a whole, the Tanana population rated their sport fishing experiences lower in quality than the other populations: only 13.4% of Tanana Valley residents fishing stocked waters reported above average or excellent experiences. Nonresidents rated their salmon fishing experience much higher than residents did. Since both groups are fishing predominantly the same waters, this may be an indication of the higher expectations residents have for a salmon fishing trip.

Discussion

The primary goals of this research were to contribute to the literature on NEV for Alaska fisheries, undertake benefit–cost applications, and inform management about the influence of policy choices on the angling public using contingent behaviour modelling. Dichotomous choice CVM in conjunction with mail survey-based data sets proved to be a successful and relatively low cost method for obtaining NEVs by management area and species. Only the most popular destinations (e.g. the Chena and Gulkana rivers) yielded sufficient responses to estimate NEVs on a per river basis. The estimated models were consistent with economic theory with the correct signs and highly significant coefficients for the price response parameter, as well as for quantity and income. The range of bids selected covered an appropriately wide range of the WTP distribution (11–88%). Standard errors were consistently within or near the objective criteria. For nine models (samples averaging 270 observations) the coefficient of variation averaged 8.2% with a range of 4–15%.

The estimated NEV for these Region III studies can be compared to the estimates from the economics literature. The values from Carson *et al.* (1987) for both residents and nonresidents combined in south-central Alaska are \$204 per day in 1987 dollars or (correcting for inflation) about \$285/day in 1997 dollars. When we convert NEVs per trip to NEVs per day, by comparison, the nonresident estimates for the current study are in the same range at \$217 per day for the grayling study and about \$300 per day for trips targeting salmon. The resident angler values from the grayling and salmon studies show NEVs around \$50 per day. These are considerably below the Carson *et al.* south-central estimate but similar to the average literature values reported for the rest of the US. For example, Walsh *et al.* (1992) in a summary of 39 coldwater fishing studies estimated a mean value per day of \$31 in 1987 dollars or about \$42 dollars in 1997 terms. The estimates for the road-accessible stocked waters in the Fairbanks area, fisheries which would be more like typical US fisheries, have per day values ranging from \$15 to \$44 – near the US average. The considerable range of values generated by this study reflects the great diversity of Alaskan fishing, from very valuable fly-out trips in wilderness settings, to road-accessible stocked sites primarily used by locals. The ability of the CVM models to correctly predict responses compares favourably with those reported in the literature. The grayling model predicts correctly the yes/no response for 77% of resident respondents and 85% of nonresidents. By comparison, the 1996 NFHS reports 22 separate state and regional sport fishing CVM models with the percent correct prediction ranging from 61% to 72%. These findings generally indicate that the CVM responses here are not random or meaningless but are consistent with economic theory.

It is anticipated that the estimated NEVs will be used in a variety of policy applications in the future. One application developed as part of these studies, the benefit–cost evaluation of the stocking program for Fairbanks-area waters, indicates that benefits exceed costs for this activity. An area for further research would be to conduct a more rigorous benefit/cost analysis for the wild stock fisheries. This would require a biological production relationship that would identify the actual marginal benefits (measured in fish stocks or angler catch rates) that are a consequence of agency research and management efforts. Additionally, one would need to estimate valuation models that included stocks or catch as variables to compute marginal values. A challenge for policy-makers is to define the cost basis, and to obtain accurate and consistent estimates of costs for purposes of program evaluation. Using the ratio of total annual NEV to expenditures as a measure of program cost efficiency, public benefits do indeed outweigh management costs for the fisheries studied, therefore meeting objectives in fisheries management plans. The extent to which the benefit/cost ratio influences program decisions is still under discussion by policy-makers. The benefit/cost ratio is only one of several considerations in program planning, and its weight in the policy-making process is likely to be variable, depending upon the importance of other influences, such as conservation or political concerns.

Overall, the greatest nonmarket benefits in Region III accrue from salmon sport fisheries. Anglers value fishing for wild species in flowing waters more than fishing stocked waters. Nonresidents value their fishing trips to Region III significantly more highly than do residents. An interesting finding is that a large portion of the total NEV for Region III sport fishing trips is from nonresidents. Fisheries management planning normally does not account for nonresident motivations or preferences; however, the revelation of the overwhelming contribution of nonresidents to net economic benefits may promote discussion by policymakers.

The contingent behaviour responses provided detailed information on angler preferences for management options. Management policies must be at least minimally supported by anglers or these clients will use the political process to lobby for changes. The policy that brings about positive changes in social and economic benefits from sport fisheries, while still achieving biological objectives, is likely to enjoy public support. Information obtained from socioeconomic research can be used to evaluate particular management policies for their influence on NEV, preferences, and fishing trip frequency. Space constraints do not allow us to summarise the many ways study findings have already been utilised by agency managers. Suffice to say that the results have already had some influence on a number of management decisions – particularly for salmon and grayling fisheries.

While implementing preferred options for management and allowing more liberal regulations is likely to result in increased angler trips, estimated changes in visits are relatively small. Changes in trip frequency would only come from a minority of the angling public. A portion of this minority stated that any increased trips to area waters from regulation changes would come at the cost of fishing trips to other waters; in other words, substitution would occur. Consistently across models, the vast majority of respondents reported that the proposed options would have no effect on the number of fishing trips they currently take. For example, in both the grayling and salmon studies a large majority (over 80%) of respondent populations would not alter the number of sport fishing trips taken regardless of restrictions or liberalisations of bag limit or bait. The ability of management to influence anglers' decisions to take fishing trips in primarily wild fisheries may be overshadowed by more significant variables such as weather, the angler's employment and economic situation, the angler's motives for initiating a trip, and other characteristics of the site besides the fishing. Accordingly, the overall impact to public welfare from the fishery-specific management options examined here may be minimal. Additionally, impacts to public welfare from implementation of preferred management options must be traded off with consideration of risks to the resource and program costs. This research confirms that anglers are a diverse group, and that in many cases there are few management options that would be preferred by a clear majority. This diversity makes optimisation difficult. Further analysis, such as segmenting anglers into groups based on fishing motivation, may reveal more obvious patterns in angler preferences for management policies.

Acknowledgements

The authors are grateful for the helpful editorial and substantive advice offered by Chuck Hollingworth, Tony Pitcher and two anonymous referees. Allen Bingham, ADF&G,

developed equations for estimating angler trips from the state-wide harvest survey. Thanks to Allen Howe, Bob Walker, Sara Case and others with ADF&G for technical support. Thanks to Mac Minard for encouragement and financial support to participate in the conference, Evaluating the Benefits of Recreational Fisheries in 1999, which led to the development of this chapter.

References

Allen, A. (1988) *Montana Bioeconomics Study: Results of the Trout Stream Angler Preference Survey*. Report for Montana Department of Fish, Wildlife, and Parks. Helena, MT.

Arrow, K., Solow, R., Learner, E., Portney, P., Radner, R. & Schuman, H. (1993) *Report of the NOAA Panel on CVM. Federal Register.* **58**(10), 4601–4614.

Belli, R.F. (1997) Reducing vote overreporting in surveys: social desirability, memory failure, and source monitoring. *Public Opinion Quarterly*, **63**(1), 90–108.

Bishop, R.C. & Heberlein, T.A. (1979) Measuring values of extra-market goods: are indirect measures biased? *American Journal of Agricultural Economics,* **61**(5), 926–930.

Bishop, R.C. & Heberlein, T.A. (1992) The contingent valuation method. In: *Natural Resource Damages: Law and Economics* (eds K. Ward & J. Duffield), pp. 281–309. John Wiley. New York.

Boyle, K.J. & Bishop, R.C. (1987) Valuing wildlife in cost-benefit analysis: a case study involving endangered species. *Water Resources Research*, **23**(5), 943–950.

Boyle, K.J., Welsh, M.P., Bishop, R.C. & Baumgartner, R.M. (1988) Analyzing the effects of Glen Canyon Dam releases on Colorado River recreation using scenarios of unexperienced flow conditions. In: *Benefits and Costs in Natural Resource Planning* (eds J. Loomis) pp. 111–130. Western Regional Research Publication W-133, Davis, CA.

Bryan, H. (1979) Leisure value systems and recreational specialization: the case of trout fishermen. *Journal of Leisure Research,* **11,** 174–187.

Cameron, T.A. (1988) A new paradigm for valuing nonmarket goods using referendum data: maximum likelihood estimation by censored logistic regression. *Journal of Environmental Economics and Management,* **15**, 355–379.

Cameron, T.A., Shaw, W.D., Ragland, S.E., Maccala, J. & Keefe, S. (1996) Using actual and contingent behaviour data with differing levels of aggregation to model recreation demand. *Journal of Agricultural and Resource Economics*, **21**(1), 130–149.

Carson, R.T., Hanemann, W.M. & Mitchell, R.C. (1986) Determining the Demand for Public Goods by Simulating Referendums at Different Tax Prices. Manuscript, University of California, San Diego.

Carson, R.T., Hanemann, W.M. & Wegge, T. (1987) *Southcentral Alaska Sport Fishing Study*. Report prepared by Jones and Stokes Associates for Alaska Department of Fish and Game, Anchorage, AK.

Clawson, M. & Knetsch, J.L. (1966) *Economics of Outdoor Recreation*. Johns Hopkins University Press, Washington, DC.

Cooper, J.C. (1993) Optimal bid selection for dichotomous choice contingent valuation surveys. *Journal of Environmental Economics and Management,* **24**, 25–40.

Cooper, J.C. (1997) Combining actual and contingent behaviour data to model farmer adoption of water-quality protection practices. *Journal of Agricultural and Resource Economics*, **22**(1), 30–43.

Davis, R.K. (1963) *The Value of Big Game Hunting in a Private Forest*. Transactions of 29th North American Wildlife & Natural Resources Conference, pp. 393–403.

Dillman, D. (1978) *Mail and Telephone Surveys*. John Wiley, New York.

Duffield, J.W. (1980) Auburn Dam: a case study in water policy and economics. *Water Resources Bulletin,* **16**(2), 226–234.

Duffield, J.W. (1984) Travel cost and contingent valuation: a comparative analysis. *Advances in Applied Microecomics,* **3**, 67–88.

Duffield, J.W. (1997) Nonmarket valuation and the courts: the case of the Exxon Valdez. *Contemporary Economic Policy,* **15**, 98–110.

Duffield, J.W. & Allen, S. (1988) *Angler Preference Study Final Economics Report*: Contingent Valuation of Montana Trout Fishing by River and Angler Subgroup. Report for Montana Department of Fish, Wildlife and Parks. Helena, MT.

Duffield, J.W. & Neher, C.J. (1996) Economics of wolf recovery in Yellowstone National Park. Transactions of North American Wildlife & Natural Resources Conference **61**, 285–292.

Duffield, J. & Patterson, D. (1991). Inference and optimal design for a welfare measure in dichotomous choice contingent valuation. *Land Economics,* **67**(2), 225–239.

Duffield, J., Loomis, J. & Brooks, R. (1987) *The Net Economic Value of Fishing in Montana*. Report for the Montana Dept. of Fish, Wildlife, and Parks. Helena, MT.

Duffield, J., Neher, C. & Brown, T. (1992) Recreation benefits of instream flow: application to Montana's Big Hole and Bitterroot Rivers. *Water Resources Research,* **28**(9), 2169–2181.

Duffield, J., Neher, C. & Merritt, M. (2000) *Effect of Proposed Changes to Rod and Reel Subsistence Harvest Regulations in the Lower Yukon/Kuskokwim Area: Surveys and Analysis*. Alaska Department of Fish and Game, Special Publication Number 00–2, Anchorage.

Duffield, J., Neher, C. & Merritt, M. (2001a) *Alaska Angler Survey: Use and Valuation Estimates for 1995, with a Focus on Tanana Valley Major Stocked Waters.* Alaska Department of Fish and Game, Special Publication Number 01–4, Anchorage.

Duffield, J., Neher, C. & Merritt, M. (2001b) *Alaska Angler Survey: Use and Valuation Estimates for 1996, with a Focus on Arctic Grayling Fisheries In Region III*. Alaska Department of Fish and Game, Special Publication Number 01–5, Anchorage.

Duffield, J., Neher, C. & Merritt, M. (2001c) *Alaska Angler Survey: Use and Valuation Estimates for 1997, with a Focus on Salmon Fisheries in Region III*. Alaska Department of Fish and Game, Special Publication Number 01–2, Anchorage.

Duffield, J., Neher, C. & Merritt, M. (2001d) *Alaska Angler Survey: Use and Valuation Estimates for 1998, with a Focus on Burbot, Pike and Lake Trout Fisheries in Region III*. Alaska Department of Fish and Game, Special Publication Number 01–3, Anchorage.

Duffield, J.W., Neher, C. Patterson, D. & Allen, S. (1990) *Instream Flows in the Missouri River Basin: a Recreation Survey and Economic Study*. Report for the Montana Department of Natural Resources and Conservation. Helena, MT.

Ferber, R. & Piskie, R. (1965) Subjective probabilities and buying intentions. *The Review of Economics and Statistics,* **47**, 322–325.

Hanemann, W.M. (1984) Welfare evaluation in contingent valuation experiments with discrete responses. *American Journal of Agricultural Economics,* **66**, 332–341.

Hanemann, W.M. (1989) Welfare evaluation in contingent valuation with discrete response: Reply. *American Journal of Agricultural Economics,* **71**, 1057–61.

Herrick, S., Strand, I. & Squires, D. (1994) Application of benefit-cost analysis to fisheries allocation decisions: the case of Alaska Walleye Pollock and Pacific Cod. *North American Journal of Fisheries Management,* **14**, 726–741.

Herriges, J.A. & Kling, C.L. (1999) *Valuing Recreation and the Environment: Revealed Preference Methods in Theory and Practice*. Edward Elgar, Northampton, MA.

Hotelling, H. (1947) *The Economics of Public Recreation*. Reprinted in Ward and Beal (2000). pp. 217–218.

Howe, A.L., Fidler, G., Bingham, A.E. & Mills, M.J. (1996) *Harvest, Catch & Participation in Alaska Sport Fisheries during 1995*. Alaska Department of Fish and Game, Fishery Data Series Number 96–32, Anchorage.

Howe, A.L., Fidler, G., Olnes, C., Bingham, A.E. & Mills, M.J. (*In press a*) Revised Edition: *Harvest, Catch, and Participation in Alaska Sport Fisheries during 1996*. Alaska Department Fish and Game, Fishery Data Series Number 97–29, (Revised) Anchorage.

Howe, A.L., Fidler, G., Olnes, C., Bingham, A.E. & Mills, M.J. (*In press b*) Revised Edition: *Harvest, Catch, and Participation in Alaska Sport Fisheries during 1997*. Alaska Department Fish and Game, Fishery Data Series Number 98–25, (Revised) Anchorage.

Howe, A.L., Walker, R.J., Olnes, C., Heineman, G. & Bingham, A.E. (*In press c*) Revised Edition: *Harvest, Catch, and Participation in Alaska Sport Fisheries during 1998*. Alaska Department of Fish and Game, Fishery Data Series No. 99–41 (Revised) Anchorage.

Just, R.E., Hueth, D.L. & Schmitz, A. (1982) *Applied Welfare Economics and Public Policy*. Prentice-Hall, Englewood Cliffs, NJ.

Krutilla, J.V., Cicchetti, C.J. & Fisher, A.C. (1972) The Economics of Environmental Preservation: a Theoretical and Empirical Analysis. *American Economic Review*, **62**, 605–619.

Loomis, J.B. (1996) Measuring the Economic Benefits of Removing Dams and Restoring the Elwah River: Results of a Contingent Valuation Survey. *Water Resources Research*, **32**, 441–447.

Lyle, J.M., Coleman, A.P.M., West, L., Campbell, D. & Henry, G.W. (2002) New large-scale survey methods for evaluating sport fisheries. In: *Recreational Fisheries: Ecological, Economic and Social Evaluation* (eds Pitcher, T.J. & Hollingworth, C.E.), Chapter 15, this volume. Blackwell Science, Oxford, UK.

Markowski, M.A., Boyle, K.J., Bishop, R.C., Larson, D. & Patterson, R. (2001) *A Cautionary Note on Interpretating Meta Analysis*. Unpublished Manuscript, Industrial Economics, Inc. Cambridge, MA.

McConnell, K.E. (1986) *The Damages to Recreational Activities from PCBs in New Bedford Harbor*. Industrial Economics, Cambridge, MA.

Mitchell, R.C. & Carson, R.T. (1989) *Using Surveys to Value Public Goods: the Contingent Valuation Method*. Resources for The Future, Washington, DC.

Mitofsky, W.J. (1996) Review: was 1996 a Worse Year for Polls than 1948? *Public Opinion Quarterly*, **62**(2), 230–249.

Morton, K.M., Adamowicz, W.L. & Boxall, P.C. (1995) Economic effects of environmental quality change on recreational hunting in Northwest Saskatchewan: a contingent behaviour analysis. *Canadian Journal of Forest Research*, **25**, 912–920.

Narayanan, R. (1986) Evaluation of recreational benefits of instream flow. *Journal of Leisure Research*, **18**(2), 116–128.

Neher, C.J. & Duffield, J.W. (1993) *Economic Analysis of the Values of Surface Uses of State Lands: Fair Market Value for Recreational Uses*. Report for Montana Department of State Lands. Helena, MT.

Park, T., Loomis, J. & Creel, M. (1989) *Confidence Intervals for Evaluating Benefit Estimates from Dichotomous Choice Contingent Valuation Survey*. Dept. of Agricultural Economics, University of Nebraska.

Pollock, K., Jones, C. & Brown, T. (1994) *Angler Survey Methods and Their Applications in Fisheries Management*. American Fisheries Society, Special Publications 25. Bethesda, MD.

Rudd, M.A., Folmer, H. & van Kooten, G.C. (2002) Economic evaluation of recreational fishery policies. In: *Recreational Fisheries: Ecological, Economic and Social Evaluation* (eds Pitcher, T.J. & Hollingworth, C.E.), Chapter 3, this volume. Blackwell Science, Oxford, UK.

SAS Institute (1988) *SAS/Stat Users Guide*. Release 6.04 Edition. Cary, NC.

Skaugstad, C., Hansen, P. & Doxey, M. (1995) *Evaluation of Stocked Game Fish in the Tanana Valley, 1994*. Alaska Dept. of Fish and Game, Fishery Data Series No. 95–20, Anchorage.

Thayer, M.A. (1981) Contingent valuation techniques for assessing environmental impacts: further evidence. *Journal of Environmental Economics and Management,* **8**(2), 27–44.

Theil, H. & Kosobud, R.F. (1968) How Informative are Consumer Buying Intentions Surveys? *The Review of Economics and Statistics,* **50**, 50–59

Traugot, M.W. & Katosh, J.O. (1979) Response validity in surveys of voting behaviour. *Public Opinion Quarterly,* **43**(3), 359–377.

U.S. Department of the Interior (1986) *Natural Resource Damage Assessment Rules and Regulations*. Federal Register **51**(148), 27725–27753(August 1).

U.S. Department of the Interior (1991) *Natural Resource Damage Assessments: Notice of Proposed Rulemaking*. Federal Register **56**(82), 19752–19753 (April 29).

U.S. Fish and Wildlife Service (1997) *National Survey of Fishing, Hunting, and Wildlife-Associated Recreation*. Washington, DC.

U.S. Fish and Wildlife Service (1998) Net Economic Values for Bass, Trout and Walleye Fishing, Deer, Elk and Moose Hunting and Wildlife Watching, 1996. *Report 96–2*. Washington, D.C.

Walsh, R.G., Johnson, D.M. & Mckean, J. (1992) Benefit transfer of outdoor recreation demand studies (1968–1988). *Water Resources Research,* **28**(3), 707–713.

Ward, F.A., Roach, B.A. & Henderson J.E. (1996) The economic value of water in recreation: evidence for the California draught. *Water Resources Research,* **32**(4), 1075–1081.

Ward, F.A. & Beal, D. (2000) *Valuing Nature with Travel Cost Models: a Manual*. Edward Elgar, Northampton.

Ward, K.M. & Duffield, J.W. (1992) *Natural Resource Damages: Law and Economics*. John Wiley, New York.

Chapter 14

Evaluating Marine Sport Fisheries in the USA

Brad Gentner and Alan Lowther

Abstract

Since 1979, the United States has been collecting data on marine recreational angling with the Marine Recreational Fishery Statistics Survey (MRFSS). This survey is designed to produce estimates of fishing catch, effort, and participation, as well as related demographic and socio-economic variables. The MRFSS consists of two independent but complementary surveys conducted annually in six two-month 'waves', all across the United States. The first part consists of an intercept survey of anglers conducted at fishing access sites. The intercept survey is designed to obtain a random sample of recreational trips for computing catch per unit effort. This sample is stratified by state, wave, mode, fishing area, catch type, and species. During the intercept interview, measuring and weighing available catch collects biological data. Disposition of catch not retained is also solicited to characterise release behaviour. The second survey component is a telephone survey of randomly contacted households in all coastal counties. This phase is used to estimate fishing effort, measured in trips, for residents of coastal counties. Overall effort, 22 518 000 trips for the North-east states in 1994 and 36 966 000 trips for the South-east states in 1997, is found by expanding the coastal effort based on a ratio obtained from the intercept survey. Combining data from the two survey components produces total catch and participation estimates.

To enable the estimation of travel cost models of recreational demand, the base MRFSS survey has been amended to include necessary data elements. Additionally, data are collected that will enable the estimation of economic impact models. In this chapter, the economic surveying strategy of the United States for marine recreational angling is presented. The discussion will focus on the key data elements we have identified for both valuation and economic impact modelling. Once the data collection strategy has been described, two regional applications of a random utility model of recreation demand are detailed. The same general random utility model was applied to data collected in the North-east and South-east United States to value access to marine angling and to develop a marginal value of catch.

Key words: marine recreational fishing, random utility models, recreational economic surveys, resource valuation, travel cost demand models.

Introduction

In 1999, 7.8 million anglers took 56.8 million recreational trips in the United States (US), exclusive of Alaska, Hawaii, and Texas (NMFS 2000). Making stock allocation and management decisions requires carefully assessing and balancing the changes in net economic benefits (value) (Box 14.1) and economic impacts (Box 14.1) stemming from these management actions across both commercial and recreational interests. Valuation of marine recreational angling requires extensive data on angling behaviour, including trip information and socioeconomic characteristics of the angler, as well as attributes of the marine recreational angling experience (fishing quality). The National Marine Fisheries Service (NMFS) has been collecting recreational catch, effort, and participation data since 1979. Socioeconomic data collection was needed to tie catch rates (fishing quality) with angler behaviour to estimate the value of marine recreational angling. Assessing economic impacts requires collecting data on anglers' trip and annual expenditures on marine sport fishing.

This chapter begins with a discussion of the history of NMFS marine recreational angling data collection and moves on to a discussion of the current data collection strategy. Next, the addition of socioeconomic and demographic questions to the base strategy will be discussed. The goal of NMFS is to collect economic data for a wide range of uses that include assessing the value and economic impact of marine angling. Two case studies are then presented

Contingent valuation = uses a survey instrument to elicit a response to a hypothetical market transaction allowing the valuation of a resource not typically traded in a market.
Economic impact = a modelling technique that details the flow of expenditures through a community. Impacts can be expressed as increments (decrements) to total output, employment, or personal income.
Effort = time spent fishing as expressed as the number of one-day trips.
Opportunity cost of time = the cost to any angler associated with the time spent travelling to an access point. Assumes that the angler derives no utility from the journey and is generated by the wage foregone to take the trip.
Participation = the number of people that have taken a fishing trip during the referent period. The MRFSS estimates participation by wave and year.
Utility = the benefit or value an individual derives from bundles of goods and services. Utility functions mathematically capture an individual's preference for various bundles of goods and services. It is assumed that consumers behave to maximise utility.
Value = net economic benefit. It is the total benefit, expressed in economic terms, to an individual or society of any good or service less the costs of obtaining that good or service (may be positive or negative).
Welfare = individual and societal well-being as expressed in economic terms.
Willingness-to-pay = the maximum sum of money an individual would be willing to pay rather than do without an increase in some good (Freeman 1993).

Box 14.1 Glossary of terms.

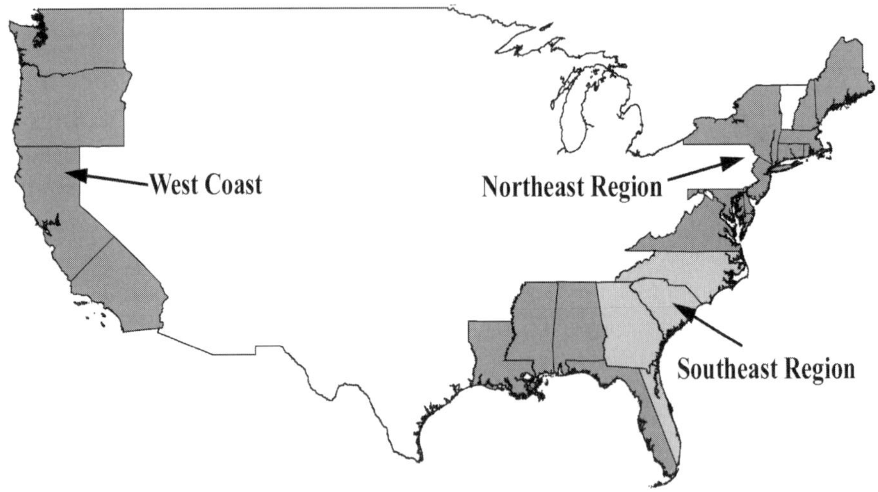

Fig. 14.1 Map of the United States divided into management regions.

involving the analysis of this data in an economic context. Specifically, we will examine the value of one-day fishing trips and the value of increasing catch rates to anglers on the East and Gulf Coasts of the USA (Fig. 14.1). Our goal is to describe the methodology of this survey and present some results that demonstrate the value of this data to the management decision process. Through an understanding of our process, it is hoped that the reader will better be able to access and use data produced by the NMFS to answer their own research questions. In addition, we hope to offer guidance to those about to embark on their own data collection effort, particularly as it applies to collecting socioeconomic data from marine anglers on a large scale. Finally, this chapter will serve as an introduction to using observed angler behaviour to value recreational angling, a nonmarket good, and will direct the reader to additional sources of information regarding valuing marine angling.

Data collection history

Recreational angling is a very important activity to the USA. Marine recreational angling is the second most popular outdoor sport nation-wide when measured by number of participants (NMFS 2000). With millions of anglers taking millions of trips in the US, this sport has profound impacts on the fisheries resource, especially for popular species. In 1998, the recreational harvest was only about 6% of the total US harvest of fish (commercial and recreational) (NMFS 2000). Although this appears to be a small impact on the marine resource overall, over half of the recreational harvest is taken from the top ten most frequently caught species of fish (Table 14.1) (NMFS 2000). For most of these species, the recreational harvest far exceeds the commercial harvest. Thus, good catch data is necessary to make resource allocations fair. In addition to the resource impacts, marine recreational angling generates high economic value for participants as well as large impacts on the economies of coastal communities and the industries associated with sport fishing. Finally, reliable catch data provided

Table 14.1 Commercial and recreational harvests of top ten recreationally caught species in 1998.

Species common name (*scientific name*)	Harvest (tonnes)	
	Recreational	Commercial
Striped bass (*Morone saxatilis*)	6107	3046
Bluefish (*Pomatomus saltatrix*)	5796	3764
Summer flounder (*Paralichthys dentatus*)	5680	6881
Dolphinfish (*Coryphaena hippurus*)	5494	416
Red drum (*Sciaenops ocellatus*)	4468	5
Spotted seatrout (*Cynoscion nebulosus*)	4329	265
King mackerel (*Scomberomorus cavalla*)	3956	2214
Atlantic croaker (*Micropogonias undulatus*)	3725	11478
Yellowtail (*Seriola lalandei*)	2585	111
Yellowfin tuna (*Thunnus albacares*)	2543	7767

through a national creel survey improves the scientific understanding of the biology of fish stocks.

Before 1955, these impacts and benefits were largely overlooked. In 1955, the United States Fish and Wildlife Service (USFWS) added questions about saltwater angling to their survey of freshwater fishing and hunting in the US. These surveys, conducted every five years, collected data on number of anglers, angler expenditures, and fishing activity level. In 1960, 1965, and 1970 adjunct surveys also collected information about catch, effort and participation.

When analysing the results of these surveys, Clark (1962), Deuel and Clark (1968), and Deuel (1973) found response bias and sampling errors. Recall bias, a bias resulting from information inaccurately recalled due to time lag, was a significant factor. In addition, because the interviewer did not inspect the catch, over- (under-) reporting of catch, by size or number, for prestigious (undesirable) species resulted in prestige bias. In addition, because of the long interval between surveys, it was impossible to detect or analyse possible seasonal variation in catch, effort, or participation (sampling error). As a result, NMFS began developing a series of specialised surveys to capture catch, effort, and participation in marine recreational angling.

In 1971, NMFS launched a pilot survey project to estimate catch, effort, and participation. Between the years of 1971 and 1978 the Marine Recreational Fishery Statistics Survey (MRFSS) tested a number of survey methodologies including mail, telephone, door-to-door, and access point (creel) surveys (Essig and Holliday 1991). In 1976, the US Congress passed the Fishery Conservation and Management Act (Public Law 94–265), mandating the collection of better marine recreation angling data, which encouraged further development of the MRFSS. After a review of the pilot methodologies, the MRFSS was pretested in 1978.

The Marine Recreational Fishery Statistics Survey (MRFSS)

By 1979, the MRFSS was fully operational in every coastal state except Alaska, Hawaii, and Texas. Alaska and Texas conduct their own surveys and attempts are made to incorporate their data in the management process at the national level and Hawaii will be included in the MRFSS for 2001. The MRFSS consists of two independent and complementary surveys. These two surveys are stratified to provide independent estimates of catch, effort, and participation across states, fishing modes, and two-month waves through each year. The fishing modes used for this stratification are shore mode, private or rental boat mode, and party or charter boat mode. This method of stratification has proven useful for developing estimates annually or seasonally and it allows individual regions easily to add sample within strata to increase the precision of the estimates.

The first survey is an intercept survey of marine anglers at fishing access sites. This survey attempts to obtain a random sample of all marine recreational fishing trips. The MRFSS maintains a list of over 6000 sites in a master site list, which is continually updated. Each of these sites is ranked by an index of relative fishing pressure by mode, month, and weekday or weekend designation. For a given date, interviewers are assigned to a specific site and to a specific mode of fishing. Interviewers are also given two adjacent, alternative sites if a minimum number of interviews cannot be obtained at the original site assignment. Sampling for private/rental and party/charter modes is conducted after the angler's fishing day has been completed. Sampling for the shore mode may be conducted when at least two-thirds of the fishing for the day has been completed and the angler then estimates total trip effort.

The intercept sampling implies a three-stage sampling framework. In stage one, a given site/day is randomly selected with probability of assignment proportional to the fishing pressure index of the site for that specific day (Cochran 1977). The second stage involves the selection of angling parties, boatloads, groups, or individuals, at the assigned site. Finally, stage three involves possible subsampling among the angling parties selected in stage two. Selection of parties (stage 2) and subsampling among parties (stage 3) is assumed random with equal probabilities. This allows the use of self-weighting estimators to obtain mean catch-per-trip estimates for each species across all strata.

This intercept survey is a creel survey primarily used to estimate mean catch-per-trip by species. Data elements collected during the base part of the intercept survey include state, county, and zip code of residence, hours fished, primary area fished, target species, gear used, and days fished in the last two and 12 months. The creel portion of the survey collects length and weight of all fish species retained by the angler and the species and disposition of all catch not retained by the angler.

For a given stratum, estimates of mean catch-per-trip multiplied by an estimate of the total number of trips (effort) equals the total catch for that stratum. The effort estimates are obtained through the second part of this survey process: the telephone survey of coastal households. Residential households are sampled randomly using the random digit dialling technique as described by Groves *et al.* (1988). All anglers in the contacted household are identified, and each is asked about their fishing activity for the previous two-month period. Multiple attempts are made to contact identified anglers. This survey is used to estimate effort by coastal residents living in households with telephones. Ratios from the intercept survey are used to correct these effort estimates to account for noncoastal residents and coastal residents who do not have telephones, as those groups are not covered in the household sampling frame. Data elements collected for this survey include the number of trips in the last 2 months and the number of trips in the last 12 months. For trips in the last 2 months, trip dates, mode, time of return, and state of access are also collected.

The entire MRFSS program costs US$7m annually and intercepts approximately 150 000 anglers in the field and 300 000 households during the telephone survey. While survey and sampling design is handled in-house, the survey implementation is competitively contracted out to survey research firms. It is these firms that recruit and train field intercept personnel and call centre personnel. Some states have decided to act as a subcontractor, using personnel from the state fish and wildlife agency to conduct the intercept interviews.

Economic add-on surveys

The collection of economic and socioeconomic data began in 1994 with the goal of collecting data for a variety of valuation models and data on angler expenditures. Because the MRFSS constitutes the best nation-wide sample frame for marine recreational angling and offers considerable savings over implementing a new program, the economic data collection was added on to the MRFSS effort. Due to the complexity and volume of data needed to meet these two goals, the valuation survey and expenditure surveys were conducted in separate years with no two regions being sampled in the same year. While there is regional variation in the survey instruments, all surveys collect a consistent set of core data elements. Each full round of economic add-ons costs approximately $150 000 per region.

Each of these two surveys involves three phases. The first phase adds a series of questions to the base MRFSS intercept survey, tying basic economic information to trip-specific catch information and behaviour. Those anglers intercepted are asked to participate in the second phase of the survey, the telephone follow-up. This follow-up survey gathers more detailed socioeconomic and preference data. Finally, the third phase adds a series of questions to the telephone survey of coastal household residents. This data is useful for creating population estimates from the intercepts and telephone follow-ups random sample of trips. It also adds the important socioeconomic and demographic variables needed to improve participation estimates and predictions of future changes in participation. Table 14.2 shows a complete timeline for the rounds of economic surveys.

Table 14.2 Recreational fisheries economic data collection timeline.

Year	Intercept add-on			Telephone follow-up			Household survey		
	NE*	SE*	WC*	NE	SE	WC	NE	SE	WC
1994	V**			V			V		
1995									
1996	F**								
1997	F	V			V			V	
1998	C**		V	E**		V	E		V
1999	F	C			E			E	
2000		F	C			E			E

* NE = Maine to Virginia; SE = North Carolina to Louisiana; WC = Washington to California.
** C = part of expenditure round but also collects core valuation data; E = expenditure survey; F = flexible questions only; V = valuation survey.

In addition to the above scheduled surveys, a small, flexible set of data elements may be added to the regional intercept surveys in years when the economic add-on is not being administered. These flexible questions have been used to gather the minimum data elements necessary to run travel cost demand models in the North-east region (Maine to Virginia, Fig. 14.1) and in the South-east region (Louisiana to North Carolina, Fig. 14.1), providing a time series of valuation information. Also, using the flexible questions to gather mailing addresses or telephone numbers from respondents willing to participate allows the use of the MRFSS sampling frame to analyse issues not covered in the standard economic add-ons. NMFS has utilised this frame a number of times to conduct additional economic surveys, including mail and other telephone surveys to explore subsistence fishing behaviour, shell fishing behaviour (not captured in the base MRFSS), and to explore new methodologies for estimating the value of recreational fishing and marine resources.

Valuation surveys

The round of valuation surveys was designed to capture the data needed to run several different models of recreation demand including travel cost models, both discrete choice and continuous trip demand models, and contingent valuation models (Box 14.1). The reader is directed either Freeman (1993) or Mitchell and Carson (1989) for more information about these and other resource valuation techniques. These surveys began in 1994 in the North-east region. The efficacy of the survey instrument was determined, design improvements were made, and the survey was continued in the South-east region in 1997. The nation-wide round of valuation surveys was completed in 1998 in the West Coast region (California to Washington). While the main focus of the intercept survey is to collect data suitable for travel cost demand models, many other data elements are collected. For example, the intercept survey collects information about anglers on overnight trips such as number of days away from residence, number of days spent fishing, lodging expenditures, and the overall purpose of the overnight trip. All intercepted anglers are asked to report their total travel-related costs for the current trip (day or overnight).

The telephone follow-up survey, which is dedicated to economic data elements, collects avidity (days spent fishing in last 2 months and other self reported measures), reasons for fishing, management preferences, motivations for fishing, willingness-to-pay (WTP, Box 14.1) for bag limit changes (contingent valuation question), boat ownership and size, ethnicity, age, gender, household size, education, employment status, flexible work schedule, wages foregone, wage rate, paid vacation, and household income. For this survey, there is both a long form and a short form. Anglers who are new to the follow-up are administered the long form while anglers who have been contacted before are given the short form, which collects only trip-specific information.

The telephone survey of coastal households is designed to collect a minimum set of socio-economic and demographic data elements needed to model participation changes using projected census demographic data. Data are collected for anyone that has fished in his or her lifetime. Respondents are asked age, gender, ethnicity, education, employment status, and household income. For those anglers who have fished in the last two or 12 months we ask how many trips were taken and whether they own a boat. This is only a subset of the data elements collected in each survey.

Expenditure surveys

With the passing of the Magnuson–Stevenson Fishery Conservation and Management Act (MSFCMA) in 1996, Congress mandated the analysis of economic impacts of management policies on fishing participants and coastal communities. As a result, NMFS set out to collect data on angler expenditures to analyse how those expenditures trickle through the economy and impact local incomes, employment, and tax revenue. Currently, expenditure data sets will allow state-level estimates for all states and possibly county-level analysis in states that pay for increased sampling levels to the base MRFSS. States that currently increase sample sizes include Maine, Connecticut, New Hampshire, Massachusetts, Maryland, Virginia, and North Carolina.

The model of choice for economic impact modelling is IMPLAN, and the data elements collected are meant to complement this analysis package. The reader is encouraged to read MIG (1997) for further information regarding this software package and I/O (input/output) modelling in general. The intercept survey collects overnight trip information, from those on overnight trips, including number of days away from residence, number of days spent fishing, lodging expenses, and the purpose of the trip. All anglers are asked travel costs, days of fishing in last 2 months, fishing ability, boat ownership, trip supply expenditures (bait, ice, refreshments, boat fees, etc.), whether they have a flexible work schedule, and wages forgone from the trip. While the intercept survey collects trip-level expenditure data a minimum set of valuation data elements is also collected.

The telephone follow-up survey is dedicated to the collection of detailed expenditure, socioeconomic, and demographic data. Expenditure data categories include semidurable goods (tackle, rods, reels, line, etc.), durable goods (motor boats and accessories, nonmotorised boats, boating electronics, mooring, boat storage, boat insurance and vehicles or homes used primarily for marine angling) and angling accessories and multipurpose items (magazines, club dues, saltwater angling specific clothing and camping gear). In addition, the expenditure survey collects a set of socioeconomic and demographic variables similar to those collected in the valuation survey.

The household telephone expenditure add-on survey is used for the same purposes as the valuation household survey. It allows the comparison of data collected from a random sample of trips (intercept survey) to data collected from a random sample of anglers during the household survey. As a result, it collects a similar set of socioeconomic and demographic elements as does the valuation household survey. Expenditure data is also well represented in this survey, with the majority of telephone follow-up expenditure categories repeated here. Again, these descriptions represent subsets of our data sets.

USA valuation case studies

Our goal from the outset was to establish values for angling access and marginal values of catch. These values would allow the assessment of the welfare (Box 14.1) effects of area closures, stock enhancements, or changes in recreational versus commercial fish stock allocations. While anglers may value other aspects of the fishing experience, such as weather, scenic beauty or companionship of other anglers, as long as these unmeasured components of the angling experience are not correlated with access and catch, a utility (Box 14.1) model is estimable.

To reach our goal, the angling experience is conceptualised as a series of nested choices. Within this framework, the first choice, to fish or undertake some other recreational activity, is made outside of the model. So, once an angler has decided to go fishing, it is hypothesised that the angler chooses a mode and species-target combination. Then a fishing site is chosen conditional on the choice of mode and species (hereinafter referred to as a fishing alternative). Because the MRFSS survey does not collect the exact area fished, a fishing site is defined as the point of fishing access. It is assumed that this choice is made to maximise an individual's utility. This structure, then, is consistent with nested random utility model (RUM) and the following presentation is adapted from the analysis of the North-east US 1994 data by Hicks *et al.* (1999a), and the analysis of the South-east US 1997 data by Haab *et al.* (2000). Both of these publications are available for downloading at http://www.st.nmfs.gov/st1/econ/pubs.html.

Within a RUM, anglers have substitute choices and can substitute away from area closures and changing catch rates by changing their choice of mode, species, or fishing site. A RUM assumes that anglers make choices that are optimal in terms of the tradeoffs between catch rates and costs. It utilises the variation in angling quality and costs across substitute fishing sites to estimate the value that anglers place on angling quality and access. Unfortunately, the treatment of all substitutes is costly from a data standpoint. The set of substitute alternatives spans all possible combinations of three modes from the MRFSS, hundreds of species, and thousands of access sites, resulting in a set of fishing alternatives in the hundreds of thousands. This large number of alternatives presents an intractable modelling problem, and, as a result, two broad aggregation strategies, across species and across sites, were used to condense the choice set into a manageable matrix of alternatives.

First, all recreationally caught species were condensed into five species groups. These species groups come directly from McConnell *et al.* (1994) and include small game fish, bottom fish, flat fish, big game fish, and all other fish (Table 14.3). This aggregation strategy works because the timing, location, and methods used to fish for each species is very similar within groups. Thus, substitution is likely within groups and unlikely across groups. For example, anglers fishing for striped bass are more likely to shift their efforts to fishing for seatrout, when striped bass fishing slows, than they are to shift their efforts to fishing for any of the tunas.

The second aggregation strategy involves reducing the number of sites. With thousands of sites on the East coast of the US and hundreds of sites in each individual state, all sites within a coastal county were aggregated into one site representing that county. Across the North-east, from Virginia to Maine, there are 63 coastal counties and therefore 63 sites. The county = site rule was not strictly followed in all cases. Some geographically diverse counties, i.e. those counties with both ocean frontage and bay frontage, were separated into two sites because of the different opportunities provided by these different types of water. Across the South-east, from North Carolina to Louisiana there are 77 coastal counties and therefore 77 sites. Seven sites (counties) did not contain sufficient observations for the model and those counties were

Table 14.3 The species groups for the Random Utility Model (RUM). The fifth species group is 'all other'.

Big game fish	Small game fish	Bottom fish	Flat fish
Thresher shark	Striped bass	Sandbar shark	Summer flounder
Shortfin mako shark	Pompano	Sand tiger shark	Winter flounder
Smooth hammerhead	Bonito	Catfish	Southern flounder
Scalloped hammerhead	Bluefish	Pollock	Sole
Wahoo	Seatrout	Sea bass	
Blue shark	Snook	Kingfish	
Dolphin (fish)	Jack	Butterfish	
White shark	Bonefish	Black drum	
Tarpon	Red drum	Perch	
Tuna	Mackerel	Groupers	
Great hammerhead	Barracuda	Snappers	
Tiger shark		Sheepshead	
Sailfish		Porgy/scup	
Marlin		Dogfish shark	
Swordfish		Smooth dog shark	
Cobia		Toadfish	
Billfish		Hake	
		Sawfish	
		Mullet	
		Nurse shark	

dropped resulting in 70 sites being used within the South-east model. As with all RUM models, defining quality across alternatives is very important when assessing alternative utility. In terms of a recreational fishing trip, expected catch is an important component of quality. Historic daily catch is used here because it is a definitive measure, it is the only measure readily available in our data, and it allows the calculation of welfare changes from changes in catch rates. There are several ways to calculate catch rates. The reader is encouraged to read Hicks *et al.* (1999a) and Haab *et al.* (2000) for discussions regarding other methods.

Deriving any demand equation requires a price. Because recreational fishing experiences are not openly traded in markets, travel cost, both the actual cost of travel plus the opportunity cost of time, is used as the price. Because it is hard to allocate travel costs to the fishing experience for anglers on multiday, multipurpose trips, only single day trips are included. As a result, the choice set is necessarily constrained by feasible, one-day, round-trip mileage. To construct travel cost, the one-way distance from the respondent's home zip code to the zip code of the site where the respondent was intercepted was multiplied by the 1994 federal travel reimbursement rate of $0.30 per mile, which includes fuel costs and vehicle depreciation. Time cost is calculated by estimating the travel time and multiplying it by the respondent's wage rate. For those anglers that did not lose income on the trip, the time cost is simply the travel time. For a more complete treatment of the calculation of travel cost see Haab *et al.* (2000). To further reduce the choice set, technically infeasible mode/species combinations were removed from the model.

With an aggregation strategy in place and the variables defined, we can move on to estimation of the nested RUM model. As a reminder, the choice of whether or not to take a fishing trip is exogenous to this model, the angler chooses the mode/species combination, and then chooses the site that maximises indirect utility from his or her set of alternatives (substitutes). Every model carries a set of implicit assumptions. Angler behaviour within this model is on a trip-by-trip basis and does not allow the angler to modify the number of trips taken each season. Therefore, each choice is independent of the next and unobservable utility is therefore independent of any other trips. These assumptions are necessary to make estimation possible. This conditional site utility and mode/species choice were modelled sequentially using limited information maximum likelihood (LIML) methods. For the specification of conditional probabilities, the reader is referred to Hicks *et al.* (1999a) and Haab *et al.* (2000). Variables used in the conditional site utility model include travel cost (TC), travel time (TT), log of the number of MRFSS intercept sites aggregated into the county site used in the model (M), and historic catch rate per trip for species group s mode m and site i ($Q_{s,m,i}$). The log of the number of sites aggregated into the county-wide site used in the model is used to account for possible aggregation bias. That is, an angler may be more likely to visit a county that contains more MRFSS intercept sites. Due to the heterogeneity across species groups, separate catch coefficients are estimated for each group within the model. This also allows for differing marginal utility of catch between species groups, which is intuitively appealing. In the North-east, the variables used to describe the mode/species choice include a variable that indicates if an individual owns a boat and selects the private rental mode (PR). The second variable (CPR) is an indicator variable that is equal to one if the mode is private rental, if the individual owns a boat, and if they are fishing in wave six, November and December, the coldest sampled wave. It is thought that CPR is likely to dampen the effect of owning a boat and choosing the private rental boat mode. Only the private rental mode alternatives explicitly contain variables in the mode/species choice model. The estimation of the models in the South-east did not use either PR or CPR.

These estimated indirect utility functions can be used to calculate value for marine anglers. Changes in quality or changes in access change an individual's indirect utility. These changes in welfare (Box 14.1) can be quantified in dollars by comparing the utility after a change in access or catch to the base utility, observed at the time of intercept. For the specification of the welfare calculations, the reader is referred to Hicks *et al.* (1999a) and Haab *et al.* (2000). Within this case study, two policy changes will be examined: elimination of all access within a state and a one-fish increase in the catch rate, which is equivalent to the marginal WTP for one fish.

North-east USA

For the intercept add-on, 33 117 surveys were attempted and 22 594 (68%) were fully completed. Of the nonrespondents, 3364 were either terminated by the respondent or terminated by the interviewer because the respondent was younger than 16 years old. Another 7151 were not usable because of item nonresponse across key variables for the analysis. For the telephone follow-up survey, all numbers were called up to four times. Attempted telephone surveys numbered 14 868, with 8226 (55%) completed correctly. Forty-five per cent of the nonresponse to this survey was due to outright refusals, wrong numbers, and numbers with no answer within four calls.

Estimates of effort will be used later in this analysis. Table 14.4 shows estimated effort by wave and state. No sampling is conducted in Wave 1 (January and February) due to cold weather and low effort. As expected, the summer Waves 3 and 4 (May to August) have the

Table 14.4 Recreational fishing effort by wave, North-east State, and subregion, in millions of anglers for 1994.

State	Wave*					
	2 (Mar. – Apr.)	3 (May – June)	4 (July – Aug.)	5 (Sept. – Oct.)	6 (Nov. – Dec.)	Total**
Virginia	0.16	0.78	1.02	0.43	0.24	2.63
Maryland	0.26	0.62	1.04	0.75	0.21	2.87
Delaware	0.02	0.17	0.29	0.13	0.08	0.69
New Jersey	0.30	1.13	2.34	1.35	0.54	5.66
New York	0.27	0.89	2.01	0.90	0.30	4.38
Connecticut	0.03	0.30	0.50	0.26	0.03	1.11
Rhode Island	0.06	0.20	0.36	0.37	0.01	1.00
Massachusetts	0.04	0.86	1.47	1.80	0.09	3.25
New Hampshire	0	0.19	0.11	0.02	NS***	0.31
Maine	0.07	0.16	0.31	0.13	NS	0.61
NE Total	1.15	5.30	9.45	5.13	1.49	22.52

* No sampling is conducted in wave 1 due to low effort.
** Numbers may not add up due to rounding.
*** NS = no sampling conducted in that wave due to low effort.

highest effort. Exploring angler trips further, 45.1% of trips in the North Atlantic (Connecticut to Maine) were in the shore mode while 47.4% of trips were in the private/rental boat mode and 7.5% were in the party or charter boat mode. In the Mid-Atlantic (Virginia to New York) region the majority of trips (56%) were in the private/rental boat mode, with 32.7% in the shore mode and 11.3% in the party or charter boat mode. In terms of catch, striped bass, scup, and bluefish were the most popular species caught in the North Atlantic, making up about 30% of the recreational catch. In the Mid-Atlantic, summer flounder, black sea bass, bluefish, and striped bass made up 37% of the total recreational catch.

Model results for the North-east are shown in Table 14.5. Overall, the model shows good fit and the all parameter estimates have the expected signs. As expected, anglers preferred less expensive sites, and those with lower travel costs and time requirements, other things being equal. In addition, anglers preferred sites with higher historic catch rates regardless of species group target.

Table 14.6 contains the welfare estimates resulting from the closure of all access by state and wave. The reader is cautioned on the use of these valuation estimates. Due to the nature of the model, these estimates are not additive across states or across species groups. For further detail see Hicks *et al.* (1999a) and Haab *et al.* (2000). These values reflect relative fishing quality and ability to choose substitutes across states. Looking at Table 14.6, Virginia and New York have the first and second highest values of access across the region. Closure of geographically large states, like New York, generates large welfare estimates because increased distances from sites increases costs. Additionally, the estimate for Virginia is high because an artificial substitution boundary has been imposed with its southern neighbour state, North Carolina. While it is likely that anglers in Virginia would include many sites in North Carolina in their choice set, North Carolina was not included in the 1994 surveying effort and, as a result, sites in North Carolina are not included in any of the choice sets in the model. Thus, when Virginia is closed, the model does not allow substitution to sites in North Carolina and anglers are modelled to travel farther than they would in reality.

Table 14.7 contains willingness to pay (WTP) for a one-fish increase in the historic catch rates. Typically anglers who target the big game species group have higher incomes, travel

Table 14.5 Model results and parameter estimates, from Hicks *et al.* (1999a).

Variable	Type of model*	Mean of variable	Parameter**	Marginal utility estimate (t-statistic)
Travel cost ($)	CSCM	61.84	$\exists_1(1\text{-}\forall)$	-0.036 (-10.46)
Travel time (hours)	CSCM	3.69	$\exists_2(1\text{-}\forall)$	-1.141 (16.12)
Ln (aggregated sites)	CSCM	3.11	$\exists_3(1\text{-}\forall)$	1.247 (33.99)
Big game catch/trip	CSCM	0.003	$\exists_4(1\text{-}\forall)$	0.974 (2.69)
Small game catch/trip	CSCM	0.39	$\exists_5(1\text{-}\forall)$	0.579 (8.68)
Bottom fish catch/trip	CSCM	0.19	$\exists_6(1\text{-}\forall)$	0.572 (100.68)
Flat fish catch/trip	CSCM	0.26	$\exists_7(1\text{-}\forall)$	0.665 (58.23)
Nonseeking catch/trip	CSCM	0.20	$\exists_8(1\text{-}\forall)$	0.324 (15.23)
π^2 (all parameters = 0)	CSCM			2780.15
Inclusive value	MSCMc	4.9	$(1\text{-}\forall)$	0.612 (19.99)
Private rental dummy	MSCM	0.15	$\exists_9$	2.490 (42.02)
Cold × Private*** rental dummy	MSCM	0.02	$\exists_{10}$	-0.553 (4.08)
π^2 (all parameters = 0)	MSCM			2172.46

* CSCM = Conditional Site Choice Model. MSCM = Mode/Species Choice Model

** For the complete model specification, including econometric equations, see Hicks *et al.* (1999a) pp. 13–18.

*** This variable represents the interaction of the private/rental mode dummy and a dummy variable for fishing during the colder seasons.

Table 14.6 Mean value of a one-day fishing trip by wave and Northeast State (in US$), from Hicks *et al.* (1999a).

State	Wave				
	3	4	5	6	Mean for all waves
Virginia	27.26	35.12	41.66	86.24	42.33
Maryland	13.63	11.53	14.06	7.36	12.09
Delaware	1.81	1.60	0.83	1.43	1.43
New Jersey	16.68	13.54	13.73	11.84	14.12
New York	20.86	20.86	21.36	24.79	21.58
Connecticut	3.48	3.29	2.71	2.54	3.07
Rhode Island	3.82	4.51	4.73	3.42	4.23
Massachusetts	8.54	8.58	9.90	5.04	8.38
New Hampshire	1.11	1.07	0.78	0.01	0.85
Maine	7.90	8.06	6.47	0.00	6.40

Table 14.7 Mean willingness to pay (WTP) per trip (in US$) for a one-fish increase in historic catch rates by Northeast State (Hicks *et al.* 1999a).

State	Big game	Small game	Bottom fish	Flat fish
Virginia	4.57	2.46	1.79	3.36
Maryland	6.51	3.44	2.44	5.30
Delaware	5.58	3.00	2.06	4.24
New Jersey	5.03	2.69	1.73	3.48
New York	4.61	2.43	1.63	3.10
Connecticut	5.99	3.29	2.25	4.43
Rhode Island	5.73	3.13	2.11	4.40
Massachusetts	5.91	3.09	2.04	4.33
New Hampshire	6.20	3.25	2.14	4.77
Maine	6.61	3.74	2.62	5.75
All states (mean)	5.39	2.89	1.97	4.01

farther in the open ocean, and have more expensive gear requirements. As expected, WTP is the highest for the big game species group. The next highest WTP is for flat fish, a group highly prized for its table quality.

It is also useful to aggregate these individual mean WTP values to the population level to look at the total welfare impacts. In order to aggregate these values, a few points are worth noting. Anglers are likely to change the number of trips they take in response to changing policies. Unfortunately, RUMs have not been definitively linked to changing demand for trips in a manner consistent with consumer demand theory. Any aggregation strategy that does not take into account utility theory is therefore *ad hoc* and the quality of any number of aggregation strategies cannot be judged theoretically. Within this analysis, it was assumed that the number of trips would not change, which is a more conservative treatment. Assuming that the number of trips does not change, one can simply multiply a population level estimate of trips for 1994, from the MRFSS, by the mean WTP values. Using the effort estimates from Table 14.4, aggregate welfare values can be calculated by multiplying the total number of trips for the North-east Region (bottom row of Table 14.4) by the values given above in Tables 14.6 and 14.7. Table 14.8 contains the aggregate WTP for access by state and wave and Table 14.9 contains the aggregate WTP for a one-fish increase in historic catch rates by species groups.

Table 14.8 Aggregate WTP for a one-day fishing trip by wave and North-east State, (Hicks *et al.* 1999a) (in millions of US$).

	Wave				
State closed	3	4	5	6	All waves
Virginia	144.4	332.0	213.8	128.6	904.7
Maryland	72.2	109.0	72.2	11.0	258.4
Delaware	9.6	15.1	4.3	2.1	30.6
New Jersey	88.3	128.0	70.5	17.7	301.8
New York	110.5	197.2	109.6	36.3	461.2
Connecticut	18.4	31.1	13.9	3.8	65.6
Rhode Island	20.2	42.6	24.3	5.1	90.4
Massachusetts	45.2	81.1	50.8	7.5	179.1
New Hampshire	5.9	10.1	4.0	0.2	18.2
Maine	41.8	76.2	33.2	0	136.8

Table 14.9 Aggregate WTP for a one-fish increase in historic catch rates, by North-east State (in millions of US$).

State	Big game	Small game	Bottom fish	Flat fish
Virginia	12.0	6.5	4.7	8.9
Maryland	18.7	9.9	7.0	15.2
Delaware	3.9	2.1	1.4	2.9
New Jersey	28.5	15.2	9.8	19.7
New York	20.2	10.6	7.1	13.6
Connecticut	6.6	3.7	2.5	4.9
Rhode Island	5.7	3.1	2.1	4.4
Massachusetts	19.2	10.0	6.6	14.0
New Hampshire	2.0	1.0	0.7	1.5
Maine	4.0	2.3	1.6	3.5
All States	120.8	64.4	43.6	88.7

South-east USA

In 1997, over 57 000 intercepts were conducted in the Southeast Region (North Carolina, South Carolina, Georgia, Florida, Alabama, Mississippi, and Louisiana) at over 1000 MRFSS intercept sites. Beyond the intercept, approximately 10 000 telephone follow-up interviews were conducted. While the survey is conducted year round in Florida, Alabama, Mississippi and Louisiana, it is not conducted in Wave 1 in North Carolina, South Carolina, and Georgia because of low effort in those colder months. After taking out missing observations and the Wave 1 interviews, 8865 useable data points are left for modelling.

Within these useable data points, 60% were in the private rental mode, 30% in the shore mode, and 10% in the party charter mode. This represents slightly more people in the private rental mode, at the expense of the shore mode, than in the North-east. The percentage of people intercepted in the party charter mode is similar between the two surveys. Thirty-two per cent of the individuals intercepted targeted fish within the small game category, 5% in the big game category, 7% in the bottom fish category, 3% in the flat fish category, and 50% of those intercepted targeted fish in the other species category or had no target. Of these intercepted anglers less than 5% were intercepted in the states of Alabama, Mississippi, and Georgia, 8% were intercepted in South Carolina, 11% were intercepted in Louisiana, 17% were intercepted in North Carolina, and more than 50% were intercepted in Florida. The three top states, in terms of effort (Table 14.10), make up over 75% of the sample. While no one state dominated the sample in the North-east, Louisiana, North Carolina, and Florida are top destinations for vacationing anglers and also support higher than average resident effort numbers. In addition, these states have large coastlines, increasing their attractiveness for in-shore fishing.

Table 14.11 displays the variable means, model parameters and estimates of goodness-of-fit criterion. Both stages of the models performed well with all the expected signs across the variables. The signs on travel cost and travel time are negative and significant suggesting that anglers choose sites that are closer (less costly) over sites that are farther away, *ceteris paribus*. The marginal utility of flat fish catch is the highest, differing from the result of Hicks *et al.* (1999a). Big game exhibits the next highest marginal utility followed by small game and

Table 14.10 Recreational fishing effort by wave, South-east State, and subregion, in millions of anglers for 1997.

	Wave						
State	1	2	3	4	5	6	Total
North Carolina	NS**	0.6	1.1	1.4	1.3	0.7	4.9
South Carolina	NS	0.1	0.4	0.5	0.3	0.2	1.6
Georgia	NS	0.1	0.1	0.2	0.1	0.1	0.6
Florida	3.5	4.0	4.9	4.9	4.0	3.5	24.7
Alabama	0.1	0.1	0.3	0.3	0.1	0.1	1.2
Mississippi	0.8	0.1	0.2	0.2	0.3	0.1	1.0
Louisiana	0.3	0.4	0.7	0.7	0.7	0.5	3.2
Total*	4.0	5.0	7.7	8.1	6.7	5.2	37.0

* Numbers may not add due to rounding
** NS = no sampling conducted in that wave due to low effort.

Table 14.11 Model results and parameter estimates, from Haab *et al.* (2000).

Variable	Type of model*	Mean of variable	Parameter**	Marginal utility estimate (t-statistic)
Travel cost ($)	CSCM	332.31	$\exists_1(1\text{-}\forall)$	-0.015 (20.6)
Travel time (hours)	CSCM	22.67	$\exists_2(1\text{-}\forall)$	-0.47 (39.6)
Ln (aggregated sites)	CSCM	2.67	$\exists_3(1\text{-}\forall)$	0.77 (35.2)
Big game catch/trip	CSCM	0.02	$\exists_4(1\text{-}\forall)$	0.32 (2.1)
Small game gatch/trip	CSCM	0.35	$\exists_5(1\text{-}\forall)$	0.15 (2.4)
Bottom fish catch/trip	CSCM	0.09	$\exists_6(1\text{-}\forall)$	0.07 (1.3)
Flat fish catch/trip	CSCM	0.01	$\exists_7(1\text{-}\forall)$	0.49 (2.9)
Non-seeking catch/trip	CSCM	0.10	$\exists_8(1\text{-}\forall)$	-0.0002 (0.05)
π^2 (all parameters = 0)	CSCM			7413.59
Inclusive value (targeted species)	MSCM	Not reported	$(1\text{-}\forall)_T$	0.53 (4.4)
Inclusive value (nontargeted species)	MSCM	Not reported	$(1\text{-}\forall)_{NT}$	0.93 (7.4)
π^2 (all parameters = 0)	MSCM			1467.77

* CSCM = Conditional Site Choice Model. MSCM = Mode/Species Choice Model.
** For the complete model specification, including econometric equations, see Haab *et al.* (2000).

then bottom fish, which is statistically no different than zero. The marginal utility catch for other species is negative and not significantly different from zero. This other species group contains individuals that have no specific species target or target species that do not fall into the species groups as defined. This mixing of objectives is thought to drive this result. The second stage parameters also provide useful insight into the behaviour of recreational anglers. The inclusive value parameter for those anglers with a species target is significant and is less than one, suggesting that the nesting structure is appropriate for targeters. However, the inclusive value for the nontargeters, while significant, is very close to one (0.93) suggesting that for nontargeters, a nested choice model is not appropriate (Haab *et al.* 2000).

The value of access was calculated first and the results are shown in Table 14.12. The reader is cautioned on the use of these valuation estimates. Due to the nature of the model, these estimates are not additive across states or across species groups. For further detail see Hicks *et al.* (1999a) and Haab *et al.* (2000). The 'All Waves' value is a weighted average across the waves, using the sample size in each wave as the weight. As an extension, Haab *et*

Table 14.12 Mean value of a one-day fishing trip by wave and South-east State (in US$), from Haab *et al.* (2000).

	Wave					
State	2	3	4	5	6	All waves
North Carolina	8.08	20.41	15.81	18.11	14.16	15.83
South Carolina	6.60	5.54	6.48	8.28	6.61	6.70
Georgia	0.97	3.75	3.30	2.51	1.92	2.58
Florida (SA)	12.65	10.36	12.38	11.27	13.93	12.01
Florida (Gulf)	56.23	44.11	42.69	43.65	44.84	45.88
Florida (all)	237.35	194.30	202.04	182.53	206.54	202.52
Alabama	1.75	1.88	1.38	1.24	1.55	1.56
Mississippi	3.46	3.46	3.49	4.03	3.66	3.63
Louisiana	8.77	11.77	13.49	11.70	12.34	11.68
Gulf Coast	86.82	79.29	82.23	82.72	81.38	82.22
South Atlantic	75.82	113.33	109.04	134.75	103.82	109.31

al. (2000) also calculated the welfare effects of closing sub-regions, like the entire Gulf Coast, South Atlantic, and all of Florida. The states with the largest values, in decreasing order, were Florida (all), Florida (Gulf), North Carolina, Florida (SA), and Louisiana. These values are high for a number of reasons. First, Florida is a very important fishing tourism destination and has a lot of coastline per unit of area. Thus, it is not surprising that the values are high in this state. In addition, closing Florida (all) produces high welfare estimates because Florida is a long peninsula surrounded on one side by the Atlantic Ocean and on the other by the Gulf of Mexico. When either one coast or the other is closed, it is a short drive to a substitute on the other side of the state, but when the entire state is closed, the nearest substitutes are a considerable distance away. In North Carolina and Louisiana, there are a number of factors contributing to the high welfare estimates. First, both states have the longest coastlines and the highest number of trips taken, which contributes to the highest number of access sites and the highest number of aggregated sites, increasing the value of fishing access in these states. These factors indicate that the model is working correctly; states with high effort should be valued higher. On the other hand, North Carolina and Louisiana share the same problem that Virginia faced in the Hicks *et al*. (1999a) model; the artificial truncation of the choice set because of a political boundary increases the distance a resident of each of these states must travel to find a substitute. In this case, Texas, Louisiana's westerly neighbour, is not included in the MRFSS survey and NMFS places its own institutional boundary between Virginia and North Carolina, as discussed before.

Table 14.13 displays the mean WTP for a one-fish increase in the catch rate, per trip. Historic catch and keep rates were used and are lower than the historic catch rates used in Hicks *et al* (1999a), especially for regulated species where illegal fish are released. Thus, these WTP values are larger than the welfare estimates calculated in Hicks *et al*. (1999a). It is interesting to note that the value of flat fish is higher than that of big game, which is the opposite of the result obtained by Hicks *et al*. (1999a). While the values are not the same for small game and bottom fish across the two analyses, the relative ranking of these two groups is the same.

As in the last case study, it is useful to aggregate these estimates up to the entire population taking trips in these regions. Table 14.10 contains MRFSS estimates of effort by wave and state for 1997. Using these estimates and multiplying by the values given in Tables 14.12 and 14.13 give us the aggregated values in Tables 14.14 and 14.15.

Table 14.13 Mean WTP for a one-fish increase in the catch and keep rate (per trip) by South-east State (in US$), from Haab *et al*. (2000).

State	Big game	Small game	Bottom fish	Flat fish
North Carolina	14.62	6.63	3.04	22.68
South Carolina	14.82	6.77	3.12	22.96
Georgia	14.44	6.41	2.98	22.14
Florida (SA)	14.63	6.60	3.01	22.47
Florida (Gulf)	15.02	6.81	3.09	23.25
Alabama	14.53	6.54	2.94	22.27
Mississippi	14.91	6.70	3.05	23.02
Louisiana	14.78	6.58	2.98	22.93
All states	14.83	6.68	3.04	22.88

Table 14.14 Aggregate WTP for a one-day fishing trip by wave and South-east State, from Haab *et al.* (2000) (in millions of US$).

	Wave					
State	2	3	4	5	6	All Waves
North Carolina	4.5	21.4	21.8	22.7	9.3	77.4
South Carolina	0.9	2.5	3.4	2.5	1.3	10.8
Georgia	0.1	0.5	0.6	0.3	0.1	1.5
Florida (SA)	22.2	22.4	24.8	21.8	24.1	115.1
Florida (Gulf)	118.4	122.2	122.4	89.1	80.1	530.9
Florida (all)	915.5	958.9	984.0	726.0	725.7	4284.0
Alabama	0.23	0.55	0.38	0.16	0.2	1.5
Mississippi	0.49	0.74	0.76	1.00	0.4	3.3
Louisiana	3.08	7.86	8.70	7.94	6.6	33.6
Gulf Coast	236.9	312.6	329.3	256.4	205.6	1340.3
South Atlantic	192.6	428.5	445.2	485.9	274.8	1820.8

Table 14.15 Aggregate WTP for a one-fish increase in historic catch rates by South-east States, from Haab *et al.* (2000) (in millions of US$).

State	Big game	Small game	Bottom fish	Flat fish
North Carolina	71.5	32.4	14.9	110.9
South Carolina	23.8	10.9	5.0	36.9
Georgia	8.3	3.7	1.7	12.8
Florida (SA)	140.2	63.3	28.9	215.3
Florida (Gulf)	173.8	78.8	35.8	269.0
Alabama	13.6	6.1	2.8	20.9
Mississippi	13.7	6.1	2.8	21.1
Louisiana	42.5	18.9	8.6	66.0
All States	488.8	220.2	100.2	754.1

Case study discussion

To summarise, recreational angling is very valuable in the USA. On the high side, if Florida were closed to recreational angling, there would be a welfare loss of $4.3bn. On the low side, closing Georgia, a state with very little coastline and low effort in marine angling, would produce a welfare loss of $1.5m. Looking at the most popular species group across both regions, small game, anglers would be willing to pay $64.4m and $220.2m across all states in the North-east and South-east, respectively, for a one-fish increase in catch rates for small game. It is important to remember that these values represent only some of the consumptive use values for the recreational use of marine resources. The total value of marine resource recreation would necessarily include nonuse values for bird watchers and other wildlife watchers, recreational boaters, bathers, and, some would argue, existence values for those who do not participate in marine recreation. It is also important to note that a nation-wide value or East and Gulf Coast wide value would be significantly higher.

It is difficult to compare the results presented here with the valuation work of others because of the aggregation across counties and the aggregation across species used in our analysis. There have been few studies in the literature that used a RUM model to value marine angling in general and even fewer that use aggregated species groups and sites aggregated to the county level. Rather than compare the results between the works of other researchers here, the reader is directed to the sources listed in Table 14.15 for similar studies and more information on valuing recreational resources.

Some caveats on the welfare estimates are in order. If a choice set for an angler is small, closing a state will have large welfare impacts for that angler. Because RUMs measure welfare relative to available substitutes, it is not equipped to handle policies eliminating all alternatives for an individual. As a result of this limitation, a total value of access, for the entire North-east or South-east region or the entire US East Coast, cannot be calculated using this specification, and the values of access presented above cannot be summed across states to reach such a value. However, as long as some alternative exists within the model, changes in welfare can be calculated.

This model is well suited to look at closures across individual counties, groups of counties, individual states, and groups of states, as shown. Because no nation-wide valuation estimates exist for saltwater angling, NMFS is currently working on a benefits transfer methodology that may allow the combination of these two regional models, even though they are estimated across multiple years. This will allow the estimation of large-scale, nation-wide, region-wide or coast-wide values. In addition, this would help resolve the artificial substitution boundary problems faced by Virginia, North Carolina, and Louisiana.

One final caveat includes potential avidity bias when using the MRFSS data. Within this methodology, a random sample of trips is being used as a random sample of anglers. This may introduce avidity bias into the welfare estimates presented here. That is, the probability of being interviewed in the random sample of trips increases with avidity. No tests were conducted to assess potential avidity bias and, as a result, no correction was made.

In conclusion, these values were generated to give big picture, aggregated values of the marine recreational fisheries on the Atlantic and Gulf Coasts of the USA and were not directed at specific policies, although these values could be used to look at policy-specific welfare impacts. In fact, region-specific and species-specific valuation estimates are possible with slight changes to the models developed here.

Availability of data and results

For more information regarding any of our data sets, surveys, or research products, please visit our web page at http://www.st.nmfs.gov/st1/econ/index.html. Here you can find all of our research output in PDF format along with a complete listing of the detailed rationales behind the surveys, and metadata. Our metadata includes survey instruments, variable names and descriptions, question justifications, and an overview of survey procedures and completion statistics, all available for downloading.

These data sets represent an excellent and flexible set of economic data on marine recreational angling suitable for market research, demographic/socioeconomic profiling, and the analysis of net economic benefits and economic impacts using a variety of techniques. These qualities, combined with the catch, effort, and participation data from the base MRFSS, allow great flexibility in model choice.

Table 14.16 Sources of other marine angling valuation research.

Reference	Description
Bockstael *et al.* (1989)	Presents welfare estimates for access and catch for the East Coast of Florida using 1988 MRFSS data.
Gautam and Steinback (1998)	Uses MRFSS data to estimate a species-specific RUM for striped bass (*Morone saxatilis*) in the North-east US.
Greene *et al.* (1997)	Uses a RUM model to develop welfare estimates of the value of catch and access for Tampa Bay, Florida using MRFSS data as well as a mail survey using the MRFSS sample frame.
Hicks *et al.* (1999b)	This reference is one of the articles in a special issue of *Marine Resource Economics* that details choice set definition in RUMs. Several applications are presented using MRFSS and other data sets.
http://www.st.nmfs.gov/st1/econ/bibliographies.html	This link points to comprehensive bibliographies on RUMs, contingent valuation, travel cost models, opportunity cost of time, and miscellaneous recreation valuation literature. These bibliographies offer both theoretical and applied work in recreation valuation with some direct applications to marine angling.
McConnell *et al.* (1994)	A pilot study for future NMFS data collection efforts. Determined the feasibility of using RUMs with data from the 1988 MRFSS.

Sources of other marine angling valuation research are given in Table 14.16.

Acknowledgements

The authors would like to thank Dr Rob Hicks, Assistant Professor, Virginia Institute of Marine Sciences, Dr Timothy Haab, Assistant Professor, Ohio State University, and Dr John Whitehead, Assistant Professor, Eastern Carolina University for their research summarised in this chapter.

References

Bockstael, N.E., McConnell, K.E., & Strand, I.E. (1989). A Random Utility Model for Sportfishing: Some Preliminary Results for Florida. *Marine Resource Economics*, **6**, 245–260.

Clark, J.R. (1962). The 1960 Salt-Water Angling Survey. *U.S. Fisheries and Wildlife Circular 153*.

Cochran, W.G. (1977). *Sampling Techniques*, 3rd edn Wiley, New York.

Deuel, D.G. (1973). The 1970 Salt-Water Angling Survey. *U.S. National Marine Fisheries Service Current Fishery Statistics 6200*.

Deuel, D.G. & Clark, J.R. (1968). The 1965 Salt-Water Angling Survey. *U.S. Bureau of Sport Fisheries and Wildlife Resources Publication 67*.

Essig, R.G. & Holliday, M.C. (1991). Development of a recreational fishing survey: the marine recreational statistics survey case study. *American Fisheries Society Symposium*, **12**, 245–254.

Freeman, A.M. III (1993). *The Measurement of Environmental and Resource Values; Theories and Methods*. Resources for the Future, Washington DC.

Gautam, A. & Steinback, B. (1998). *Valuation of recreational fisheries in the North-east US. Striped Bass: a case Study*. In: *Recreational Fisheries; Social, Economic and Management Aspects* (eds Hickley., P. & Tompkins, H.), Chapter 23. Blackwell Science. Oxford, UK. pp. 165–183.

Greene, G., Moss, C.B. & Spreen, T.H. (1997). The Demand for Recreational Fishing in Tampa Bay, Florida: A Random Utility Approach. *Marine Resource Economics,* **12**, 293–305.

Groves, R.M., Biemer, P.P., Lyberg, L.E., Massey, J.T., Nichols, W.L. II & Waksberg, J. (1988). *Telephone Survey Methodology*, Wiley, New York.

Haab, T., Whitehead, J. & McConnell, T. (2000). The Economic Value of Marine Recreational Fishing in the Southeast United States: 1997 Southeast Economic Data Analysis. *Final Report for NMFS Contract No. 40WCNF802079*, National Marine Fisheries Service, Southeast Regional Office, St. Petersburg, FL. Available at http://www.st.nmfs.gov/st1/econ/pubs.html, last accessed May 14, 2001.

Hicks, R., Steinback, S., Gautam, A. & Thunberg, E. (1999a). Volume II: The Economic Value of New England and Mid-Atlantic Sportfishing in 1994. *NOAA Tech Memo No. NMFS-F/SPO-38*. Available at http://www.st.nmfs.gov/st1/econ/pubs.html, last accessed May 14, 2001.

Hicks, R.L., Gautam, A.B., Van Voorhees, D., Osborn, M. & Gentner, B. (1999b). Thalassorama: An Introduction to the NMFS Marine Recreational Fisheries Statistics Survey with an Emphasis on Economic Valuation. *Marine Resource Economics*, **14**, 375–385.

McConnell, K.E., Strand, I.E., Jr., Valdes, S.K. & Weninger, Q.R. (1994). *The Economic Value of Mid and South Atlantic Sportfishing*. Volume 2 of a report on Cooperative Agreement #CR811043 01 0 between the University of Maryland, the Environmental Protection Agency, the National Marine Fisheries Service & the National Oceanic and Atmospheric Administration.

MIG (1997). *IMPLAN Pro: Users Guide, Analysis Guide, and Data Guide.* 2nd printing. Minnesota IMPLAN Group, Minnesota.

Mitchell, R.C. & Carson, R.T. (1989). *Using Surveys to Value Public Goods: The Contingent Valuation Method.* Resources for the Future, Washington DC.

NMFS (2000). Fisheries Statistics and Economics Division. *Marine Recreational Fisheries Statistical Survey Real Time Data Queries*. Available at http://www.st.nmfs.gov/st1/recreational/database/queries/index.html, last accessed May 14, 2001.

Chapter 15

New Large-Scale Survey Methods for Evaluating Sport Fisheries

Jeremy M. Lyle, Anne P.M. Coleman, Laurie West, David Campbell and Gary W. Henry

Abstract

This paper describes a methodology that has been applied in large-scale regional surveys and recently a national survey of recreational fishing in Australia. The methodology has been developed in response to the need to provide reliable and detailed catch, effort and economic information whilst recognising logistic and financial constraints. The result has been a multifaceted design, the primary component of which is an integrated telephone and diary survey.

Using general population (or licence list) sampling, an initial screening interview is employed to collect profiling information and identify intending fishers. This is followed by a diary survey in which responsibility for data collection rests with survey interviewers. Diarists are contacted regularly by telephone throughout the diary period and the interviewers record any details of fishing or expenditure activity since last contact. The frequency of this contact is tailored to the needs and behaviour (fishing avidity) of the diarist. Diary response rates of greater than 95% have been achieved across a range of surveys using this methodology.

An important contributor to this success has been the underlying design philosophy to minimise respondent burden and recall bias. Yet, simplicity for the respondent translates to substantial responsibilities on the part of the interviewer, where in turn, careful staff recruitment, training and management are vital. Additionally, a rigorous approach to other design aspects has resulted in a range of quality control and validation measures to address various response biases and other sources of nonsample error.

The efficacy of the telephone-diary methodology has been demonstrated in the Australian studies reviewed and indicates its wider applicability and policy relevance to recreational fisheries elsewhere.

Key words: large-scale recreational fishing surveys, recreational fishing in Australia, survey methodology, telephone-diary surveys.

Introduction

Recreational fishing is a popular pastime in Australia with about one-quarter of the population of about 19 million people believed to go fishing at least once each year. A wide diversity of marine and freshwater fishing opportunities are available, spanning tropical and temperate latitudes. The size of the recreational sector and the need to manage for resource sustainability places an obligation on government to understand and manage the impacts of all sectors of the fishing industry, while ensuring public benefits from the resource are maximised.

Although catch, value and fishery status information are generally available for the commercial sector (e.g. Australian Bureau of Agricultural and Resource Economics 2000, Caton and McLoughlin 2000), equivalent and ongoing broad-scale data sets do not exist in Australia for the recreational fishery. The comparatively high cost of recreational fisheries research, together with limited research budgets, has resulted in a general paucity of information for this sector.

The need for national fishery statistics for the recreational sector was recognised in the early 1980s, with the conduct of a modest, Commonwealth government-funded survey in 1984. The survey involved personal interviews of about 2400 households and provided national estimates of participation rates, fisher demographics and economic impacts (PA Management Consultants 1984). Recreational effort and harvest levels were not assessed. The survey found that about 3.75 million adults (26% participation rate) fished in the preceding 12 months, expending over A$2bn in 1983/84 on fishing and related equipment. However, as this survey was restricted to capital cities and major provincial towns, it was unclear as to how representative the sample was of the population as a whole. Furthermore, the inclusion of nonfishing elements limited the utility of the economic impact assessment [see Kearney 2002 for a discussion of the difficulty of establishing an accurate participation figure – eds].

Inferences about the magnitude of the recreational harvest in Australia can be made from household seafood consumption surveys. For instance, in 1990/91 an estimated 24 000 tonnes of recreationally caught fish was consumed nationally (PA Consulting 1992). Another study estimated that 31 000 tonnes of seafood was 'home produced' (i.e. caught recreationally) in Australia in 1991/92 (Australian Bureau of Statistics 1994). Since these estimates were based on recalled consumption over periods of 3 or 12 months, respectively, their accuracy is uncertain. Despite these shortcomings, these surveys did highlight the importance of the fishery at a time when recreational impacts on resources were largely not recognised or simply overlooked.

Since the 1980s, most Australian fishery agencies have conducted biological and small-scale on-site (creel) surveys of recreational fisheries (McGlennon 1995). General population surveys have also been conducted at least once in each of the States and Territories. Their focus has been limited to broad profiling of the recreational fishery in terms of participation, demographics, key target species and broad effort levels (reviewed by McIlgorm and Pepperell 1999). Where detailed information about recreational catch and effort has been required, it has generally been for specific sites or bodies of water (Caputi 1976, Henry 1984, Battaglene 1985, West and Gordon 1994, Griffin 1995). Although relatively broad-scale on-site (creel) surveys have been conducted in Australia, they have tended to be restricted to regional studies with limited coverage of fishing activities (e.g. McGlennon and Kinloch 1997, Steffe *et al.* 1996, Sumner and Williamson 1999).

With growing awareness of the significance of the recreational fishery in the early 1990s, a national policy for recreational fishing was developed in Australia. The policy was released in 1994 and endorsed the principle that 'fisheries management decisions should be based on sound information including fish biology, fishing activity, catches and economic and social values of recreational fishing' (National Recreational Fisheries Working Group 1994). The policy recommended that a national survey of recreational fishing be undertaken once every five years. Recognition was also given to public concern over the then poor quality of data on recreational fishing.

In this chapter, we describe an innovative methodology that has been applied in recent large-scale regional and national studies of recreational fishing in Australia. The methodology has been developed to address a broad range of policy issues including resource sharing and allocation, resource assessment and sustainability, economic impacts (expenditure and investment), effectiveness of government programs and awareness of fisheries regulations.

Survey design considerations

A number of factors need to be considered in evaluating potential design options for recreational fishing surveys. They include the scale of the survey (spatial and temporal), survey objectives (what information is required and how it will be used) and, importantly, the available budget. Recreational fishing surveys are to some extent about compromises that seek to balance accuracy and precision against logistics and cost. This is particularly the case in a country of Australia's substantial landmass (almost 8 million km^2), small population size and limited public funding for national projects.

In order to achieve comprehensive coverage of recreational fishing and economic activity at national or even state levels in Australia, conventional on-site survey methods would be logistically very difficult and cost prohibitive to implement as the primary source of data collection. As a consequence, large-scale surveys will, by necessity, have some reliance on off-site survey techniques. Available methodologies can collect data 'retrospectively' (recall surveys) or 'prospectively' (diary surveys) and can employ, singly or in combination, mail, telephone or face-to-face interviewing techniques (Pollock *et al.* 1994).

Recall surveys by mail and telephone are relatively cost-effective, simple to administer and suited to demographic and socioeconomic profiling of fishers, in addition to assessing attitudes and awareness. Door-to-door surveys can provide similar information but are labour intensive and more costly. However, where detailed information about levels of harvest, effort and economic activity are required, recall surveys are of limited applicability due largely to the impact of recall biases, since respondents are often expected to recall activity for periods of up to 12 months (Pollock *et al.* 1994). Recall or memory bias is a complex issue that is influenced not only by the length of the recall period but by the frequency of participation (Thompson and Hubert 1990, Fisher *et al.* 1991, Tarrant and Manfredo 1993, Tarrant *et al.* 1993, Connelly and Brown 1995). As a general rule, surveys with recall periods of 2 or more months produce significant over-estimates of effort and catch and under-estimates of expenditure (Pollock *et al.* 1994). Mail, telephone and door-to-door surveys can also suffer from other sample and nonsample errors, such as incomplete coverage and nonresponse, which if not minimised or calibrated for, can introduce significant biases (Brown 1991, Pollock *et al.* 1994).

Notwithstanding the issue of recall bias, several examples exist where large-scale surveys using door-to-door, telephone and mail survey techniques have been conducted to collect effort, catch and economic data. The National Survey of Fishing, Hunting and Wildlife-Associated Recreation in the United States of America (Grambsch and Fisher 1991) and the Survey of Recreational Fishing in Canada (Economic and Policy Analysis Directorate 1997) have been conducted and repeated over many years, providing national and regional statistics about participation, effort, harvest, expenditure and investment in recreational fishing. Key information about effort, harvest and economic activity is currently based on 4 and 12-month recall periods for the American and Canadian studies, respectively. Although warning of the impacts of recall bias in angler surveys, Brown (1991) noted that recall based estimates may have value in trend and comparative analyses, an objective of these national surveys.

A complex methodology using both on- and off-site techniques has been developed for the Marine Recreational Fishery Statistics Survey in the USA. This survey is designed to provide detailed effort, catch and biological data for the marine fishery to support resource assessment and management. The key elements are a general population telephone survey (administered in coastal counties) to determine effort and an on-site access point survey to determine catch rates and collect biological data (Essig and Holliday 1991). Catch is then estimated as the product of effort and catch rate. In this study, recreational fishers who are contacted in the telephone survey are asked to recall their fishing effort (days fished) over the 2 months prior to interview. Potentially, the influence of recall issues, even for this comparatively short period, will be to over-estimate effort and thus influence the accuracy of catch estimation [compare Gentner and Lowther 2002, who provide a detailed account of a US survey method – eds].

Recall problems can be minimised by the application of logbooks or diaries, since fishing details are recorded soon after the activity has been completed. Diary surveys are relatively cost-effective to administer and can provide considerable detail about catch, effort and expenditure. Significantly, diary surveys enable coverage of the full range of fishing activities, including both day- and night-time fishing and boat and shore-based fishing, weighted in proportion to their occurrence within the sample population. Typically, diary respondents are expected to document their own fishing activities and mail back completed records, usually with minimal direct or ongoing contact between respondents and the survey agency. In practice this often results in varying levels of respondent comprehension and compliance with survey instructions. Reported information may be ambiguous, incomplete or missing and respondents often fail to return completed records or drop out of the survey altogether (Pollock *et al.* 1994, Connelly and Brown 1996). Nonresponse can also be a major problem and introduce biases since nonrespondents may differ in their fishing characteristics to respondents (Connelly and Brown 1996). Additional problems with conventional mail-back diary surveys arise when respondents have to be reminded to send in their fishing information. Often this information was never put into diaries, so if collected at all, it must be done retrospectively, often several months after the activity occurred. Other concerns with self-reported data relate to exaggeration of catches (prestige bias), rounding of responses to numbers ending in zero or five (digit bias) and misidentification of fish species (Essig and Holliday 1991, Pollock *et al.* 1994).

Mail-back diary surveys have been conducted at regional and national levels in New Zealand (Bell *et al.* 1993, Teirney *et al.* 1997, Bradford 1998) and recently in state-wide surveys of recreational fishing in Queensland, Australia (Higgs 1999). In these studies a random telephone survey was used to identify recreational fishers who were then invited to participate in a 12-month diary survey. At 3-monthly intervals, diarists were sent a reminder to submit their trip records for the period. Follow-up telephone calls were used to contact diarists who did not respond to the reminder letter. Significantly, in the Queensland study, 34–48% of diarists who reported fishing activity in each quarter did so through follow-up telephone interviews rather than in the diary (Higgs 1999).

An alternative survey approach involving a longitudinal telephone survey and a *de facto* diary has been described by Weithman (1991) for the collection of state-wide catch and effort information in Missouri. The survey was conducted over a 6-year period during the 1980s and was comprised of three 2-year segments. A sample of resident and nonresident licensed fishers was contacted in an initial telephone interview, with the primary aim being to enlist fishers to participate in the survey for the following 2 years. Respondents who agreed to participate were sent a list of instructions and data record forms (the diary) and were asked to keep detailed records of when and where they went fishing and what they caught. They were then contacted every 1–3 months by telephone and asked to provide information about individual fishing trips over the telephone. The frequency of contact was proportional to the amount of fishing by individual respondents. From Weithman's account it is unclear as to how much information was recorded on the data record forms and how much was based on recalled activity. However, by comparison with conventional mail-back diary surveys, response rates were very high, averaging 90% for the first year and 80% for the entire 2-year program, suggesting considerable potential for the combined telephone and diary survey method in recreational fishing surveys. In this regard, the approach addressed some major problems associated with telephone and conventional mail-back diary surveys, namely recall bias and nonresponse. It is surprising then, that this approach appears to have received little attention in recreational fisheries research.

Survey context with recent Australian studies

In the early 1990s the Northern Territory government identified the need for comprehensive and reliable recreational fishing data, including catch, effort and expenditure, for the entire Territory. A survey design consultant (Kewagama Research) was commissioned to review and recommend possible data collection methods. As a large region (over 1 million km^2 in area and 11 000 km of coastline) with a small resident population (fewer than 200 000 people) and a large number of visitors (over 900 000 annually), a particularly innovative and cost-effective approach was required. After extensive development work (including two pilot tests), a multifaceted survey design – the central component being an integrated telephone and diary survey – was implemented in late 1994. The survey provided a comprehensive assessment of the recreational fishery for 1995 (Coleman 1998).

The publication of the National Policy on Recreational Fishing, the success of the Northern Territory Survey and the release of a substantial amount of Commonwealth government funding into a trust for the conduct of community projects (Natural Heritage Trust) renewed the impetus for the development of an Australia-wide recreational fishing survey. A steering committee comprising representatives of each of the government fisheries agencies and key stakeholder groups was established to determine the preferred methodology and the process to complete the development and planning for a national survey. A feasibility study was conducted and a general population telephone screening survey and telephone-diary survey, similar to that used in the Northern Territory, was recommended as the most appropriate and cost-effective technique for gathering reliable national data on recreational fishing (Kewagama Research 1998). Through an extensive developmental phase, involving each of the Commonwealth, State and Territory fishery agencies, the survey design was refined and tested. The National Survey was implemented in early 2000 as a series of concurrent state-wide surveys, conducted and managed in-house by each fisheries agency, but co-ordinated nationally. The National Survey will provide a comprehensive description of the recreational fishery over a 12-month period, with data collection completed mid-2001.

While the National Survey was being developed, the telephone-diary methodology was applied successfully in state-wide surveys of licensed recreational fishers in South Australia (McGlennon 1999) and Tasmania (Lyle 2000), providing further evidence of its efficacy.

The primary focus of this chapter is a description of the methodological approach applied to the Northern Territory and Australian National Surveys. They both represent general population surveys with fundamentally similar objectives, namely to describe the characteristics of recreational fishers (participation rates, sociodemographics); evaluate effort and catch by species, method and region; assess economic impacts in terms of investment and expenditure associated with fishing; and evaluate awareness and attitudes to fishing-related matters. All saltwater and freshwater fishing activities were within the scope of these surveys.

Methodology

The surveys comprised three primary components: (1) an initial screening survey, designed to identify fishing households, collect basic profiling information and invite recreational fishers to participate in a follow-up diary survey; (2) the diary survey, in which fishing and expenditure activity was monitored over a period of time through regular telephone contact by survey interviewers; and (3) an attitudinal survey administered as a final telephone interview at the completion of the diary survey.

Sample design

The survey design for general population sampling has been based on single-stage cluster sampling (Thompson 1992), where the primary sampling unit was the household (chosen by random sampling) and the secondary unit was the recreational fisher within the household, with all fishers in the household included in the sample. Cluster designs are recommended where there is no frame listing of all elements or where a frame listing is prohibitively expensive to obtain (Schaeffer *et al.* 1996). Since a listing of recreational fishers was not available in Australia, random sampling of households has been adopted as a feasible means of screening the resident population. In this situation, the major advantages of cluster sampling over simple random sampling (subsampling of fishers within a household) are the provision of correct weightings to both single and multiple fisher-households and a cost benefit in providing multiple fisher data through a single (screening survey) contact.

Resident sampling

The basis for sampling residents was the 'white pages' telephone directory (electronic version), which was used as a proxy for private dwelling listings at state and national levels. Available census data indicated that about 98% of the resident population resided in private dwellings, such as houses, units and apartments, as opposed to nonprivate dwellings such as hotels, hospitals, jails, military establishments (Australian Bureau of Statistics 1996). The use of directory lists enabled obvious business numbers and multiple household listings to be filtered out and the sample population stratified into regions consistent with those used by the Australian Bureau of Statistics to report statistical data.

Stratified random sampling was used in the selection of telephone numbers. For the National Survey, sample allocation by state was designed to provide general consistency in terms of the level of precision for state-wide estimates of participation, effort, catch rates and total harvest (target levels were less than 5% relative standard error). Assumed values of regional participation rates, mean effort, catch rates and response rates were used to model the effects of stratification and sample size on the precision of estimates. As a general principle, lower sampling rates were applied to the most densely populated strata.

Nonresident sampling

In the National Survey, interstate activity by residents of other states was addressed directly since any such fishing and economic activity by respondents was collected as a routine component of the diary survey. For state-wide studies such as the Northern Territory Survey, however, an alternative sampling frame was required to sample nonresidents. This was achieved through a survey of nonprivate dwellings, namely hotels, motels, guesthouses and caravan parks – but not jails, nursing homes or staff quarters. Nonprivate dwellings were stratified by region and establishment type in order to align with Northern Territory Tourism Commission data and stratified random samples of nonprivate dwellings selected. By necessity, the definition of a 'household' within nonprivate dwellings varied and could be a room, caravan, tent or bed, depending how the establishment was organised. Systematic random sampling (Pollock *et al.* 1994) was used to select the household within the establishment, the 'skip' interval determined by dividing the number of occupied sites or rooms by the required number of interviews. Unlike screening survey interviews for residents, visitor interviews were conducted face-to-face. In the process of sampling nonprivate dwellings, international as well as interstate visitors were encountered.

Coverage of international visiting fishers was not feasible in the National Survey through the telephone screening so an alternative data source, the ongoing quarterly International Visitor Survey (conducted by the Australian Bureau of Tourism Research), has been used to provide basic information about overseas fishers. This survey involves face-to-face interviews conducted in international departure lounges at the major Australian airports. Information about recreational fishing is limited to whether the respondent had fished or not during his/her visit. For the National Survey, therefore, information about overseas visitor fishing activity was restricted to participation and demographic profiling data (country of origin, age and gender).

Screening survey

The household screening survey was administered as a structured interview in which demographic data, such as age and gender of all household members, boat ownership and details of past and likelihood of future recreational fishing activity for all members over 5 years of age were established. For fisher-households, additional information including labour force status, education and ethnicity were collected. All household members identified as intending fishers (likely to go recreational fishing in the following 12 months) were invited to participate in the diary survey. Respondents who agreed to participate received a diary kit, comprising a covering letter (to establish survey legitimacy), a show card of the most common species/species groups relevant to their region and a diary for each intending fisher.

Diary survey

After receiving the diary kit, data requirements were explained to respondents in a brief interview and the next contact arranged. Unlike mail-back diary surveys, the diary was employed more as a 'memory jogger' than a logbook, and significantly, responsibility for data collection rested with survey interviewers. Regular telephone contact was maintained with diarists in order to collect details of any fishing or related expenditure that had occurred since the last contact. The level of fishing activity determined the frequency of such interviews but, as a general rule, respondents were contacted at least once a month, even if no fishing was planned. Respondents were not required to return any documentation to the survey office.

The use of interviewers in this manner allowed for greater detail and reporting precision than could be achieved in a self-administered diary, with interviewers being able to immediately clarify any misunderstandings or apparent reporting errors and therefore ensure consistency and completeness. Importantly, respondents were only required to record basic information in their diaries, for example start and finish times for fishing, catch and release details by species and expenditure by item category. More detailed data, such as species targets (if any), fishing platform (boat or shore), water body type (freshwater, marine, estuary, offshore, etc.), were collected and recorded during the telephone interview. This in turn, provided for greater data utility, for example a day's fishing effort can be apportioned between target fisheries and expenditure can be defined in terms of where it occurred and level of attribution to recreational fishing.

A conservative diary period of just 4 months was employed in the Northern Territory Survey because of concern over possible effects of nonresponse arising from respondent fatigue. Subsequent surveys of licensed recreational fishers employing this methodology produced negligible evidence of respondent fatigue for diary periods of 6–8 months (McGlennon 1999, Lyle 2000). This experience, combined with a pilot test, justified the implementation of a single 12-month diary wave for the National Survey.

Attitudinal survey

The attitudinal survey comprised a telephone interview at the end of the diary survey in which respondents 15 years of age and older were asked a range of questions aimed at assessing awareness and attitudes in relation to resource and management issues. Questionnaires were designed to address specific issues of relevance to the agency or agencies involved in the survey.

Quality control and validation

To optimise overall data quality, a range of quality control measures have been employed through all phases of survey development and implementation. In the design phase particular attention was given to the precise wording and order of survey questions, as was the minimisation of respondent burden, which in turn, minimises nonresponse bias and other sources of nonsample error. Careful recruitment, training and supervision of interviewers, coupled with follow-up re-interviews, ensured interviewer reliability and integrity. Also, systematic clerical and computer-based editing checks were applied to all completed data forms.

Validation work associated with the surveys was focused on three main sources of information: (1) concurrent creel surveys to provide species identification assessment, behavioural comparisons between telephone and nontelephone owners, catch rate comparisons and size composition information; (2) specifically designed components within the survey itself, including nonresponse and nonintending fisher follow-up surveys; and (3) secondary data sets such as population census data. The major issues for validation related to sample coverage and representation, and data-reporting quality.

Sample coverage and representation

The use of telephone directory listings meant that the sample did not provide coverage of households without a telephone or with an unlisted number. Smith (1983) warned that while telephone ownership rates may be high in developed countries, ownership rates may be low among particular socioeconomic groups and these groups may behave differently in terms of fishing activity to the wider population. For commercial and privacy reasons, the exact proportion of Australian households with a telephone listing is unavailable, but household telephone ownership rates are known to be high.

Two primary strategies were employed to assess coverage and representation of the telephone screening survey. Firstly, because basic demographic profiling was assessed for all households (rather than just fishing households), survey data could be benchmarked in terms of age, sex and household size against Australian Bureau of Statistics resident population data. A strong alignment between census and sample data was confirmed for the Northern Territory Survey (Coleman 1998) and similar validation will be undertaken for the National Survey. Secondly, to assess whether nontelephone owners have different fishing behaviour from telephone owners, respondents from concurrent creel surveys were questioned about telephone ownership. In the Northern Territory Survey, no significant differences were detected between resident nontelephone and telephone owners in terms of the number of days fished previously or in catch rates from Darwin Harbour, a region that accounted for almost half of the annual fishing effort in the Northern Territory (Coleman 1998).

Data reporting quality

A number of biases and other factors have the potential to impact on data quality, particularly where the information is self-reported. They include nonresponse bias, recall bias, prestige bias, behavioural shifts and catch misidentification. A variety of validation and design elements have been employed to address these data quality issues.

Nonresponse bias

The overall design approach resulted in very high response rates to all facets of the surveys, thereby minimising the impacts of nonresponse. Response profiles for screening and diary survey components are summarised in Table 15.1, along with results of other Australian surveys that have used the same basic methodology. Overall, screening survey response rates were in the range of 86–91%, with refusals generally accounting for less than 5% of the sample. A rigorous approach was applied to minimise nonresponse, especially from noncontacts. For instance, in the resident screening, multiple call-backs (generally a minimum of 10) were made at different times of the day and week over a period of several weeks to establish contact.

Diary acceptance or uptake rates for eligible respondents were also high, greater than 85% and, with the exception of the nonresident component of the Northern Territory Survey (where respondent 'tracking' was inherently difficult), full participation in the diary survey component exceeded 95% for a range of studies and diary duration (Table 15.1). While final survey response rates for the National Survey were not available at the time of writing, preliminary assessments indicate a full diary participation rate of at least 93%. The primary cause of nonresponse in each of the surveys has been loss of contact through disconnected telephones rather than withdrawal from the survey. Strategies such as the recording of alternative contact details, follow-up letters and electronic white page searches assisted in respondent tracking.

These response rates are not only evidence of the likely insignificance of nonresponse bias but also represent performance indicators for the methodology. With the possible exception of the Missouri study (Weithman 1991), these diary response rates are exceptional by comparison with alternative mail-back diary surveys, where diary response rates below 50% are common (Sztramko *et al.* 1991, Anderson and Thompson 1991, Tarrant *et al.* 1993, Connelly and Brown 1996).

Further, in relation to nonresponse, subsamples of noncontacts and refusals in the screening survey have been successfully resurveyed in the National Survey, to assess whether these groups differed in terms of demographic characteristics and fishing participation to respondents. Results of this follow-up work have yet to be fully evaluated, but will be incorporated in data analysis as adjustment factors, where appropriate.

Recall bias

The potential for recall bias was minimised by the frequency of contact with respondents and the use of the diaries. By regularly contacting respondents throughout the diary period, routinely within 2 weeks of any activity, details of activity not recorded in the diaries were

Table 15.1 Response rates for recent Australian recreational fishing surveys using the telephone-diary survey methodology.

Sample frame	Survey area	Survey duration	Screening survey			Diary survey				Source
			Total sample	Response rate (%)	Refusal rate (%)	Noncontacts (%)	Diary period (months)	Diary acceptance rate (%)	Final diary response rate (%)	
Resident population	Northern Territory	Oct 94–Feb 96	3 776*	86	6	8	4	90	98	Coleman (1998)
Nonresident visitors	Northern Territory	Jan 95–Dec 95	2 572*	87	5	8	4	94	80	Coleman (1998)
Licensed fishers	Tasmania	Dec 96–Apr 98	2 038	88	1	11**	6	97	96	Lyle (2000)
Licensed fishers	South Australia	Oct 98–May 99	330	89	<1	11**	8	98	97	McGlennon (1999)
Resident population	National	May 00–Apr 01	37 248*	91	4	5	12	85	93	Authors' unpublished data

* 'Household' based

** Includes noncontacts and sample loss

obtained with minimal concern in relation to recall bias.

Prestige bias

By design, reporting accuracy was optimised by a range of measures that relied on the rapport that existed between the respondent and interviewer. Put simply, the respondent was routinely brought into the survey process in terms of understanding objectives and, for example, that poor or zero catches were common and very important to the study. In addition, comparisons between diary and creel surveys in terms of catch rates were routinely undertaken. In the Northern Territory Survey, no differences between diary and creel surveys were found for overall catch rates or for catch rates from various types of targeted fishing for Darwin Harbour (Coleman 1998).

Digit bias is, to some extent, linked to prestige bias in that catches may be inflated through the effects of rounding up. No evidence of digit bias emerged in the Northern Territory Survey (A.P.M. Coleman, unpublished data) and preliminary examinations of catch distributions for the National Survey (J.M. Lyle, unpublished data).

Behavioural shifts

Involvement in a survey has the potential to condition respondent behaviour. This can occur through sensitisation, where respondents may be prompted to go fishing more often, or on the other hand experience fatigue, where they may actually go fishing (or report activity) less often, or drop out altogether (Connelly and Brown 1996). While such behavioural shifts are difficult to quantify, all respondents were routinely briefed in terms of not going fishing any more or less often than they would normally have done within the survey period. This aspect was neutrally reinforced throughout the survey by interviewers, especially during periods of nil or low activity. The fact that a relatively high proportion of diarists ultimately did not do any fishing during the survey tends to support this observation. For example, just 38% of the 1735 resident fishers in the Northern Territory Survey reported fishing activity within their designated 4-month diary period (Coleman 1998).

Related to behavioural change is the possibility that respondents identified as nonintending fishers at the time of the initial screening may have changed their minds and actually gone fishing during the diary period. In order to assess the extent of this, a sample of nonfisher households was interviewed again at the end of the diary period to establish whether they had actually fished. For instance, in the Northern Territory, around 10% of nonintending fisher households were recontacted, with less than 2% of household members reporting fishing activity (Coleman 1998). This information can then be incorporated as an adjustment to the actual participation levels determined from respondent behaviour during the diary period.

Species identification and fish sizes

Some studies have established that recreational fishers cannot reliably identify species and incorrectly estimate weights and lengths (Essig and Holliday 1991). Clearly, as diary information is self-reported it is important that respondents correctly identify their catch. To assist with this, respondents were provided with show cards of the main species (or species groups) and interviewers routinely referred respondents to these cards where uncertainties in fish identification arose. Design of these show cards and the species classification system generally was an important part of survey development. Although excellent reporting precision can often be achieved at the species level, species groupings are sometimes required (namely, where fishers cannot reasonably delineate particular species, even using show cards).

The concurrent creel surveys were designed to evaluate the identification skills of fishers, in terms of the level of detail required by the diary survey. If required, however, additional detail for grouped species (i.e. relative occurrence in catches of the various species within a grouping) can be obtained from these surveys. For the Northern Territory Survey, Coleman (1998) established that correct identification occurred in 98% of cases. Similar validation work has been undertaken as part of the National Survey.

Recognising the problems with estimates of fish size and weight, diarists were only required to report catches in terms of numbers. Creel survey data were used to assess mean lengths and weights, the latter being applied to convert catch numbers to weights thus enabling comparisons with commercial fisheries production levels.

Data outputs and utility

Outputs

The methodology has been developed to provide detailed information at a number of levels, namely the household, the fisher, the fishing activity and the related expenditure, each of which may be cross-linked for analytical purposes. Expansion of survey data to produce total population estimates was achieved using latest available information from the Australian Bureau of Statistics, with expansions conducted at the household level for expenditure and the person level for fishing activity. While conventional expansion processes were employed, several adjustment factors were also included to account for (albeit) very minor effects of nonresponse and other factors as determined through various follow-up surveys.

For analytical purposes, the household and fisher can be described in terms of key socio-economic and demographic characteristics as well as region of residence or 'home economic zone'.

Fishing behaviour can be analysed in terms of catch (harvest and release), effort (hours, days fished, etc.) and catch rates (per unit of effort), and further described in terms of locality or fishing region, water body characteristics, fishing platform, target species or fishery and fishing method/gear. These 'splits' provide a basis for increased descriptive and analytic detail. For instance, hours fished and catch can be dissected by target species with an additional capability to assess the relative impacts of different components within the recreational sector, such as boat versus shore fishing or resident versus nonresident fishers.

Fishing-related expenditure can occur independently of, or in association with, recreational fishing activity. For example, the purchase of fishing tackle may have occurred during a lunchtime break at work or during a trip involving recreational fishing. Expenditure was described in terms of expenditure items (such as fishing tackle, boat maintenance, boat fuel, boat hire/charter, bait, accommodation), the region or economic zone in which it occurred, and the level of attribution to recreational fishing. Since trips involving recreational fishing may involve interests and activities in addition to recreational fishing, inclusion of all trip expenditure can be misleading. To overcome this, diarists were asked to identify the proportion of any reported expenditure that they would attribute directly to recreational fishing (Campbell *et al.* in press). Further, in order to isolate fishing trip-related expenditure from everyday expenditure, selected items such as food and drink were only included when these items were purchased away from home. By definition, for the National Survey such expenditure was recorded when it occurred more than 40 km from home.

To permit the aggregation of data at state or national levels, economic zones have been selected to align with Australian Bureau of Statistics reporting areas (i.e. intrastate divisions) while fishing regions (i.e. discrete bodies of water such as lakes, rivers or estuaries, inland areas or areas incorporating coastal and adjacent offshore waters) were also defined to fall within state boundaries. An advantage of the national approach has been to enable regional aggregations of data that reflect the distribution of key fisheries, rather than being constrained by jurisdictional boundaries.

While large-scale studies provide substantial detail about most fishing activities, those that are relatively uncommon (or undertaken by a minority of fishers) may require supplementary sampling. For specific activities, supplementary general population sampling or, where available, use of alternative sampling frames such as licence lists, may be applied to increase the level of coverage. In the National Survey, for example, both types of supplementary sampling have been undertaken to assess the recreational rock lobster fisheries in several states.

Utility

While the Northern Territory and National Surveys were largely a response to a need for effort and catch information for use in resource assessment, sharing and allocation, their data utility extends beyond these policy issues. A particular benefit from the methodology is the richness of the data, where the many links that can be made between data groupings make it possible to assess the use of recreational fish resources according to:

- Catch and characteristics of the catch
- Spatial and temporal patterns in catch and effort
- How the catch was taken, including gear used, platform and effort
- Characteristics of the fisher, including home location, socioeconomic and demographic characteristics, and fishing behaviour
- Goods and services used in relation to fishing including fishing gear, travel, accommodation, food and drink
- Regional expenditure patterns and the significance of visitor versus resident expenditure within regions
- Relationships between awareness and attitudes and fishing behaviour

Increasingly, the issue of bycatch in fisheries is being highlighted in relation to resource utilisation and sustainability. By linking information on targeting and catch, the extent of bycatch in the recreational fishery can be quantified (McGlennon and Lyle 1999).

The survey data can be used to evaluate impacts of management decisions or options based on levels of awareness or attitudes to regulations such as seasonal closures, size and bag limits, and to model the potential effects of such options on fisher behaviour and harvest (Porch and Fox 1991).

Survey outputs have relevance to researchers using on-site survey techniques, particularly in relation to issues of survey coverage. For instance, quantifying shore-based and night-time fishing is usually difficult for creel surveys and as a consequence, these activities may be ignored or only partially assessed (Pollock *et al.* 1994). Through the use of the telephone-diary methodology, the relative importance of these activities can be assessed against activities that are more readily evaluated using creel survey techniques, such as daytime boat fishing.

Expenditure data have been collected primarily to observe the amount of expenditure attributable to recreational fishing within an economic zone and the items and services concerned. Such information will be useful in assessing the total and relative impacts of recreational fishing on regional economic activity. This in turn may affect investment in things such as road access, boat ramps, accommodation and fishing equipment, the provision of such recreational fishing inputs involving both public administrators and private enterprise.

Expenditure data can also be used to estimate the value that recreational fishers place on fish species or groupings of fish using econometric models such as a random utility model (Haab and Hicks 1999). The derived results here are an important input to the estimation of optimal allocation of fish resources between competing uses. However, reallocation of fish resources between competing users on the basis of long-term economic criteria ignores short-term secondary impacts, such as those on local providers of complementary goods and services, who are likely to have a changed customer base. These impacts can be examined using the National Survey data set (Alward *et al.* 1992). Equally, care is required in the use of expenditure data as an indicator of short-term impacts, as people are often able to adjust quickly to changes in resource allocation (Campbell *et al.* 2000).

Ultimately, the benefits arising from any survey will depend on a number of factors. These include the social impacts resulting from the uses of the data insofar as there are social cost savings from resource management decisions that optimise the long-term health of the fisheries. At a more basic level there is of course the issue of the provision of higher quality data at equivalent or less cost than alternative methodologies. A difficulty in assessing cost savings and benefits in recreational fishing surveys is that most of the impact occurs outside of the market, and cannot be directly assessed from market choices and 'willingness to pay'. As discussed, important to the benefits of the methodology applied in the Northern Territory and National Surveys are data quality and reliability. While the question of costs is discussed below, the breadth of policy issues to which the data might be applied indicates the possible benefits obtainable from the surveys.

Survey costs

The total budget associated with the implementation of the 2000/01 National Survey was just under A$2.8m (approximately US$1.5m, May 2001), inclusive of all operating, salary and travel costs for all survey components. Prior to implementation, however, an extensive development phase was undertaken, addressing factors such as output specifications, sampling strategy, statistical design and analysis, data management, survey documentation and implementation strategy (Survey Development Working Group 2000). The total cost of this phase was approximately A$475 000.

The survey implementation budget was constructed around four major activities, namely final survey development (accounting for 10% of the budget), recruitment and training (11%), data collection (69%) and analysis and reporting (10%). Of particular interest are the field costs (specifically interviewer and telephone expenses) associated with the telephone-diary component. Based on the initial sample of 37 000 households, actual response rates and a 95% final diary participation rate, the field costs for screening and diary components were A$690 000. This represents an anticipated 17 500 'fisher' years of data, which cost A$40 per fisher year or A$79 per fisher-household year. Significantly, because there are inefficiencies in general population sampling in comparison to using licence lists, the effective cost of locating a fisher or fisher-household was disproportionately high. For instance, if screening costs are discounted, the cost per fisher year is reduced to A$30 or A$59 per fisher-household year.

Alternatively, costs can be expressed in terms of the commitment in interviewer time per fisher-household. On average each fisher-household accounted for 2.8 hours of interviewer time over the 12-month diary period. For the majority of respondents the actual time commitment was much lower, since most did not fish very often and interviews were therefore very brief.

Other field survey costs associated with the National Survey included A$220 000 for on-site validation surveys and A$16 000 for nonresponse and nonintending fisher follow-up.

The lack of comparability in detail and quality of data complicate any survey cost comparisons between the present methodology and the alternative mail-back diary technique. However, any apparent cost savings of mail-back diary surveys can be very much eroded by the need for reminder and nonresponse follow-up, in attempting to assess and calibrate for associated bias effects and ultimately, the cost of uncertain data quality.

Conclusions

The methodology employed in these large-scale Australian studies has been developed in direct response to policy requirements, specifically the need for robust and reliable recreational fishing catch, effort and economic data. Central to this outcome is the underlying design philosophy that focuses on minimising respondent burden and addresses response biases and other sources of nonsample error. While methodological similarities exist between these studies, each survey has been specifically tailored to needs, as is evident in the development of a nonresident sampling component for the Northern Territory Survey.

The telephone-diary methodology represents an important innovation in recreational fisheries research and the efficacy of this approach has been clearly demonstrated in these Australian studies. Between the initial application of the method in the Northern Territory and the National Survey, the methodology has been refined from a conservative 4-month to a 12-month diary period. This development has resulted in obvious cost savings in requiring only a single screening survey to provide annual data.

The overall success of the methodology in achieving high response rates and data reporting quality has been achieved through a comprehensive respondent management process involving highly-trained survey interviewers. Interviewers are, in effect, able to personalise the conduct of the survey to match the behaviour of individual respondents, yet at the same time ensure consistency in data collection. Routinely, the same interviewer was used to contact individual respondents throughout all phases of the study, thus developing a strong rapport with respondents. This clearly places considerable responsibilities on the interviewer and thus the processes of interviewer recruitment and training, along with on-going interviewer management, represent critical elements to the overall survey success.

The telephone-diary methodology not only represents a cost-effective means for conducting large-scale recreational fishing surveys, but it also has the advantage of collecting all substantive data using a single methodology and therefore problems with complemented surveys, such as data comparability links, are effectively avoided. Based on our experiences in Australia, the telephone-diary methodology has widespread potential for use in recreational fisheries research, in particular in countries with high telephone ownership rates. However, the increasing trend towards personal mobile telephone ownership, especially among certain demographic groups, will create future challenges in ensuring representative sample coverage.

Acknowledgements

The Northern Territory government funded the Northern Territory Survey while the National Survey was funded jointly by the National Heritage Trust, the Fisheries Research and Development Corporation and the State and Territory governments of Australia.

We gratefully acknowledge our colleagues in each of the Australian fisheries agencies for their contributions to the development and implementation of the survey methodology. We are particularly indebted to Murray Johns, David McGlennon, Cheryl Munro, Shirley Munro, Dennis Reid and Russell Willing for their inputs to this process. We also thank Colin Buxton and two anonymous referees for useful comments on previous drafts of this paper.

References

Alward, G.S., Workman, W.G. & Maki, W.R. (1992) Regional economic impact analysis for Alaskan wildlife resources. In: *Valuing Wildlife Resources in Alaska* (eds Peterson, G.L. *et al.*), pp. 61–86, Westview Press, Boulder, CO.

Anderson, L.E. & Thompson, P.C. (1991) Development and implementation of the angler diary monitoring program for Great Bear Lake, Northwest Territories. *American Fisheries Society Symposium* **12**, 457–475.

Australian Bureau of Agricultural and Resource Economics (2000) *Australian Fisheries Statistics 1999*. Australian Bureau of Agricultural and Resource Economics, Canberra.

Australian Bureau of Statistics (1994) *Home production of selected foodstuffs, Australia, year ended April 1992*. Australian Bureau of Statistics, Canberra. Catalogue No. 7110.0.

Australian Bureau of Statistics (1996) *Estimated resident population by gender and age: States and Territories of Australia*. Australian Bureau of Statistics, Canberra. Catalogue No. 3230.0.

Battaglene, S.C. (1985) A creel survey of the Lake Keepit recreational fishery. *Australian Zoologist* **21**, 565–578.

Bell, J.D., Bell, S.M. & Teirney, L.D. (1993) Results of the 1991–92 marine recreational fishing catch and effort survey MAF Fisheries South Region. *New Zealand Fisheries Data Report No.* **39**, 79p.

Bradford, E. (1998) National marine recreational fishing survey 1996: scaling the diary survey results to give the total recreational harvest. *National Institute of Water and Atmospheric Research Technical Report* **17**, 33p.

Brown, T.L. (1991) Use and abuse of mail surveys in fisheries management. *American Fisheries Society Symposium* **12**, 255–261.

Campbell, D., Brown, D. & Battaglene, T. (2000) Individual transferable catch quotas: Australian experience in the southern bluefin tuna fishery. *Marine Policy* **24**, 109–117.

Campbell, D., West, L., Lyle, J., *et al.* (in press) *The Australian national recreational fishing survey: 2000–01*. Proceedings of the Tenth Conference of the International Institute of Fisheries Economics and Trade, Oregon State University, Corvallis, OR.

Caputi, N. (1976) Creel census of amateur line fishers in the Blackwood River estuary, Western Australia, during 1974–75. *Australian Journal of Marine and Freshwater Research* **27**, 583–593.

Caton, A. & McLoughlin, K. (eds) (2000) *Fishery Status Reports 1999: Resource Assessment of Australian Commonwealth Fisheries*. Bureau of Rural Resources, Canberra.

Coleman, A.P.M. (1998) Fishcount: A survey of recreational fishing in the Northern Territory. *Fishery Report No.* 43. Northern Territory Department of Primary Industry and Fisheries, Australia.

Connelly, N.A. & Brown, T.L. (1995) Use of angler diaries to examine biases associated with 12-month recall on mail questionnaires. *Transactions of the American Fisheries Society* **124**, 413–422.

Connelly, N.A. & Brown, T.L. (1996) Using diaries to estimate fishing effort and fish consumption: A contemporary assessment. *Human Dimensions in Wildlife* **1**, 22–34.

Economic and Policy Analysis Directorate (1997) 1995 Survey of Recreational Fishing in Canada. *Economic and Commercial Analysis Report No.* 154.

Essig, R.J. & Holliday, M.C. (1991) Development of a recreational fishing survey: The Marine Recreational Fishery Statistics Survey case study. *American Fisheries Society Symposium* **12**, 245–254.

Fisher, W.L., Grambsch, A.E., Eisenhower, D.L. & Morganstein, D.R. (1991) Length of recall and accuracy of estimates from the National Survey of Fishing, Hunting, and Wildlife-Associated Recreation. *American Fisheries Society Symposium 12*, 367–374.

Gentner, B. & Lowther, A. (2002) Evaluating Marine Sport Fisheries in the USA. In: *Recreational Fisheries: Ecological, Economic and Social Evaluation* (eds Pitcher, T.J. & Hollingworth, C.E.), Chapter 14, this volume. Blackwell Science, Oxford, UK.

Grambsch, A.E. & Fisher, W.L. (1991) Improving surveys: Lessons from the National Survey. *American Fisheries Society Symposium 12*, 5–12.

Griffin, R.K. (1995) Recreational fishing surveys in the Northern Territory – 1978 to 1993. In: *Recreational fishing: what's the catch?* (ed Hancock, D.A.) pp.113–119. Australian Society for Fish Biology, Canberra.

Haab, T.C. & Hicks, R. (1999) Choice set considerations in models of recreational demand: History and current state of the art. *Marine Resource Economics* **14**, 271–282.

Henry, G.W. (1984) Commercial and recreational fishing in Sydney estuary. *Fishery Bulletin 1*. Department of Agriculture, New South Wales.

Higgs, J. (1999) Experimental recreational catch estimates for Queensland residents. *RFISH Technical Report 2.* Queensland Fisheries Management Authority, Australia.

Kearney, R.E. (2002) Recreational fishing: value is in the eye of the beholder. In: *Recreational Fisheries: Ecological, Economic and Social Evaluation* (eds Pitcher, T.J. & Hollingworth, C.E.), Chapter 2, this volume. Blackwell Science, Oxford, UK.

Kewagama Research (1998) *National Recreational Fishing Survey – Feasibility Study*. Research report to Department of Primary Industries and Energy, Canberra.

Lyle, J.M. (2000) *Assessment of the licensed recreational fishery of Tasmania (Phase 2)*. Final report to Fisheries Research and Development Corporation, Project 1996/161.

McGlennon, D. (1995) A review of recreational fishing surveys in Australia. In: *Recreational fishing: what's the catch?* (ed Hancock, D.A.) pp. 61–67. Australian Society for Fish Biology, Canberra.

McGlennon, D. (1999) *Survey of the licensed recreational rock lobster fishery in South Australia 1998/99*. South Australian Research and Development Institute Final Report.

McGlennon D. & Kinloch, M.A. (1997) *Resource allocation in the South Australian marine scalefish fishery*. Final Report to Fisheries Research and Development Corporation, Project 93/249.

McGlennon, D. & Lyle, J.M. (1999) Bycatch in Australian recreational fisheries: Is it an issue? In: *Establishing meaningful targets for bycatch reduction in Australian fisheries* (eds Buxton, C.D. & Eayrs, S.E.) pp. 61–67. Australian Society for Fish Biology, Sydney.

McIlgorm, A. & Pepperell, J.G. (1999) *A national review of the recreational fishing sector*: A report to Agriculture, Fisheries and Forestry – Australia. Dominion Consulting Pty Ltd.

National Recreational Fisheries Working Group (1994) *Recreational Fishing in Australia: A National Policy*. Department of Primary Industries and Energy, Canberra.

PA Consulting (1992) *National Seafood Consumption Study: Summary Report*. Prepared for Fisheries Research and Development Corporation.

PA Management Consultants (1984) *National survey of participation in recreational fishing*. Report to the Australian Recreational Fishing Confederation, No. 1.

Pollock, K.H., Jones, C.M. & Brown, T.L. (1994) *Angler Survey Methods and Their Implications in Fisheries Management*. American Fisheries Society Special Publication 25. American Fisheries Society, Bethesda MD.

Porch, C.E. & Fox, W.W. (1991) Evaluating the potential effects of a daily bag limit from the observed frequency distribution of catch per fisher. *American Fisheries Society Symposium 12*: 435–456.

Schaeffer, R.L., Mendenhall, W. & Ott, R.L. (1996) *Elementary Survey Sampling, Fifth edition*. Duxbury Press, Belmont California.

Smith, C.L. (1983) Evaluating human factors. In: *Fisheries techniques* (eds Neilsen, L.A. & Johnson, D.L.), pp. 431–445. American Fisheries Society, Bethesda MD.

Steffe, A.S., Murphy, J.J., Chapman, D.J., Tarlinton, B.E., Gordon, G.N.G. & Grinberg, A. (1996) *An assessment of the impact of offshore recreational fishing in New South Wales waters on the management of commercial fisheries*. Final report to Fisheries Research and Development Corporation, Project 94/053.

Sumner, N.R. & Williamson, P.C. (1999) A 12-month survey of coastal recreational boat fishing between Augusta and Kalbarri on the west coast of Western Australia during 1996–97. *Fisheries Research Report No. 117* (Fisheries Western Australia).

Survey Development Working Group (2000) Development of the National Recreational and Indigenous Fishing Survey. Final report to the National Heritage Trust and Fisheries Research and Development Corporation, FRDC Project No. 98/196. *New South Wales Fisheries Report Series No. 23* (Vol.1).

Sztramko, L.K., Dunlop, W.I., Powell, S.W. & Sutherland, R.G. (1991) Applications and benefits of an angler diary program on Lake Erie. *American Fisheries Society Symposium 12*: 520–528.

Tarrant, M.A. & Manfredo, M.J. (1993) Digit preference, recall bias, and nonresponse bias in self-reports of angling participation. *Leisure Sciences* **15**, 231–238.

Tarrant, M.A., Manfredo, M.J., Bayley, P.B. & Hess, R. (1993) Effects of recall bias and nonresponse bias on self-report estimates of angling participation. *North American Journal of Fisheries Management* **13**, 217–222.

Teirney, L.D., Kilner, A.R., Millar, A.R., Bradford, E. & Bell, J.D. (1997) Estimates of recreational harvest from 1991–92 to 1993–94. *New Zealand Fisheries Assessment Research Document 97/15*.

Thompson, S.K. (1992) *Sampling*. Wiley, New York.

Thompson, T. & Hubert, W.A. (1990) Influence of survey method on estimates of statewide fishing activity. *North American Journal of Fisheries Management* **10**, 111–113.

Weithman, A.S. (1991) Telephone survey preferred in collecting angler data statewide. *American Fisheries Society Symposium* **12**, 271–280.

West, R.J. & Gordon, G.N.G. (1994) Commercial and recreational harvest of fish from two Australian coastal rivers. *Australian Journal of Marine and Freshwater Research* **45**, 1259–1279.

Chapter 16
Evaluating the Marine Recreational Fishery in South Africa

Marc H. Griffiths and Steve J. Lamberth

Abstract

The South African marine linefishery targets almost 200 species, of which 31 contribute substantially to catches. User groups may be broadly divided into recreational, commercial and subsistence components. The commercial component consists of about 18 600 participants (2581 vessels of 5–15 m long), which target both pelagic and demersal species beyond the surf-zone. The recreational component may be divided into estuarine anglers (72 000), shore anglers (412 000), spearfishers (7000) and a recreational boat-based sector (12 000 participants). Subsistence linefishing is largely limited to estuarine and shore-based activities in the Transkei and KwaZulu-Natal. The linefishery, excluding the estuarine component, is estimated to provide employment for approximately 132 000 people, and to contribute about R2.2bn (US$0.3bn) to the South African GDP.

Although the commercial component is responsible for 79% of the estimated total catch, the recreational component provides 81% of the employment and generates 82% of the revenue. Species targeted by the linefishery display diverse life-history strategies, including tactics – long life-spans (>20 years), estuarine dependence, sex change and aggregating behaviour – that cause populations to be particularly vulnerable to over-fishing. Owing to poor enforcement and ineffective regulations many species are now severely over-exploited. Estuarine and shore anglers have been responsible for over-fishing estuarine dependant and surf-zone species whereas boat-based recreationals have, together with their commercial counterparts, contributed to the demise of bottom-dwelling offshore species, particularly in KwaZulu-Natal.

The imminent transfer of about 1703 part-time commercial vessels to the recreational sector is anticipated to increase the impact of the recreational boat-based sector in the Cape regions, particularly if regulations are too liberal or compliance is not improved. Several initiatives have recently been developed to rebuild depleted linefish stocks: a new Linefish Management Protocol, designed to monitor stocks and adjust regulations accordingly; programmes to improve monitoring and compliance; and downsizing the commercial fleet. The net effect of these initiatives remains to be tested. Circumstantial evidence suggests that the depletion of stocks of large predators may have had substantial ecosystem effects, but remains to be quantified.

Key words: biological assessment, compliance, impacts, linefishing, management protocol, monitoring, netfishing, regulations, socio-economics, spearfishing.

Introduction

South Africa has a 3000 km coastline that, incorporating the southernmost tip of the continent, is bordered by both the Atlantic and Indian Oceans (Fig. 16.1). The highly productive Benguela upwelling ecosystem along the west coast, and the influence of the warm Agulhas Current along the east coast, have together ensured great diversity and abundance of marine fishes. Considering that the country has one of the highest coastal population densities in Africa – 81 people per km^2, compared with the average African density of 55 per km^2 – it is little wonder that fishing is a popular pursuit.

The linefishery, including recreational, commercial and subsistence components, exploits over 200 demersal and pelagic fishes, of which 95 are regarded as important (Van der Elst and Adkin 1991) [the South African linefishery uses rod, line, hook and bait – eds]. The commercial component is boat-based and operates on the continental shelf in depths of 5–200 m. The recreational component consists of several sectors, including estuarine anglers, who fish from boats or the shore; rock and surf anglers; a boat-based sector which operates in a similar environment to the commercial component; and a recreational spearfishing sector operating from boats and the shore. Subsistence linefishing is currently limited to estuarine and shore-based activities in the northern regions of the east coast.

Until 1985, when commercial permits were allocated for the first time (only for offshore boat-based activity), there was no clear distinction between commercial and recreational fishers. Fishers operated largely without restriction, with many selling catch surplus to their own

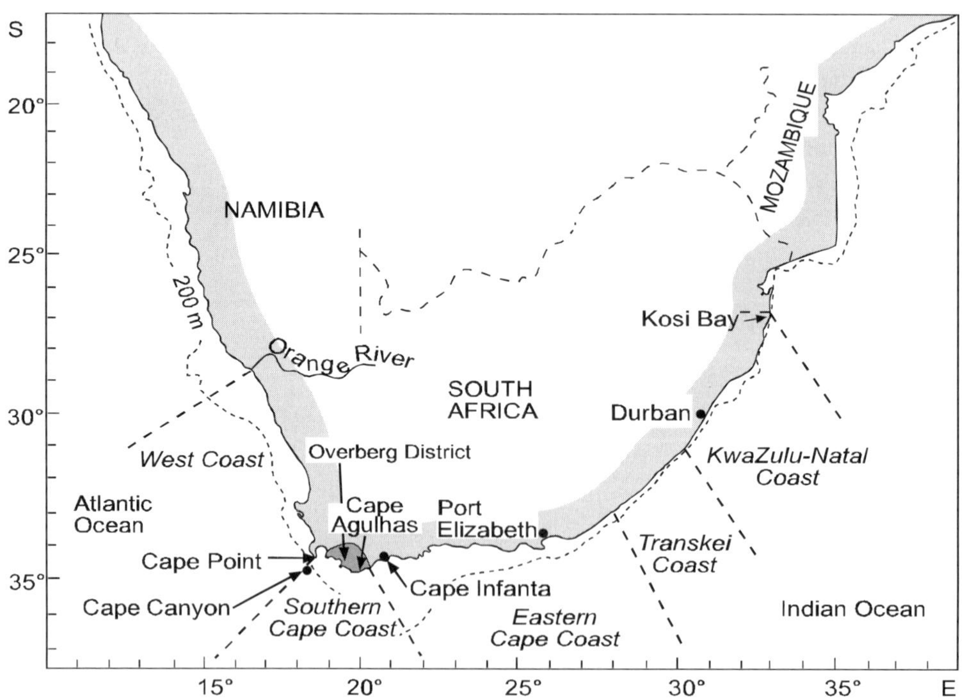

Fig. 16.1 Map of South Africa showing regions and geographical features mentioned in the text.

consumption. Marine recreational angling permits were not required until 1999. Recreational anglers are, for the purposes of this chapter, therefore defined as those without commercial permits (1985–1998) or those with recreational permits (1999 onwards). Subsistence fishers were formally recognised as a sector apart from the recreational sector in 1998. They are defined as poor people who personally harvest marine resources as a source of food or to sell them to meet basic needs or food security. Subsistence permits are yet to be allocated for the linefish resource.

Historically, subsistence linefishing can be traced back to the fishing activities of indigenous Khoi who used hooks made of animal teeth, bones, and thorns, thousands of years ago, and to European seafarers in the 1500s (Muller 1997, Gore 2000). Commercial activities were initiated after the Dutch colonised the Cape in 1652, but due to various restrictions, the fishery was slow to develop, despite an abundance of fish (Thompson 1913). When the British captured the Cape Colony in 1795, all fishing restrictions were removed, and by the mid-1800s the linefishery had become a thriving industry. It is generally accepted that recreational fishing was introduced by British settlers who arrived in 1820 (Van der Elst 1989). These early anglers originally cast handlines (of hemp, gut or braided flax) from the shore, but later used cane rods and wooden centre-pin reels. Boat-based recreational fishing was, however, not introduced until after World War II (Horne 1974). Two types of vessel were and still are used – the large striker or tuna boats, which are harbour based, and trailable ski-boats (4–8 m vessels with semidisplacement hulls and outboard engines, launched off slipways and beaches).

Given the diversity of user groups and species targeted, and the large number of participants, the management of the linefishery is extremely complex. In addition to these complexities, misconceptions of the potential impact of linefishing, and focus on landed catch value as a means of prioritisation, has led to the linefishery being neglected in favour of large commercial fisheries, e.g. demersal trawl and pelagic purse-seine. Notwithstanding this setback, substantial progress has been made during the last 10 years and the management of the linefishery is currently being transformed. This chapter describes the marine recreational linefishery in South Africa with evaluations of (1) past and future management procedures, (2) impact on resources, (3) impact on commercial and subsistence sectors, and (4) possible ecosystem effects.

Management of the South African marine linefishery

Monitoring the recreational fishery

Past

Long-term trends in catch, effort and size composition are essential for stock assessment and therefore for the effective management of a fishery. Owing largely to the diverse nature of the South African linefishery, existing catch and effort data are patchily distributed in space, time and across user groups. Current data sources include club logbooks, competition results, voluntary catch cards, and observations made by enforcement officers whilst on patrol or at limited access points. Unfortunately, the collection of recreational data has been neglected in most regions and, owing to biases associated with competition data and voluntary returns, the existing data are generally unreliable (Mann-Lang 1996). As a result of these, and of the data limitations associated with the compulsory returns collected from the commercial component (Penney 1997, Sauer *et al.* 1997), it had not been possible to quantify the catches made by each sector or to make inter-sector comparisons, even for specific regions.

The national survey

A 2-year national survey was, owing to the limitations of existing data, initiated in 1994 to evaluate the catches, socioeconomic aspects and attitudes of the main sectors participating in the linefishery. An additional aim of the survey was to evaluate prospective methods for the future monitoring of the fishery. Survey design included the division of the coastline into five regions (Fig. 16.1) with the following techniques – aerial surveys, roving creel surveys, access point surveys and interviews/questionnaires – employed in each (Sauer *et al.* 1997; Brouwer *et al.* 1997). Due to logistical constraints and the quasi-independent status of the Transkei at the time, estuaries and the Transkei region were excluded from the initial survey. A survey of the Transkei commenced in 1997, but estuaries are yet to be addressed on a national level.

Although the survey provided much useful information, there were some limitations due largely to insufficient manpower (one fieldworker per 750 km of coastline) and the snapshot nature of the exercise. The vast areas covered meant that the sampling frequencies at specific sites were often insufficient to estimate catches of migratory species that are characterised by large catches made over short and infrequent time periods. Similarly, sampling was also insufficient in gauging the effort expended by commercial and recreational ski-boat fishers whose nomadic nature is dictated by the movements of these migratory species.

In terms of the total number of participants the shore-angling sector is by far the largest, followed by commercial boat, recreational boat and spearfishing (Table 16.1). Recreational linefishing sectors are estimated to be growing at a rate of 2% (McGrath *et al.* 1997) and the spearfishery at 6% per annum (Mann *et al.* 1997). Demand for access to the resource is therefore expected to increase, accordingly. Demographic characteristics of the participants are given in Table 16.2.

Recreational anglers were predominately middle-aged males, with 16–20 years of experience (Table 16.2). Compared to shore anglers (17%), a large proportion of boat anglers fell under organised angling with 48% belonging to an angling club. Economic contribution and

Table 16.1 Estimated total catch and number of participants and boats in the four main sectors of the linefishery (1994–1996). Commercial data from National Marine Linefish System of Marine and Coastal Management.

	Participants	Boats	Annual catch (tonnes)	Source*
Rock and surf	412 000		2 836	1, 5
Commercial boat	18 583	2 581	16 671	3
Recreational boat	12 054	3 444	1 460	2, 5
Spearfishing	7 000		214	4

* Sources: (1) Brouwer *et al.* (1997); (2) Sauer *et al.* (1997); (3) NMLS; (4) Mann *et al.* (1997); (5) McGrath *et al.* (1997)

Table 16.2 Characteristics of South African (excluding the Transkei) shore-anglers and recreational skiboat operators (condensed from data in McGrath *et al.* 1997).

Characteristics of anglers	Shore-anglers	Skiboat-anglers
Male (%)	98	99
Mean age (years)	40	41
Mean years fished	21	16
Club member (%)	17	48

Table 16.3 Economic value of the South African linefishery, excluding estuaries and the entire Transkei region (from McGrath *et al.* 1997).

Sector	Rand (million)	US$ (million)	Percentage	Number employed
Shore angling	1653	212	76	99 180
Boat fishing				
Recreational	128	17	6	7680
Commercial	386	49	18	24 700
Total	2167	278		131 560

employment figures followed similar trends to that of the numbers participating (Table 16.3), but in the case of total catch, highest values were estimated for the commercial component, followed, in descending order, by rock and surf, recreational boat, and spearfishing (Table 16.1). Owing to large temporally related fluctuations in the availability of the dominant linefish species, and the low boat-based sampling frequency of the 1994 national survey (Sauer *et al.* 1997), the estimates of boat-based catch are regarded as imprecise.

Future

As a result of the limitations associated with catch returns (voluntary or compulsory) and snapshot surveys (see above), future catch and effort (as well as biological) data will be collected by means of observers (Griffiths *et al.* 1999). Following the successful completion of a pilot phase, an observer programme covering all major boat-based access points is currently being implemented; recreational boats in the Cape provinces are therefore being monitored for the first time. The intention is to expand this programme to include data for shore anglers (including the estuarine sector) collected during enforcement patrols and access point inspections (see coastal trusts below).

Marine inspectors in KwaZulu-Natal currently collect catch and effort data from between 110 000 and 120 000 angler outings each year (1996–1999) while on shore patrol. The data are held on the National Marine Linefish System (NMLS) – a national database created by the Linefish Section of Marine and Coastal Management in 1984 to store and process all available information related to the monitoring of the linefishery. In addition, telephone surveys, facilitated by the introduction of a marine angling permit in 1999, will soon be used to provide independent catch estimates as well as socioeconomic information.

Biological assessment and regulations

Past

Owing to complexities accruing from the multispecies, multiuser nature of the fishery, it is managed on an effort rather than on a Total Allowable Catch (TAC) basis. The general philosophy is to control fishing effort within the commercial and recreational sectors by input (number of participants) and output (e.g. bag and size limits) controls, respectively. In spite of the 200-year history of the fishery, the levels of funding and effort historically channelled into linefish research and management have been inadequate; the basic life histories of many species have only recently been understood, and for many others remain unknown. The main reason for this is that commercial landed value has been used as the main criterion when prioritising resources for research. Socioeconomic factors (e.g. employment opportunities and community dependence) and hence the true value of recreational fishing have unfortunately been ignored. As a result of this approach research and monitoring of the linefishery was neglected in favour of valuable commercial resources such as hake, rock-lobster and pelagic fish.

Biological studies were initiated on a few important boat-caught sea breams (Sparidae) in the 1960s, as a result of growing concern for the linefish resource (Ahrens 1964, Nepgen 1977). A decade passed, however, before the fishery received research attention that was consonant with its status. Apart from the pelagic nomads, the remaining species are typically long lived (15–42 years) and are predictably located (resident or coastal migrant), making them particularly vulnerable to over-exploitation (Griffiths 2000). In addition to this, some species change sex when they attain a predetermined size, and others depend on estuaries during certain phases of the lifecycle.

The first attempt at managing linefish resources was marked by the introduction of minimum size limits for selected species in 1940. In turn, owing to the absence of life-history information, these regulations were determined on a fairly arbitrary basis. With the exception of a closed season for elf (*Pomatomus saltatrix*) in KwaZulu-Natal, and snoek (*Thyrsites atun*) in the Cape, no other restrictions were promulgated until a comprehensive management framework was introduced for the linefishery early in 1985. Included in this framework was the introduction of a two-tiered licence system for commercial fishers (full-time [A-category] and part-time [B category]); freezing of commercial effort at the 1985 level; introduction of revised minimum size limits equal to the sizes at maturity; division of linefish species into categories based on their perceived vulnerability to exploitation, with associated bag limits for commercial and recreational fishers; and the introduction of national closed seasons for certain species.

One of the categories consisted of a group of species designated for recreational use, and as such were prohibited from sale. The main objective behind this initiative was to steer fishing effort away from 'vulnerable' reef fishes to the more 'resilient' shoaling species. This was the first attempt at managing all of the important linefish species using a single management framework. Key points were the use of rules which could be applied to all species (e.g. minimum size limit equal to size at maturity) and the tailoring of regulations to suit particular species groups, rather than attempting to develop individual regulations for each species.

Although species categories and size limits have been revised (in 1992 and again in 1999), the management framework, in essence, has remained unchanged. Owing to a lack of biological and fisheries data, the level of protection afforded to each species has depended largely on subjective perceptions of its vulnerability to exploitation, rather than on quantitative evaluations. In the absence of long-term catch and effort data, management decisions were also undoubtedly influenced by the 'shifting base-line syndrome': wherein successive generations of fisheries scientists, managers or user groups lose sight of the state of resources prior to their involvement; earlier accounts of abundance perceived as unverified anecdotes (Pauly 1995). Furthermore, the absence of clear management guidelines and the existence of strong lobby groups have necessitated considerable compromise between managers and fishers regarding the implementation of management action for certain species.

Future

Analysis of historical data and recent stock assessments indicate that many linefishes, including so-called resilient species, are severely over-exploited (Griffiths 1997a,b,c, 2000, Hutton *et al.* 2001). In addition, bag frequencies from research surveys reveal that many of the current regulations are failing to limit recreational catch (Attwood and Bennett 1995, Griffiths 1997a). As a result of these findings a new Linefish Management Protocol (LMP) was recently developed for the linefishery in which management plans are required for each species (Griffiths *et al.* 1999). Key features of the LMP include:

- The type of data required to assess individual stocks
- The quantitative approach to be used (owing to a lack of accurate catch and effort series, spawner biomass per recruit models are the default)
- Appropriate biological reference points (spawner biomass-per-recruit ratios of 25% and 40% accepted as the default limit and target reference points, respectively – historical reduction in CPUE and sex ratio of sex-changing species were also accepted as interim stock status indicators in the absence of stock assessment)
- Long-term goals for each stock
- Management action that will be taken to achieve these goals
- User group involvement in the development of regulations (and combinations thereof) to achieve the biological reference points

Application of the LMP, to those species for which sufficient data are available, has resulted in considerable revision of both size and bag limits for recreational fishers. Necessary modifications to fishing mortality (to attain biological reference points) are translated into recreational bag limits by analysis of surveyed bag frequencies (Attwood and Bennett 1995, Griffiths 1997a). Bag limit reductions of more than 80% are required to achieve target levels of fishing mortality for many species (Table 16.4).

Because of the large number of species targeted by the South African linefishery, it is not possible to assign the same management priority to each one. Prioritisation for research and assessment is further complicated by the multisectoral nature of the fishery as each sector has its own suite of important species and different management actions may be required for the same species. For these reasons the frequency with which stocks are reviewed as well as

Table 16.4 Summary of stock status and the species-specific management regulations emanating from the application of the new Linefish Management Protocol to species for which sufficient scientific information existed. Daily bag limits were generally determined for recreational anglers (based on bag frequency analysis), but those with an asterisk apply to commercial fishers as well.

Family	*Species*	Common name	Stock status	Assessment method	Daily bag	Minimum size (cm)
Carangidae	*Seriola lalandii*	Yellowtail	Optimally exploited	VPA(40%), *CPUE*(60%)	10→10	None
Dichistiidae	*Dichistius capensis*	Galjoen	Collapsed	SB/R (17%), *CPUE*(20)	5→*2	35→35
Gempylidae	*Thyrsites atun*	Snoek		*CPUE* (58%)		
Merluccidae	*Merluccius capensis*	Hake	Optimally exploited	SB/R+APM	5→5	None
Pomatomidae	*Pomatomus saltatrix*	Elf/shad	Over-exploited	SB/R (34%)	5→4	30→30
Sciaenidae	*Argyrosomus inodorus*	Silver kob	Collapsed	SB/R (6–9%),*CPUE* (4–9%)	5→5	40→50
	Argyrosomus japonicus	Dusky kob	Collapsed	SB/R (2%)	5→*1	40→60
	Argyrosomus thorpei	Squaretail kob	Collapsed	*CPUE* (5%)	5→5	35→40
	Atractoscion aequidens	Geelbek	Collapsed	SB/R (5%) VPA (7%) *CPUE* (3%)	10→2	60→60
Scombridae	*Scomberomorus commerson*	King mackerel	Over-exploited	SB/R (33%)	10→10	None
	Scomberomorus plurilineatus	Queen mackerel	Optimally exploited	SB/R (50%)	10→10	None
	Thunnus alalunga	Longfin tuna	Optimally exploited	APM (20%)	10→5	None
Serranidae	*Epinephelus andersoni*	Catface rockcod	Collapsed	SB/R (19%)	5→5	40→50
	Epinephelus marginatus	Yellowbelly rockcod	Collapsed	SB/R (24%)	5→1	40→60
Sparidae	*Chrysoblephus anglicus*	Englishman	Collapsed	*CPUE* (?)	5→1	None→40
	Chrysoblephus cristiceps	Dageraad	Collapsed	SB/R (3%), *CPUE* (6%)	5→*1	30→40
	Chrysoblephus gibbiceps	Red stumpnose	Collapsed	*CPUE* (1–5%)	5→1	30→30
	Chrysoblephus laticeps	Roman	Collapsed	*CPUE* (5–17%)	5→2	30→30
	Chrysoblephus puniceus	Slinger	Collapsed	SB/R (15%), *CPUE* (14%)	5→5	25→25
	Cymatoceps nasutus	Poenskop	Collapsed	*CPUE* (20%)	2→*1	50→50
	Lithognathus lithognathus	White steenbras	Collapsed	SB/R (6%)	5→*1	60→60
	Petrus rupestris	Red steenbras	Collapsed	*CPUE* (0.2%)	2→*1	40→60
	Polysteganus praeorbitalis	Scotsman	Collapsed	*CPUE* (35%)	5→1	30→40
	Polysteganus undulosus	Seventy-four	Collapsed	SB/R(< 5%),CPUE (10%)	0→0	Moratorium
	Pterogymnus laniarius	Panga	Underexploited	APM(67%)	10→10	None
	Sarpa salpa	Strepie	Underexploited	SB/R(60%)	0→10	Stat. Quo
	Sparodon durbanensis	Musselcracker	Collapsed	*CPUE*(20%)	5→*1	60→60

SB/R = spawner biomass per recruit APM = age structured production model, VPA = Virtual Population Analysis, *CPUE* = Catch per unit effort (a status indicator). Numbers in parentheses in the 'Assessment Method' column indicate current spawner biomass per recruit ratios, spawner biomass (VPA or APM) and catch per unit of effort as a percentage of pristine or historical levels. Ranges are given where calculations were made for separates stocks or regions. Data sources: Griffiths (2000), Mann (2000) and Web.

Table 16.5 The top 30 species for conservation, all fishing sectors and for conservation and sectors combined as determined using Multicriteria Decision Analysis.

Rank	Conservation	All sectors	Conservation and all sectors
1	*Epinephelus albomarginatus*	*Pomatomus saltatrix*	*Pomatomus saltatrix*
2	*Polysteganus praeorbitalis*	*Sarpa salpa*	*Dichistius capensis*
3	*Chrysoblephus puniceus*	*Dichistius capensis*	*Diplodus sargusl*
4	*Epinephelus andersoni*	*Thyrsites atun*	*Argyrosomus japonicus*
5	*Argyrosomus* spp.	*Diplodus sargus*	*Chrysoblephus puniceus*
6	*Chrysoblephus lophus*	*Argyrosomus japonicus*	*Argyrosomus* spp.
7	*Chrysoblephus cristiceps*	*Thunnus alalunga*	*Epinephelus albomarginatus*
8	*Polysteganus undulosus*	*Pomadasys olivaceum*	*Thyrsites atun*
9	*Chrysoblephus gibbiceps*	*Pomadasys commersonni*	*Polysteganus praeorbitalis*
10	*Lithognathus lithognathus*	*Lithognathus lithognathus*	*Lithognathus lithognathus*
11	*Argyrosomus thorpei*	*Liza richardsonii*	*Epinephelus andersoni*
12	*Chrysoblephus anglicus*	*Pachymetopon grande*	*Argyrosomus inodorus*
13	*Petrus rupestris*	*Argyrosomus inodorus*	*Sarpa salpa*
14	*Argyrosomus japonicus*	*Pachymetopon blochii*	*Chrysoblephus cristiceps*
15	*Argyrosomus inodorus*	*Chrysoblephus laticeps*	*Chrysoblephus lophus*
16	*Epinephelus marginatus*	*Seriola lalandii*	*Polysteganus undulosus*
17	*Argyrops spinifer*	*Argyrosomus* spp.	*Pachymetopon grande*
18	*Cymatoceps nasutus*	*Neoscorpis lithophilus*	*Chrysoblephus gibbiceps*
19	*Epinephelus tukula*	*Rhabosargus globiceps*	*Rhabdosargus globiceps*
20	*Diplodus cervinus*	*Rhabosargus sarba*	*Argyrosomus thorpei*
21	Serranidae spp.	*Merluccius capensis*	*Pachymetopon blochii*
22	*Rhabdosargus globiceps*	*Atractoscion aequidens*	*Petrus rupestris*
23	*Promicrops lanceolatus*	*Argyrozona argyrozona*	*Chrysoblephus anglicus*
24	*Epinephelus malabricus*	*Scomberomorus commerson*	*Atractoscion aequidens*
25	*Dichistius capensis*	*Lichia amia*	*Cymatoceps nasutus*
26	*Porcostoma dentate*	*Sparodon durbanensis*	*Epinephelus marginatus*
27	*Umbrina* spp.	*Thunnus* spp.	*Diplodus cervinus*
28	*Cephalopholis sonnerati*	*Rhabdosargus holubi*	*Liza richardsonii*
29	*Atractoscion aequidens*	*Chrysoblephus puniceus*	*Argyrops spinifer*
30	*Epinephelus albomarginatus*	*Cheimerius nufar*	Serranidae spp.

research priorities is being determined using a Multiple Criteria Decision Analysis (MCDA) approach. The methodology incorporates scoring priority alternatives (fish) on the basis of a number of criteria that indicate species importance, including catches, socioeconomic value, existing indications of stock depletion, extent of biological knowledge and endemism (see Lamberth and Joubert 1999). This approach assesses the relative need for remedial measures per species, and serves as a forecast of fisher reaction, thereby defining appropriate action that is acceptable to both fishers and management. High priority species of conservation and sector importance are listed in Table 16.5.

Institutional framework and the decision-making process

According to South Africa's new Marine Living Resources Act (Act No. 18 of 1998) marine-living resources are the heritage of all South Africans; but the central government is responsible for their management. Management recommendations are submitted by the Chief Directorate of Marine and Coastal Management to the Minister of Environmental Affairs and

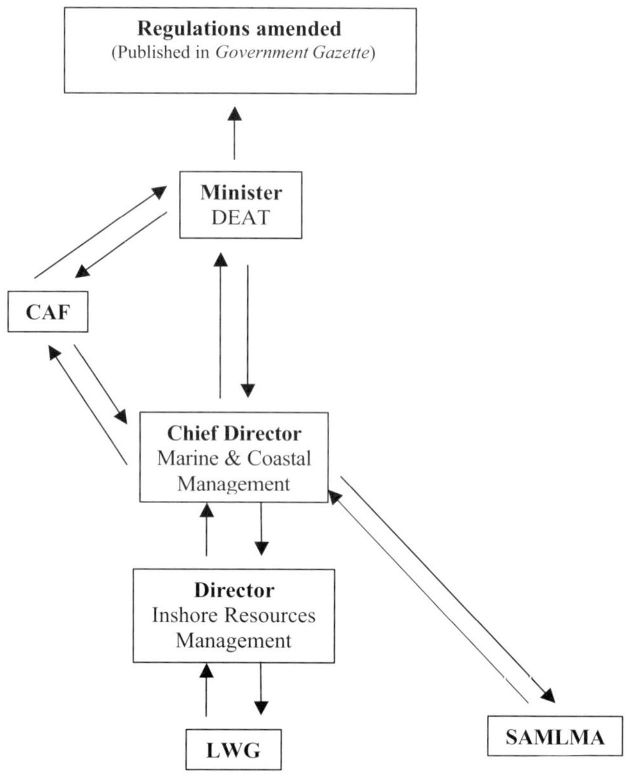

Fig. 16.2 Key role players and flow of information involved in the formulation of management regulations for the South African linefishery. LWG = Linefish Working Group, SAMLMA = South African Marine Linefish Management Association, CAF = Consultative Advisory Forum, DEAT = Department of Environmental Affairs and Tourism.

Tourism for approval and promulgation. Owing to its multispecies, multiuser nature, and also the large area of operation, there are three key role players involved with the management of the linefishery (Fig. 16.2).

The Linefish Working Group (LWG) of MCM, consisting of eight scientists from both MCM and other institutions (involved with linefish research), is tasked with the formulation and scientific evaluation of linefish management recommendations. Where necessary, fisheries managers are also invited to participate. Recommendations generated by the LWG are usually submitted to the South African Marine Linefish Management Association (SAMLMA), consisting of scientists (including MCM scientists) and representatives from all commercial and recreational fishing sectors, for further evaluation and input. It is at this stage that user groups have the opportunity to determine the most acceptable combination of regulations (designed to attain the target reference point). Conversely recommendations generated by SAMLMA (initially tabled by scientists or fishing sectors) are channelled (via the Chief Director of MCM) to the LWG, prior to ministerial submission. Although draft management regulations generally pass through the Consultative Advisory Forum (CAF) – a body comprised of 18 fishery representatives and scientists (external to MCM), established to provide advice to the minister – the CAF may autonomously address issues concerning the management of marine-living resources, but in such cases scientific input is usually requested from MCM. It is also possible for user groups to independently submit recommendations to the Chief Director/Minister, but these are generally evaluated by the LWG, and if necessary, SAMLMA. The current linefish management system is therefore both flexible and involves input from a wide variety of sources; as a result it embraces elements of both 'consultative' and 'co-operative' comanagement (*sensu* Sen and Nielsen 1996).

Compliance

Notwithstanding the new management plan or the efficacy of the regulations emanating from it, the success of fisheries management depends heavily on the degree of user compliance, which is often a function of enforcement level (Hemming and Pierce 1997). The national survey reveals that although most anglers agreed with management regulations, a substantial number had 'broken the rules' at some stage, and knowledge of regulations was generally poor (Table 16.6). Both knowledge of regulations and frequency of transgressions were clearly correlated with the degree of enforcement, directly in the former and inversely in the latter (Tables 16.6 and 16.7).

Table 16.6 Opinions, compliance and knowledge of South African linefish regulations by recreational users. Shore-angler data from Brouwer *et al.* (1997) and McDonald *et al.* (1999) and skiboat data from Sauer *et al.* (1997) and Fennessey *et al.* (1999). All values are frequency (%).

Region and Parameters	Shore-anglers			Skiboat-anglers		
	Agree	Obey	Knowledge	Agree	Obey	Knowledge
West Coast						
Size limit	81	61	35	83	67	36
Bag limit	76	56	43	62	50	50
Closed season	82	60	36	75	55	51
Reserves	86	88		88	90	
Southern Cape Coast						
Size limit	84	78	43	86	43	38
Bag limit	75	71	45	57	86	62
Closed season	76	71	33	57	43	14
Reserves	83	87		100	71	
Eastern Cape Coast						
Size limit	91	50	35	82	30	38
Bag limit	80	70	36	62	56	62
Closed season	84	75	45	90	79	14
Reserves	86	91		93	84	
Transkei						
Size limit	87	49	16	86	50	31
Bag limit	75	58	21	79	64	56
Closed season	72	71	30	84	73	64
Reserves	72	94		90	85	
KwaZulu-Natal						
Size limit	93	69	51	89	64	47
Bag limit	76	75	61	79	70	68
Closed season	72	79	72	87	79	92
Reserves	83	97		97	98	

Table 16.7 The average number of inspections per recreational vessel per year (Sauer *et al.* 1997, Fennessy *et al.* 1999), and the percentage of shore anglers inspected by compliance officers over the 12 months prior to their being interviewed (Brouwer *et al.* 1997). The number of compliance officers per kilometre of coastline is also given.

	Shore anglers % inspected	Skiboat angler Number of inspections	Inspectors/km
West Coast	1.39	4	0.06
Southern Cape	0.63	1	0.06
Eastern Cape	0.75	2.7	0.03
Transkei	Not available	1.5	0
KwaZulu-Natal	10.9	3	0.53

The number of enforcement officers per km of coastline was, in ascending order, 0 in the Transkei, 0.03–0.06 in the Cape regions and 0.53 in KwaZulu-Natal (Table 16.7). Compounding this low enforcement effort, as much as 80% of the manpower in the Cape regions to the west of Port Elizabeth is expended on issues related to the control of quotas for other fisheries, thereby dramatically reducing the capacity for the enforcement of linefish regulations. The possibility of increasing the number of inspectors is, owing to political dictates, extremely slim at this stage (Mr P. Goosen, Deputy Director of Compliance, Offshore, personal communication).

The management regulation most supported and most adhered to during the 1994/5 survey was closed areas (Table 16.6). The higher degree of compliance is, however, more likely to be a function of the fact that no-take Marine Protected Areas have defined boundaries, which are easier to police, than to angler choice *per se*. In spite of the illegal nature of the activity, 38% of spearfishers (Mann *et al.* 1997) and 40% of the recreational boat anglers (Sauer *et al.* 1997) interviewed, admitted to selling their catch. With the imminent reduction in the number of commercial fishing rights, it is anticipated that most of those displaced will, having the vessels and other necessary equipment, retain access to the resource via the recreational route. Failure by this group to comply with recreational regulations may sabotage the stock rebuilding process in the absence of better enforcement.

On the brighter side, a new approach has been adopted by the Monitoring, Control and Surveillance Directorate of MCM, with emphasis on improving compliance rather than merely apprehending offenders once the damage has already been done (M. Kroese, Head Marine Conservation Inspector, Marine and Coastal Management). Activities associated with quota control are to be contracted out, thus freeing the current officers for increased patrol based/linefish associated activities with the added commitment of user group education. Other recent initiatives include the involvement of the South African Navy and the Police Water Wing in sea-based patrols, and intended partnerships with municipalities and district councils for the monitoring and compliance of shore anglers (*sensu* Taylor 1999). The highly successful pilot phase of the partnership programme was conducted in the Overberg District (Fig. 16.1). A trust, generated from monies from the Overberg District Council and Marine Living Resources Fund (the MLRF is generated from fishing levies and permit fees and is administered by the Consultative Advisory Forum, a statutory body advising the Minister of Environmental Affairs and Tourism), is used to employ monitors at key coastal access points, who record shore-based activities, including catch and effort data. These monitors will also fulfil an educational role.

Cost of management

Recreational fishing permits were introduced for marine linefish in 1998 in accordance with the 'user pays' principle. Prior to that, costs of research and compliance were subsidised by levies from larger commercial fisheries and tax-payers. Unfortunately, problems experienced with the data system used by the agency that sell the permits have precluded a meaningful balance sheet for the recreational fishery at this stage.

Impacts of the recreational linefishery

Potential impacts on other linefishing sectors: intersector competition

In spite of the limitations of recent catch and effort data, estimates of the regional catch composition of each sector are considered to be accurate. The catch contributions of the ten most important species for each sector in each region are given in Table 16.8. The ten most important species for each of seven sectors comprised only 31 species, suggesting that overlap does exist between some sectors. This has resulted in commercial fishing (including boat-based, trawl and net sectors) being most often blamed by recreational anglers for perceived catch declines (Lamberth *et al.* 1994, Sauer *et al.* 1997, Brouwer *et al.* 1997).

Nevertheless, the more sophisticated cluster analysis of the catch contribution by 180 species, using the Bray-Curtis measure of similarity, reveals some interesting comparisons (Fig. 16.3). Firstly, the catch composition of 'inshore' (shore-angling, seine and gillnet and estuary), 'offshore' (commercial and recreational boat) and spear fisheries were fairly dissimilar, indicating that the conflict between these major groups is largely a result of perceived rather than actual competition. Moreover, stock assessments reveal that South African recreational shore anglers have been directly responsible for the depletion of several species, including representatives of the Sparidae (Bennett 1993), Dichistiidae (Bennett 1988) and Sciaenidae (Griffiths 1997a).

Given that highest similarities in catch composition are found between boat-based commercial and recreational sectors (almost 100% in some regions), one would anticipate the greatest potential for recreational impact here. Indeed analysis of sectoral catch contribution reveals that recreational boat anglers have contributed substantially (20–80%) to the catches of at least nine collapsed stocks (Table 16.9), all of which occur in KwaZulu-Natal. The reason for this geographic phenomenon is that recreational effort, as a proportion of the total boat-based effort, increases steadily from west to east. Of the total number of boat days monitored, recreational vessels accounted for about 4% on the west coast, 6% in the Southern Cape, 42% in the Eastern Cape, 55% in the Transkei and 94% in KwaZulu-Natal (Sauer *et al.* 1997, Fennessy *et al.* 1999). This trend is largely a function of the fact that most recreationals along the south and west coasts obtained commercial permits (to subsidise their sport) when they were first issued in 1985, thereby becoming classified as part-time commercials (Griffiths 2000).

Table 16.8 Percentage catch contribution (by mass) of the ten most important species in the recreational shore, boat and spearfishing sectors and commercial boat and beach-seine and gill-net sectors. Included are the seven most important species for recreational angling and commercial gillnets in estuaries. Regions are: W = western Cape, S = southern Cape, E = eastern Cape, T = Transkei and K = KwaZulu-Natal.
Data sources: Baird and Pradervand 1999, Baird *et al.* 1996, Beckley and Fennessy 1996, Beckley *et al.* 2000, Brouwer 1996, Brouwer *et al.* 1997, Cowley 2000, Guastella 1994, Hutchings and Lamberth 1999, Kyle 1995, Kyle 2000, Kyle 1996, Kyle 1999, Lamberth 2000a, Lamberth 2000b, Lamberth *et al.* 1994, Lamberth 1996, Lamberth *et al.* 1997, Lamberth and Joubert 1999, Lechanteur 2000, Mann 1994, Mann 1995, Mann 1996, Mann *et al.* 1998, Mann *et al.* 1997, Mann *et al.* 1996, Marais and Baird 1980, McDonald *et al.* 1998, Pradervand and Baird in press, Sauer and Erasmus 1996, Sauer *et al.* 1997.

		Recreational shore angling					Recreational boat angling			
		W	S	E	T	K	W	S	E	K
Callorhynchidae	*Callorhynchus capensis*		0.1	<0.1			<0.1			
Carangidae	*Lichia amia*			0.6	0.5	0.7		<0.1	0.3	
	Seriola lalandii			0.1			1.1	20.3		
Dichistiidae	*Dichistius capensis*	29.8	13.7	1.1	1.0		<0.1	<0.1		
Gempylidae	*Thyrsites atun*						58.4	20.7		
Haemulidae	*Pomadasys commersonni*		0.3	0.7	4.1	1.1				0.1
	Pomadasys olivaceum		1.7	12.8	2.5	3.1		<0.1	<0.1	0.1
Merluccidae	*Merluccius capensis*						0.2	0.3	23.0	
Mugilidae	*Liza richardsonii*	9.5	1.6	5.8			<0.1	<0.1		
Oplegnathidae	*Oplegnathus conwayi*									
Pomatomidae	*Pomatomus saltatrix*	0.1	52.3	21.7	19.0	24.9	<0.1	1.3	0.4	1.2
Sciaenidae	*Argyrosomus inodorus*	1.7	4.6				<0.1	7.3	18.4	
	Argyrosomus japonicus			2.9	3.5	1.0				0.6
	Atractoscion aequidens						<0.1	9.1	13.4	3.3
	Umbrina spp.		1.4	0.2			<0.1	<0.1	<0.1	
Scombridae	*Scomberomorus commerson*									33.2
	Scomberomorus plurilineatus									2.0
	Thunnus alalunga						28.4	0.3		
	Thunnus albacares						0.7	0.4	10.4	18.9
Scorpidae	*Neoscorpis lithophilus*			1.6	14.5	4.3				
Sparidae	*Argyrozona argyrozona*						<0.1	14.4	2.2	
	Cheimerius nufar			0.7		0.1		0.2	3.2	2.4
	Chrysoblephus laticeps			0.2			<0.1	1.7	2.5	
	Chrysoblephus puniceus							<0.1		3.8
	Diplodus sargus	1.0	1.5	7.8	15.9	8.9	<0.1	<0.1	<0.1	<0.1
	Gymnocrotaphus curvidens		0.1	<0.1			<0.1	<0.1		
	Lithognathus lithognathus	0.5	4.7	1.4	0.3		<0.1	<0.1		
	Pachymetopon blochii	12.5	0.6				3.8	3.5		
	Pachymetopon grande			4.0	13.3	0.1				<0.1
	Sarpa salpa	0.3	3.3	23.7	10.9	44.3	<0.1	<0.1		
	Sparodon durbanensis		0.8	0.8	0.4			<0.1		

It is also perhaps noteworthy that most of the damage and stock collapse occurred during the 1960s and early 1970s, a time when there was no clear distinction between commercial and recreational fishers, and when most of the latter sold surplus catch (Griffiths 2000). Owing to the poor status of many linefish stocks, the fishery is now officially regarded as being in a state of emergency (*Government Gazette* No. 21949, Notice 4727 of 2000) in terms of section 16 of the Marine Living Resources Act. Consequences of this will be a dramatic reduction (66%) in the number of commercial participants and more stringent regulations for recreationals (Table 16.4). The 1994/5 ratios of commercial to recreational effort will

Table 16.8 *(Continued).*

		Recreational Spearfishing			Commercial boat fishing				
		W	**S&E**	**K**	**W**	**S**	**E**	**T**	**K**
Callorhynchidae	*Callorhynchus capensis*				<0.1		<0.1		
Carangidae	*Lichia amia*		0.6	4.7		<0.1	<0.1	<0.1	<0.1
	Seriola lalandii	4.3	1.9	0.3	1.1	20.3	1.1	2.1	0.4
Dichistiidae	*Dichistius capensis*	6.1	6.7	0.3	<0.1	<0.1	<0.1		<0.1
Gempylidae	*Thyrsites atun*				58.3	20.7	1.8	0.6	0.1
Haemulidae	*Pomadasys commersonni*						<0.1		0.1
	Pomadasys olivaceum					<0.1	<0.1		
Merluccidae	*Merluccius capensis*				0.2	0.3	30.2		
Mugilidae	*Liza richardsonii*				<0.1	<0.1			
Oplegnathidae	*Oplegnathus conwayi*		15.4	3.0			<0.1		
Pomatomidae	*Pomatomus saltatrix*				<0.1	1.3	0.2	<0.1	<0.1
Sciaenidae	*Argyrosomus inodorus*	<0.1			<0.1	7.3	7.4		
	Argyrosomus japonicus						7.4	8.0	6.0
	Atractoscion aequidens				<0.1	9.1	3.7	11.3	15.0
	Umbrina spp.		4.8	3.8	<0.1	<0.1	<0.1		<0.1
Scombridae	*Scomberomorus commerson*			17.0		<0.1		0.3	3.4
	Scomberomorus plurilineatus			31.4					0.5
	Thunnus alalunga				28.3	0.3	<0.1	0.1	<0.1
	Thunnus albacares				0.7	0.4	0.3	0.1	1.3
Scorpidae	*Neoscorpis lithophilus*								<0.1
Sparidae	*Argyrozona argyrozona*				<0.1	14.4	13.7	0.7	<0.1
	Cheimerius nufar					0.2	1.0	4.1	8.2
	Chrysoblephus laticeps	<0.1	19.3		<0.1	1.7	1.0	0.1	<0.1
	Chrysoblephus puniceus					<0.1	<0.1	22.1	31.1
	Diplodus sargus	0.1	1.0		<0.1	<0.1	<0.1		<0.1
	Gymnocrotaphus curvidens		10.6		<0.1	<0.1	<0.1		
	Lithognathus lithognathus	<0.1			<0.1	<0.1	<0.1		
	Pachymetopon blochii	88.1	6.7		3.8	3.5	<0.1		<0.1
	Pachymetopon grande		6.7	9.8			<0.1	0.3	0.1
	Sarpa salpa				<0.1	<0.1	<0.1		<0.1
	Sparodon durbanensis		2.9	2.8		<0.1	<0.1	<0.1	<0.1

therefore change substantially as excommercial fishers (predominantly part-time operators) enter the recreational sector (2001). In addition to this one-off boost, the recreational linefishing sectors are estimated to be growing at a rate of 2% per annum (McGrath *et al.* 1997). Considering the status of the linefish resource, the need for effective regulations for recreational fishers and good compliance has never been so great.

Apart from directly impacting on the resource, recreational boat-based fishers may also affect the livelihoods of their commercial counterparts by depressing the market price through the illegal sale of fish (Anon in *Fishing Industry News* 2000), restricting access to the sea by crowding slipways, directly competing for aggregated fish or causing shoals to sound or stop feeding.

Boat-based recreational anglers may be subdivided into those that operate further offshore targeting oceanic migrants – tuna and swordfish in the Cape regions and marlin in KwaZulu-Natal – and those that target both coastal migrants and bottom dwellers in the inshore environment. Having discussed the impacts of the latter group on the commercial sector, it is necessary to comment on the former as well. Although the commercial linefishery does not target

Table 16.8 *(Continued).*

		Commercial seine and gillnet			Recreational estuary angling				Commercial estuary gillnet		
		W	S	K	W	S	E	K	W	S	K
Callorhynchidae	*Callorhynchus capensis*	13.7	0.3								
Carangidae	*Lichia amia*	<0.1	<0.1	0.7		3.9	1.8		0.1	3.2	
	Seriola lalandii		15.1								
Dichistiidae	*Dichistius capensis*	1.0	<0.1								
Gempylidae	*Thyrsites atun*	0.1	<0.1								
Haemulidae	*Pomadasys commersonni*		<0.1	7.4		62.0	32.8	17.8		10.2	7.7
	Pomadasys olivaceum		<0.1	0.1							
Merluccidae	*Merluccius capensis*	1.0									
Mugilidae	*Liza richardsonii*	78.9	73.8		2.0	0.1			86.0	48.5	
Oplegnathidae	*Oplegnathus conwayi*										
Pomatomidae	*Pomatomus saltatrix*	1.0	0.9	0.9	92.1	0.1	0.7	0.6	8.0	0.3	
Sciaenidae	*Argyrosomus inodorus*	1.0	0.9								
	Argyrosomus japonicus		0.2	15.7		7.8	50.7	63.5		2.7	19.5
	Atractoscion aequidens	<0.1	<0.1								
	Umbrina spp.	<0.1	0.8								
Scombridae	*Scomberomorus commerson*										
	Scomberomorus plurilineatus			0.1							
	Thunnus alalunga										
	Thunnus albacares										
Scorpidae	*Neoscorpis lithophilus*										
Sparidae	*Argyrozona argyrozona*										
	Cheimerius nufar										
	Chrysoblephus laticeps										
	Chrysoblephus puniceus										
	Diplodus sargus	<0.1	0.2	0.1		0.7	0.1			0.2	
	Gymnocrotaphus curvidens										
	Lithognathus lithognathus	0.1	2.7		0.7	14.0	2.0		<0.1	1.8	
	Pachymetopon blochii	0.1									
	Pachymetopon grande										
	Sarpa salpa	0.1	0.2	0.1			<0.1			0.1	
	Sparodon durbanensis		<0.1			0.6	0.1			0.1	

billfishes, swordfish are caught by domestic longliners operating west of Cape Infanta while both swordfish and marlin are taken by foreign longliners in the South African EEZ, as their activities extend further north along the east coast. Eighty Japanese and 23 Taiwanese longliners had permits to operate within the South African EEZ in 2000; foreign vessels are, however, being gradually phased out as local expertise/capacity is developed.

South Africa is world renowned for its large swordfish which, until 1997, was prohibited from sale. In 1997 an experimental tuna longline fishery was initiated (Penney and Griffiths 1999). However, owing both to lack of fisher experience and relative abundance, swordfish 'bycatch' comprised 70% of the total landed catch up to January 1999 (Kroese 1999). Commensurate with this development there has been a dramatic drop-off in the catch rate of swordfish during an annual linefishing competition, the last two years producing no swordfish at all (Fig. 16.4). Not surprisingly this situation has led to considerable user conflict.

The South African sportfish sector targeting large gamefish is, owing to relatively small catches and to a catch-and-release policy, unlikely to have a substantial impact on the longline fishery. Approximately 37% of the 187 swordfish caught from 1990 to 1998 were released

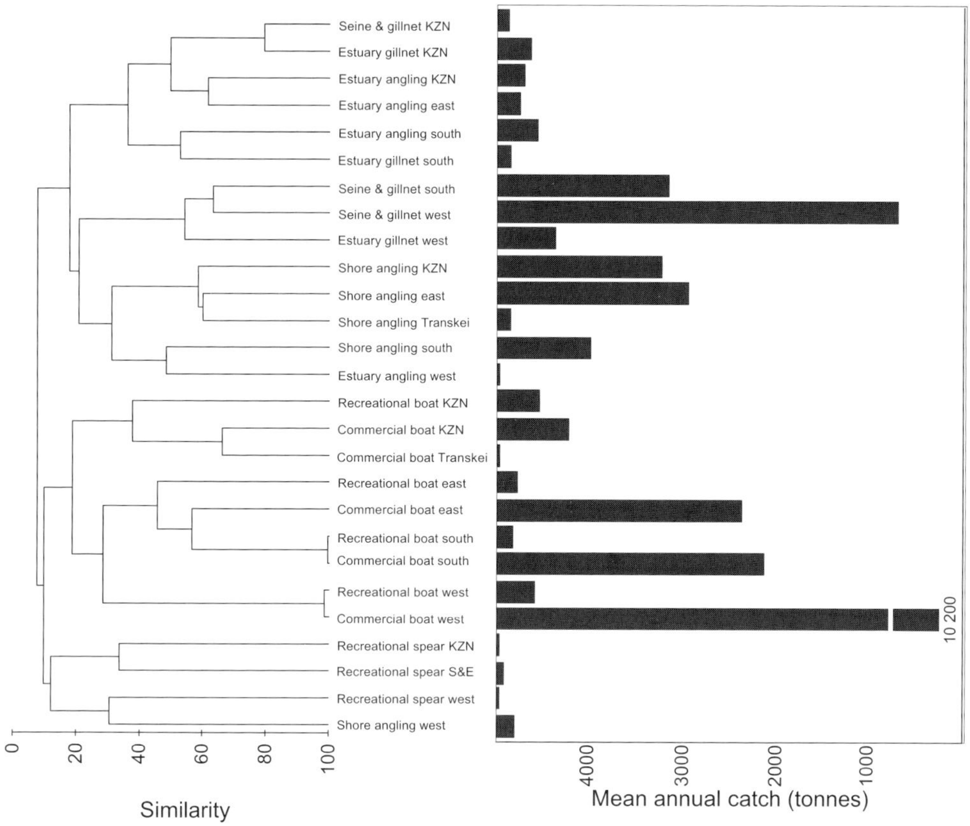

Fig. 16.3 Cluster analysis of the catch contribution by region and fishing sector using the Bray-Curtis measure of similarity, and the total catch estimated for each sector (bars). In the absence of an appropriate survey, the estuarine angling estimates of total catch are tentative.

(unpublished records of the South African Deep Sea Angling Association). Similarly 65% of marlin caught during 1999 were released (unpublished records of the Marlin Club of South Africa). The total longline catch (dressed weight) of swordfish recorded within the South African EEZ from January 1997 to December 2000 was 1670 tonnes (domestic = 781 tonnes and foreign = 889 tonnes) or 3200 individual swordfish (unpublished records, Marine and Coastal Management). Although a return to decommercialisation has been strongly mooted by both the local recreational sector and the international Broadbill Foundation, this is unlikely to occur for the following reasons: the swordfish bycatch is unavoidable and owing to its migratory nature, the South African stock is also targeted by foreign vessels outside of the EEZ. Apart from the Asian tuna fleets which frequent the southern hemisphere, European longliners have recently moved southwards following the over-exploitation of the northern Atlantic swordfish stock. There is as yet no international TAC for Indian Ocean swordfish. The South African experimental longline fleet is currently limited to a swordfish catch of 3500 tonnes (dressed weight): 1500 tonnes west of Cape Agulhas and 2000 tonnes east of this point.

Subsistence linefishers are difficult to identify as, with the exceptions of the traditional trap and spear fisheries in Kosi Bay (Fig. 16.1), they are dispersed within the recreational

Table 16.9 Percentage contribution of recreational and commercial fishing sectors to the total catches of 19 species regarded to be over-exploited or collapsed (see Table 16.4). Vulnerability characteristics are estuarine dependency (E), sex change (S), coastal migrations (C), relatively high age at maturity (M), longevity (L) and residency (R). In the absence of a suitable survey, estuarine catches were not included in the analysis. The recreational impact on estuarine dependant species such as *Argyrosomus japonicus* and *Lithognathus lithognathus* would therefore have been underestimated here.

Common name	Species	Vulnerability characteristics	Recreational shore	Recreational boat	Recreational spearfishing	Commercial boat	Commercial seine and gillnet
Dusky kob	*Argyrosomus japonicus*	E,C,M,L,R	27.7	0.8		58.7	12.8
Dageraad	*Chrysoblephus cristiceps*	S,C,M,L,R		17.9		82.1	
Geelbek	*Atractoscion aequidens*	C,M,L,R		11.6		88.3	0.1
Seventy-four	*Polysteganus undulosus*	S,C,M,L,R		44.3		55.7	
White steenbras	*Lithognathus lithognathus*	E,C,M,L,R	57.7	0.0	0.0	0.6	41.6
Silver kob	*Argyrosomus inodorus*	C,L	8.4	9.9	0.0	71.5	10.2
King soldierbream	*Argyrops spinifer*	S,C,M,L		73.4		26.6	
Englishman	*Chrysoblephus anglicus*	S,M,L		15.1	0.8	84.1	
False Englishman	*Chrysoblephus lophus*	C,M,L		32.8		67.2	
Slinger	*Chrysoblephus puniceus*	S,C,M,L,R		6.8		93.2	
White-edged rockcod	*Epinephelus albomarginatus*	S,M,L,R		38.9		61.1	
Red steenbras	*Petrus rupestris*	C,M,L,R		22.4	1.2	76.4	
Scotsman	*Polysteganus praeorbitalis*	S,M,L,R	8.0	24.4		67.7	
Catface rockcod	*Epinephelus andersoni*	S,M,L,R	7.8	82.1		10.1	
Poenskop	*Cymatoceps nasutus*	S,C,M,L,R	34.0	0.0	12.3	53.7	
Galjoen	*Dichistius capensis*	M,L,R	81.7	0.0	2.5	0.1	15.7
Squaretail kob	*Argyrosomus thorpei*	L,R	1.4	24.4		74.1	
Yellowbelly rockcod	*Epinephelus marginatus*	S,M,L,R	62.2	21.7		16.2	
Red stumpnose	*Chrysoblephus gibbiceps*	M,L,R		15.3	10.6	74.1	

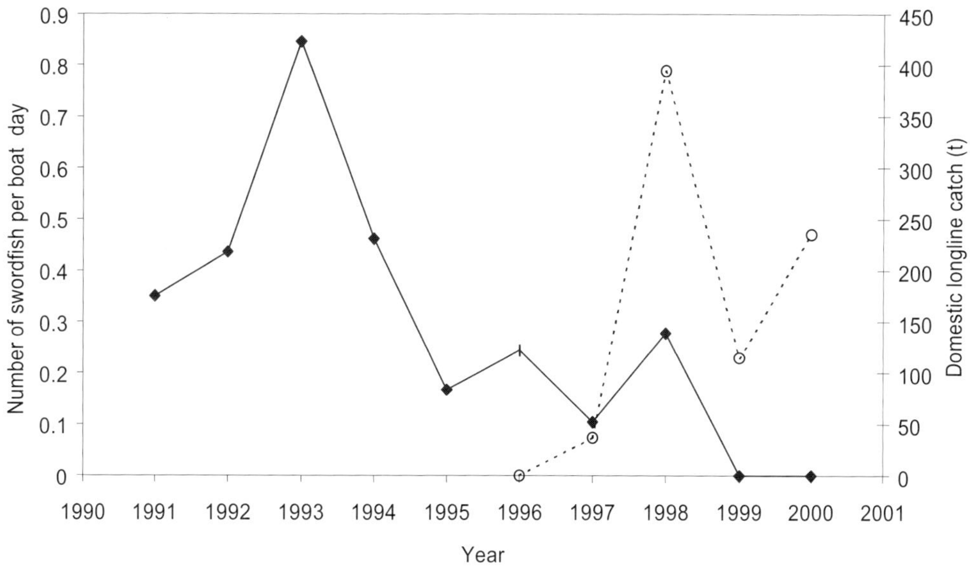

Fig. 16.4 Annual catch rates of swordfish by recreational anglers (diamonds) participating in the Gordons Bay Broadbill Classic (GBBC) angling competition, and catches made by the domestic longline fleet (circles) in the South African EEZ. The GBBC is fished in an area known as the Cape Canyon (see Fig. 16.1) in May of every year.

shore and estuary fisheries and amongst the poorer participants of the commercial line and net fisheries. As a result, subsistence permits are yet to be allocated for the linefish resource. At least 25 000 South African households in poverty (excluding the Transkei) rely on shore angling as a significant contribution to their protein supply (McGrath *et al.* 1997). A recent survey revealed that subsistence linefishing is largely restricted to shore-based and estuarine activities in the Transkei and KwaZulu-Natal. Thirty-one per cent of shore anglers interviewed in the former region fished for food and occasionally sold their catch (McDonald *et al.* 1999).

The existence of a subsistence component within the recreational shore fishery is further highlighted by the appearance of small-bodied species such as *Pomadasys olivaceum* and *Sarpa salpa* within the high priority list of this sector (Table 16.8). Although the catching of small species by subsistence fishers is expected to minimise conflict with recreational sportsfishers who target the larger species, subsistence fishers also retain the juveniles of these larger species, most of which are below the minimum legal size. This once again accentuates the need for improved enforcement and compliance. Subsistence linefishing permits will be allocated for the first time in 2002; and will be allocated only for estuarine and shore-based activity.

Ecological impacts

It is often contended that recreational angling sensitises the public to the environment and related issues such as pollution and habitat degradation (Kearney 1999). While this may be true, this component of the linefishery has been directly responsible for over-fishing several populations/species of inshore predator in South Africa, and has contributed to the demise of several others. Although there has been no empirical research on the ecosystems effects of linefishing in South Africa, most of the depleted species are important predators of both benthic and pelagic organisms (teleosts and large invertebrates) (Griffiths 2000) and their removal is likely to have precipitated substantial changes to the species diversity and the trophic pathways (of pelagic and demersal food webs), of shallow (< 100 m) subtidal ecosystems. Surveys in the southern Cape revealed that not only were predators more abundant in a marine reserve (Buxton and Smale 1989), but there were also differences in the familial composition of cryptic fishes on exploited and unexploited reefs (Burger 1990). The exploitation of predators of large invertebrates (including a large sparid fish) was recently shown to change the community structure and reduce both primary and secondary production in a temperate-reef ecosystem off New Zealand (Babcock *et al.* 2000). Bennett and Attwood (1989) showed recovery of several South African shore-angling species following the establishment of a marine reserve, a no-take MPA within which commercial and recreational exploitation is prohibited by legislation.

Sardine, *Sardinops sajax*, and anchovy, *Engraulis japonicus*, are important food of pelagic and several reef associated linefishes (Griffiths 2000). These two forage species spawn on the Agulhas Bank and utilise the Benguela current to transport eggs and larvae onto nutrient-rich West Coast nursery areas (Shelton and Hutchings 1990, Armstrong *et al.* 1987, Hampton 1992, Barange *et al.* 1998, 1999). Thereafter the juvenile sardine and anchovy return to the Agulhas Bank, where they were spawned; biologically pumping energy/carbon from the highly productive west coast upwelling system onto the eastern seaboard. This process presumably allows warm-temperate reef-ecosystems to support larger shoals of piscivores than the reefs *per se* could sustain. It is therefore not incidental that the most targeted reef-associated species have been those that feed on clupeoids, i.e. carpenter, silver kob, geelbek and seventy-four. The extent to which these linefishes may have fertilised reefs (nitrogen excretion and faecal pellets) is unclear, but it is possible that their demise has concomitantly reduced the links of reef ecosystems with the pelagic foodweb. Additional research, particularly ecosystem modelling, is deemed necessary to provide clearer understanding of ecological impacts.

Conclusions

South Africa has a large, multisectoral marine recreational linefishery with almost half a million participants. Although the commercial sector is responsible for 79% of the estimated total catch, the recreational component provides 81% of the employment [about 1.5% of the adult population of South Africa – eds] and generates 82% of the revenue. Appreciation of the socioeconomic value of marine recreational angling, and the fact that the fishery now generates substantial revenue through the new permit system (an anticipated R16m based on current participation), will hopefully secure the funding and research attention that it deserves.

On the downside recreational angling in South Africa has, as a result of both poor compliance and an inadequate management framework, directly resulted in the over-fishing of several populations/species of inshore predator and has contributed to the demise of several others. Not only are catch rates now well below optimal sustainable levels, but ecosystem effects are also anticipated. Although the implementation of the new management protocol developed for the linefishery should result in more appropriate regulations for recreational anglers, surveys have shown that dramatically improved enforcement and compliance are prerequisites to effective stock recovery programmes. Whilst initiatives to improve monitoring and compliance are encouraging/laudable, the net effect of these is yet to be evaluated.

In addition, modelling the ecosystem effects of linefishing is required. In order to further appreciate the value of recreational angling in South Africa, it is also necessary to balance the costs of management against revenue generated (including permit fees), so that this can be compared with the balance sheet for the commercial component.

Acknowledgements

We thank Paul Cowley of the J.L.B. Smith Institute of Ichthyology in Grahamstown and Bruce Mann of the Oceanographic Research Institute, Durban for unpublished information on the catch composition of anglers in Eastern Cape and KwaZulu-Natal estuaries; the South African Deep Sea Angling Association and the Gordons Bay Boat Angling Club for billfish recreational catch statistics; and Andy Cockcroft, Colin Attwood and two anonymous reviewers for constructive comment on earlier drafts of the manuscript.

References

Ahrens, R.L. (1964) A preliminary report on the biology of the seventy-four *Polysteganus undulosus* (Regan) in Natal waters. MSc thesis, University of Natal, Durban,

Anon (2000) Linefishers call for intervention. *Fishing Industry News*, Southern Africa. December 2000, **6.**

Armstrong, M.J., Berruti, A. & Colclough, J. (1987) Pilchard distribution in South African waters, 1983–1985. *South African Journal of Marine Science* **5**: 871–886.

Attwood, C.G. & Bennett, B.A. (1995) A procedure for setting daily bag limits on the recreational shore-fishery of the south-western Cape, South Africa. *South African Journal of Marine Science* **15**, 241–251.

Babcock, R.C., Kelly, S., Shears, N.T., Walker, J.W. & Willis, T.J. (2000) Changes in community structure in temperate marine reserves. *Marine Ecology Progress Series* **198**, 125–134.

Baird, D. & Pradervand, P. (1999) Assessment of the recreational angling fishery in Eastern Cape estuaries. In: *Proceedings of the Third South African Marine Linefish Symposium. South African Network for Coastal and Oceanic Research, Occasional Report* 5 (ed B.Q. Mann), pp. 73–76.

Baird, D. Marais, J.F.K. & Daniel, C. (1996) Exploitation and conservation of angling fish species in selected South African estuaries. *Aquatic Conservation: Marine and fresh water ecosystems* **6(4)**, 319–330.

Barange, M., Pillar, S.C. & Hampton, I. (1998) Distribution patterns, stock size and life-history strategies of Cape horse mackerel *Trachurus trachurus capensis*, based on bottom trawl and acoustic surveys. *South African Journal of Marine Science* **19**, 433–447.

Barange, M., Hampton, I. & Roel, B.A. (1999) Acoustic surveys of anchovy and sardine on the South African continental shelf in the 1990s. *South African Journal of Marine Science* **21**, 367–391.

Beckley, L.E. & Fennessy, S.T. (1996) The beach-seine fishery off Durban, KwaZulu-Natal. *South African Journal of Zoology* **31**(4), 186–192.

Beckley, L.E., Mann, B.Q., Robertson, W.D., Fielding, P.J. & van der Elst, R.P. (2000) *Subsistence fishery resource assessment: Port Edward to Kosi Bay*. SA Association for Marine Biological Research Unpublished Report No. 193.

Bennett, B.A. (1988) Some considerations for the management in South Africa of galjoen, *Coracinus capensis* (Cuvier), an important shore-angling species off the south-western Cape. *South African Journal of Marine Science* **6**, 133–142.

Bennett, B.A. (1993) The fishery for white steenbras *Lithognathus lithognathus* off the Cape coast, South Africa, with some consideration for its management. *South African Journal of Marine Science* **13**, 1–14.

Bennett, B.A. & Attwood, C.G. (1993) Shore-angling catches in the De Hoop nature reserve, South Africa, and further evidence for the protective value of marine reserves. *South African Journal of Marine Science* **13**, 213–222.

Brouwer, S.L. (1996) *Evaluation of participation in, and management of, the South African Linefishery in the south-east Cape*. Final Report to the Sea Fisheries Research Institute, Department of Environment Affairs and Tourism, South Africa.

Brouwer, S.L., Mann, B.Q., Lamberth, S.J., Sauer, W.H.H. & Erasmus, C. (1997) A survey of the South African shore-angling fishery. *South African Journal of Marine Science* **18**, 165–178.

Burger, L.F. (1990) *The distribution patterns and community structure of the Tsitsikamma rocky littoral ichthyofauna.* MSc Thesis, Rhodes University.

Buxton, C.D. & Smale, M.J. (1989) Abundance and distribution patterns of three temperate marine reef fish (Teleostei: Sparidae) in exploited and unexploited areas off the southern Cape coast. *Journal of Applied Ecology* **26**, 441–451.

Cowley, P.D. (2000) *Utilisation of living marine resources in the estuaries between Port Elizabeth and East London, Eastern Cape Province*. Unpublished report, JLB Smith Institute of Ichthyology, Grahamstown.

Fennessey, S.T., McDonald, A. & Mann, B.Q. (1999) The Transkei skiboat fishery. In: *Proceedings of the third Southern African marine linefish symposium, Arniston, 28 April-1 May 1999.* (ed B.Q. Mann) *South African Network for Coastal and Oceanic Research, Occasional Report* 5, 63–68.

Gore, R. (2000) People like us. *National Geographic* **198**(1), 91–117.

Griffiths, M.H. (1997a) Management of the South African dusky kob *Argyrosomus japonicus* (Sciaenidae) based on per-recruit models. *South African Journal of Marine Science* **18**, 213–228.

Griffiths, M.H. (1997b) The application of per-recruit models to *Argyrosomus indorus*, an important South African sciaenid fish. *Fisheries Research* **30**, 103–115.

Griffiths, M.H. (1997c) Towards a management plan for the South African linefishery: objectives and strategies. In: *Management and monitoring of the South African marine linefishery* (Proceedings of a linefish workshop held at the Tsitsikamma National Park, 28–29 July 1997) (eds A.J. Penney, M.H. Griffiths & C.G. Attwood). *South African Network for Coastal and Oceanic Research, Occasional Report* 5, 3–14.

Griffiths, M.H. (2000) Long-term trends in catch and effort of commercial linefish off South Africa's Cape Province: snapshots of the 20th Century. *South African Journal of Marine Science* **22**, 81–110.

Griffiths, M.H., Attwood, C. & Thompson, R. (1999) New management protocol for the South African linefishery. In: *Proceedings of the third Southern African marine linefish symposium, Arniston, 28 April–1 May 1999* (ed B.Q. Mann). *South African Network for Coastal and Oceanic Research, Occasional Report* 5, 145–146.

Guastella, L.A.-M. (1994) A quantitative assessment of recreational angling in Durban harbour, South Africa. *South African Journal of Marine Science* **14**, 187–203.

Hampton, I. (1992) The role of acoustic surveys in the assessment of pelagic fish resources on the South African continental shelf. *South African Journal of Marine Science* **12**, 1031–1050.

Hemming, B. & Pierce, B.E. (1997) Fisheries enforcement: our last fisheries management frontier. In: *Developing and sustaining world fisheries resources, the state of science and management.* (Proceedings of the 2nd world fisheries congress, Developing and sustaining world fisheries resources, the state of science and management, Brisbane Australia, 28 July – 2 August 1996) (eds D.A. Hancock, D.C. Smith, A. Grant and J.P. Beumer), pp. 675–678. CSIRO, Australia.

Horne, C. (1974) *Game Fishing Transformed.* Don Nelson, Cape Town, 163pp.

Hutchings, K. &. Lamberth, S.J. (1999) *Management and socio-economic considerations for the drift and beach-seine net fisheries on the West and southwestern coasts of South Africa.* Final Report for Marine and Coastal Management, Department of Environmental Affairs and Tourism, South Africa.

Hutton, T., Griffiths, M.H., Sumaila, U.R. & Pitcher, T.J. (2001) Co-operative versus non-cooperative management of shared linefish stocks in South Africa: an assessment of alternative management strategies for geelbek (*Atractoscion aequidens*, Sciaenidae). *Fisheries Research* (in press).

Kearney, B. (1999) Evaluating recreational fishing: managing perceptions and/or reality. In: *Evaluating the benefits of recreational fisheries* (Papers, Discussion and Issues: a Conference held at the UBC Fisheries Center, June 1999) (ed T.J. Pitcher). *Fisheries Center Research Reports* **7**(2), 9–14.

Kroese, M. (1999) An overview of swordfish catches in the South African experimental pelagic longline fishery with emphasis on the South-Western Atlantic Ocean. In: *Proceedings of the third Southern African marine linefish symposium, Arniston, 28 April–1 May 1999* (ed B.Q. Mann). *South African Network for Coastal and Oceanic Research, Occasional Report* 5, 88–90.

Kyle, R. (1995) Wise use of wetlands by rural indigenous people. The Kosi Bay Nature Reserve: a case study. In: *Wetlands of South Africa* (ed Gowan, G.I.). pp. 273–291. Department of Environment Affairs and Tourism, Pretoria.

Kyle, R. (1996) *Kosi Bay Nhlange gillnetting project. A comprehensive report on the results for 1993, 1994 and 1995 and a general discussion on the project.* Unpublished report, KwaZulu Department of Nature Conservation, Pietermaritzburg.

Kyle, R. (1999) Gillnetting in nature reserves: a case study from the Kosi Lakes, South Africa. *Biological Conservation* **88,** 183–192.

Kyle, R. (2000) *Report on Estuarine gillnetting.* Report to the Subsistence Fisheries Task Group, Marine and Coastal Management, 7pp.

Lamberth, S.J. (1996) *The distribution of total catch and effort between all sectors of the linefishery in the southwestern Cape.* (Regional program to evaluate participation in, and management of, the linefishery in the SW Cape, SE Cape and KwaZulu-Natal). Final Report to the Sea Fisheries Research Institute.

Lamberth, S.J. (2000a) *Current fishing activities in estuaries from the Orange River to Cape Point.* Report to the Subsistence Fisheries Task Group, Marine and Coastal Management.

Lamberth, S.J. (2000b) *Current fishing activities in estuaries from Cape Point to Mossel Bay.* Report to the Subsistence Fisheries Task Group, Marine and Coastal Management,

Lamberth, S.J., Bennett, B.A. & Clark, B.M. (1994) Catch composition of the commercial beach-seine fishery in False Bay, South Africa. *South African Journal of Marine Science* **14,** 69–78.

Lamberth, S.J, Sauer, W.H.H., Mann, B.Q., Brouwer, S.L., Clark, B.M. & Erasmus, C. (1997) The status of the South African beach-seine and gill-net fisheries. *South African Journal of Marine Science* **18**, 195–202.

Lamberth, S.J. & Joubert, A. (1999) Prioritizing linefish species for research and management: a first attempt. In: *Proceedings of the third Southern African marine linefish symposium, Arniston, 28 April–1 May 1999* (ed B.Q. Mann). *South African Network for Coastal and Oceanic Research, Occasional Report* 5, 130–137.

Lechanteur, Y.A.R.G. (2000) The ecology and management of reef fishes in False Bay, southwestern Cape, South Africa. PhD thesis. University of Cape Town, South Africa,

Mann, B.Q. (1994) *Multiple usage of St Lucia's fish resources*. Unpublished report to Southern African Nature Foundation.

Mann, B.Q. (1995) Quantification of illicit fish harvesting in the Lake St Lucia Game Reserve, South Africa. *Biological Conservation* **74**, 107–113.

Mann, B.Q. (1996) *Implementation and assessment of the Lake St Lucia subsistence gill net fishery: a case study*. Unpublished Report Oceanographic Research Institute 128, 1–14.

Mann, B.Q. (2000) South African marine linefish status reports. *Oceanographic Research Institute, South Africa, Special Publication* 7.

Mann, B.Q., Fennessy, S.T. & Beckley, L.E. (1998) *Evaluation of linefishery participation and management in the Transkei region of the eastern Cape (Port Edward to Coffee Bay)*. Final report for the Sea Fisheries Research Institute, Department of Environment Affairs and Tourism, South Africa.

Mann, B.Q., Scott, G.M., Mann-Lang, J.B. *et al.* (1997) An evaluation of participation in and management of the South African spearfishery. *South African Journal of Marine Science* **18**, 179–191.

Mann, B.Q., Beckley, L.E. & van der Elst, R.P. (1996) *Evaluation of linefishery participation and management along the KwaZulu-Natal coast*. Unpublished Report Oceanographic Research Institute **134**, 1–17.

Mann-Lang, J.B. (1996) *National Marine Linefish System recreational data – a critical review*. Unpublished Report, Oceanographic Research Institute, South Africa 136.

Marais, J.F.K. & Baird, D. (1980) Analysis of anglers' catches in the Swartkops estuary. *South African Journal of Zoology* **15**, 61–65.

McDonald, A., Sauer, W.H.H., Mann, B.Q. & Hecht, T. (1998) *Evaluation of the linefishery in the Transkei region of the eastern Cape (Kei Mouth to Coffee Bay)*. Final report for the Sea Fisheries Research Institute, Department of Environment Affairs and Tourism, South Africa.

McDonald, A., Mann, B.Q. & Sauer, W.H.H. (1999) Evaluation of the shore fishery in the Transkei region of the Eastern Cape (Kei Mouth to Port Edward). In: *Proceedings of the third Southern African marine linefish symposium, Arniston, 28 April–1 May 1999* (B.Q. Mann ed.) *South African Network for Coastal and Oceanic Research, Occasional Report* 5, 69–72.

McGrath, M.D., Horner, C.C.M., Brouwer, S.L., *et al.* (1997) An economic valuation of the South African linefishery. *South African Journal of Marine Science* **18**, 203–211.

Muller, C.F.J. (1997) South African fisheries through the centuries. *Fishing Industry Handbook, South Africa, Namibia and Mocambique* **25**, 204–215.

Nepgen, C.S. De V. (1977) The biology of the hottentot *Pachymetopon blochii* (val.) and the silverfish *Argyrozona argyrozona* along the Cape south-west coast. *Investigational Reports of the Sea Fisheries Branch of South Africa* 105.

Pauly, D. (1995) Anecdotes and the shifting baseline syndrome in fisheries. *Trends in Ecology and Evolution* **10**, 430.

Penney, A.J. (1997) The National Marine Linefish System: a decade in review. In: *Management and monitoring of the South African marine linefishery* (Proceedings of a linefish workshop held at the Tsitsikamma National Park, 28–29 July 1997) (eds A.J. Penney, M.H. Griffiths and C.G. Attwood), pp. 23–50. *South African Network for Coastal and Oceanic Research, Occasional Report* 5.

Penney, A.J. & Griffiths, M.H. (1999) A first description of the developing South African pelagic longline fishery. (SCRS/98/144) *ICCAT Coll. Vol. Sci. Pap.* **49**(1), 162–173.

Pollock, K.H., Jones, C.M. & Brown, T.L. (1994) Angler survey methods and their application in fisheries management. *Am. Fish. Soc. Symp. 25*: 3.

Pradervand, P. & Baird, D. (in press) The Eastern Cape estuarine linefishery: trends in catch and effort. *South African Journal of Marine Science.*

Sauer, W.H.H. & Erasmus, C. (1996) *Evaluation of the line and net fisheries along the West Coast of South Africa.* Final report for the Sea Fisheries Research Institute, Department of Environment Affairs and Tourism, South Africa.

Sauer, W.H.H., Penney, A.J., Erasmus, C., *et al.* (1997) An evaluation of attitudes and responses to monitoring and management measures for the South African boat-based linefishery. *South African Journal of Marine Science* **18**, 147–163.

Sen, S. & Nielsen, J.R. (1996) Fisheries co-management: a comparative analysis. *Marine Policy* **20**, 405–418.

Shelton, P.A. & Hutchings, L. (1990) Ocean stability and anchovy spawning in the southern Benguela Current region. *Fishery Bulletin US*, 88(2), 323–38

Taylor, V. (1999) Coastal trusts as a possible solution to the conservation of our coastal resources. In: *Proceedings of the third Southern African marine linefish symposium, Arniston, 28 April–1 May 1999* (eds B.Q. Mann). pp. 119–125. *South African Network for Coastal and Oceanic Research, Occasional Report* 5.

Thompson, W.W. (1913) *The sea fisheries of the Cape Colony, from Van Riebeeck's day to the eve of the Union, with a chapter on trout and other freshwater fishes.* Maskew Miller, Cape Town.

Van der Elst, R.P. (1989) Marine recreational angling in South Africa. In: *Oceans of Life off Southern Africa* (eds A.I.L. Payne & R.J.M. Crawford), pp. 164–176.Vlaeberg Publishers, Cape Town.

Van der Elst, R.P. & Adkin, F. (1991) Marine linefish: Priority species and research objectives in southern Africa. *Oceanographic Research Institute Special Publication, South Africa* **1**, 1–132.

Chapter 17

The Next Chapter: Multicultural and Cross-Disciplinary Progress in Evaluating Recreational Fisheries

Øystein Aas

Abstract

Recreational fishing is becoming increasingly important world-wide, and there is a growing demand for multidisciplinary research aimed at assisting science-based fisheries management. This book presents different theoretical perspectives and empirical data on recreational fisheries, and represents one of the first serious attempts to gather such research from different continents and cultures. In this closing comment I discuss some challenges for such research, including:

- Cultural and regional differences in practices, resources and knowledge
- The reductionist tradition, which is reflected throughout science-based natural resources management
- Research biased by concentrating on salmonids and other high-profile fishes
- Poorly developed scholarly communication between Europe and North America

Such challenges – together with other trends such as broader involvement in the decision-making process, increased demand for high-quality research and more sophisticated methods – must be addressed by researchers and managers in the fisheries community if progress is to be made. Future themes for research include provision of high-quality fishing experiences, the effects of increased tourism, conflict among user groups, and recreational fishing by children, elderly persons, indigenous people, and women.

Key words: cultural differences, multidisciplinary research, recreational fisheries, research strategies.

Introduction

Recreational fisheries are becoming increasingly important around the world, primarily in developed countries, but also in developing countries. The growth in developing nations is not only a result of increased interest in tourism, but is also an emerging and growing leisure activity for residents in these countries. Both within and among regions, world fisheries are characterised by immense biological, geographical, economic, social, political and cultural heterogeneity.

Parallel with the growth of recreational fishing as sport in western countries from the beginning of the 1800s (Lowerson 1989), research into different aspects of recreational fisheries has gradually grown and evolved over two centuries. Growth in reported research has been exponential from the 1960s onwards (Socioeconomics Section, American Fisheries Society 1993). At the beginning of the third century in the modern history of recreational fishing, the state-of-the-art includes three major applied research directions each of which has the goal of enhancing the quality of recreational fisheries:

- Fisheries biology and ecology (Siegler and Siegler 1990)
- Fisheries economics (Anderson 1986)
- Social aspects of recreational fisheries (Ditton 1996)

In each of these fields, but especially in the biological sciences, research has become increasingly specialised, resulting in many new areas of inquiries and subdirections.

This book focuses on the different values, including benefits and losses, associated with recreational fishing. The primary focus is on contributions founded in fisheries economics and in human dimensions. However, some of the studies focus on ecological aspects and on new management instruments, while some also bring results from creel surveys, or 'surrogate biology' (Brown 1987). However, much research on sport fisheries is necessarily multidisciplinary, and attempts to cross the gulf between the biological and social sciences (Aas and Ditton 1998).

This chapter will try to summarise some of the issues raised in the book and to identify areas that must be addressed if progress is to be made in science-based recreational fisheries management. Furthermore, some ideas for future research needs will be suggested. The author would like to emphasise his background in the social aspects of recreational fisheries.

Understanding cultural differences in recreational fisheries on a global scale

A major contribution of this compilation is its attempt to gather research into recreational fisheries from different continents, cultures and different research traditions. Several contributions from nonwestern countries focus on the need to assess recreational, subsistence and commercial fishing as a whole, whether for ecological, cultural, social or economic reasons (Griffiths and Lamberth 2002, Sumaila 2002), while other authors present new and maybe provocative developments in the management of angling in western countries, for instance Policansky (2002) in his discussion of catch-and-release. This difference in focus illustrates how greatly the context and the environment for recreational fisheries may vary.

The challenges in Policansky's (2002) and Cox and Walters' (2002) setting, north-western North American coldwater recreational fisheries, and in the multiple-use marine fisheries in Namibia and South Africa described by Sumaila (2002) and Griffiths and Lamberth (2002), will obviously differ. Developing countries in Africa and Asia might face serious challenges related to lack of appropriate legal instruments, the early phase of establishing management institutions and lack of knowledge in many areas. Major issues might include how to develop recreational fisheries while at the same time enhancing the role of fisheries in food provision, combating severe poaching or establishing sustainable industries without harming biological resources. In western countries, challenges might relate to conflicts between anglers and animal rights groups (Aas *et al.* 2002, Policansky 2002), postindustrial river restoration issues and harvest regulations so strong that consumption of fish is impossible. Obviously, the issues faced by politicians, managers and researchers will differ around the world. However, the task for research in the area of recreational fisheries remains much the same, regardless of geographical location and issue at hand, namely to guide management or policy decisions and to provide for 'optimal' solutions in each given situation.

The matter of definitions

Many definitions exist for recreational fishing, and applying one, generic definition in these very different settings and contexts might be difficult, and maybe not a constructive task at all. Definitions might have different purposes, as some might be for legal uses, some for research, and others for local management purposes. In addition, the surroundings and the culture in which the activity takes place may differ. Most of the definitions of recreational fishing focus on the type of fishing activity or fishing gear used. Very often, recreational or sport fisheries are synonymous with angling, meaning fishing with rod, line and hook. However, it is widely accepted that recreational fishing includes the use of other fishing methods as well (Hickley 1998).

A scientific definition of recreational fishing originating in the leisure sciences is 'any form of fishing done during free time (as opposed to working time), that is subjectively defined by the individual as being leisure' (cf. Neulinger 1974, Iso-Ahola 1980). Such a definition of recreational fishing has its basis in the western tradition, which has drawn a strict distinction between work and leisure since the industrial revolution. Therefore, this definition soon becomes problematic in cultures with another cultural or economic framework, for instance, in many developing countries, among people working off the land (farmers, foresters), and among indigenous people (Booth and Kessler 1996). Such a definition also might become increasingly problematic in postindustrial western countries if the distinction between work and leisure is reduced (Sylvester 1999). The term 'subsistence' fishing is often used to describe a third type of fishery that is not commercial and not recreational, but where fishing is used as a means to supplement incomes and direct consumption of food (Glass *et al.* 1995). In many areas the distinction between subsistence and recreation is unclear, and especially in developing countries and rural districts it is necessary to consider subsistence and recreational fishing together (Aas and Skurdal 1996, Duffield *et al.* 2002).

In other areas where native subsistence fisheries and nonresident tourist anglers utilise the same resource, we have sometimes seen severe conflicts, some including the use of violence (Lyman 1999). Such conflicts may often introduce many elements, among them cultural, economic and access issues.

This issue of definitions illustrates the importance of developing and testing models and tools aimed at assisting a fair allocation of fishery resources among recreational, subsistence and commercial users (see Sumaila 2002), as well as the importance of research into cultural and motivational differences between stakeholders and user groups (Aas *et al.* 2002, Calvert 2002).

Our biased research tradition

The short history of scientific fisheries management has not differed from that of natural resources management in general. In the western world, there is growing acceptance of the view that natural resource management traditionally has been dominated by a reductionist, 'maximum sustainable yield' paradigm, suggesting that humans can improve the performance of forests, rangelands, rivers and lakes (Booth and Kessler 1996). In such a view, living systems are orderly, predictable and exhibit properties that are nothing more than the sum of their individual parts. According to Bottom (1997), the history of modern fisheries conservation, and especially hatcheries, is an excellent example of the strong agricultural philosophy embedded in resource management from the 1800s. Such a philosophy opposes much of what we now know to be central characteristics of living ecosystems. Ecosystems are extremely complex, dynamic, stochastic and exhibit large variations in production and annual 'yield'. Yet they seem remarkably resilient to perturbation, and they cannot be explained as the sum of single components.

The gap between the agricultural tradition and new ecological insights continues to challenge the field of recreational fisheries management, as it challenges other areas of natural resources management. In the areas of economic and human dimensions research dominating this book, it is easy to identify the kinship to the reductionist research tradition most often associated with natural sciences. Psychology-inspired researchers split up the holistic, complex recreational fishing experience into single pieces labelled escape, relaxation, nature experiences, companionship, catching fish, and challenge/competition (Fedler and Ditton 1994, Calvert 2002). Economists put a single monetary value on the fishing trip of a lifetime, which to the subject may be immaterial (Rudd *et al.* 2002, Duffield *et al.* 2002). Therefore, the western, mechanistic view still dominates much of the socioeconomic research into recreational fisheries. This is a challenge to most of the human dimensions and the economic research in recreational fisheries. For instance, the stochastic, often strong variation in resources between seasons, must be better addressed in economic models.

A critical view of the traditions of scientific fisheries management and research can also be stimulated by unprejudiced and reflective discussions about the definitions of recreational and subsistence fisheries. Considering the perspectives of indigenous people and their world views is also valuable, both to the discussions on the definitions issue, the tradition/paradigm issue and to the evolving ecosystems paradigm in resource management. Generally, there is acceptance that indigenous people set humans within the context of land over time to a greater extent than in the traditional western technological-reductionist view (Booth and Kessler 1996). The discussion above also illustrates the need for reflection among fisheries experts about their world view. Different research traditions exist in the natural sciences, in social sciences and in the humanities. The quantitative versus the qualitative methods approach (Smith 1983), the reductionist versus holistic view, discussions about what are data and what are results, all illustrate the need among experts in this area to be conscious of the different traditions that exist between disciplines that might meet when investigating recreational fisheries. Adopting humanistic research traditions, with their different scholarly traditions, can only strengthen and supplement existing socioeconomic research (Driver *et al.* 1999).

Improving scholarly communication between Europe and North America

Much research in recreational fisheries has had a strong bias towards North American and European problems and perspectives (Orbach 1980). It is from these geographic areas that most existing knowledge originates. Research in 'new geographical areas' must build upon previous research in recreational fisheries, and at the same time acknowledge the specific cultural, political and natural setting and surroundings for their study area. Since the evaluation of recreational fisheries covers many different disciplines, and can benefit from applied research in many other topics (i.e. leisure and recreation, tourism, ecosystems management), the task of being aware of the state-of-the-art can be a challenging one.

Despite the huge challenges to communicating on a global scale, there are severe problems facing communication between western countries, even those who all use the English language. This book confirms that the situation in Europe and North America regarding management of, and research into, recreational fisheries, differs in several ways. Participation in recreational fisheries varies strongly among countries in the western world (Pinter and Wolos 1998, Steffens and Winkel 2002, Lyons *et al.* 2002, Gentner and Lowther 2002, Toivonen 2002). Rates of participation are much higher in Scandinavia and North America (20% and higher) than in mainland Europe and Great Britain (normally less than 10% of the adult population). This difference exists despite variations within each group in the extent to which fisheries in each country depend on marine or freshwater resources. Some observers believe there exists a strong difference between the two groups in that central European recreational fisheries are very specialised and partly oriented towards competitions (Lowerson 1989), while North American and Scandinavian recreational fisheries constitute a much more family-oriented activity.

It is also obvious that the amount, type and specialisation in research on recreational fisheries differs between Europe and North America. North American fisheries managers and researchers seem to have a somewhat longer tradition in research aimed at monitoring and valuing recreational fisheries than Europe (see for instance, Gentner and Lowther 2002, Thunberg and Milon 2002). As a consequence, the empirical basis seems to be far more complete and developed on the western side of 'the Atlantic divide'. However, many research efforts in Europe seem to lack information about the body of knowledge in North America (Aas and Ditton 1998), and do not always build upon this rich empirical literature, which is of much value if one country plans to increase its knowledge of angling. On the other side, from a quick look at North American social and economic scientists' references, they generally lack knowledge of the few but relevant published works in Europe. Essentially, much economic and social research in recreational fisheries appears parochial, with a strong national orientation and relatively small frames of reference in terms of theory, concepts and empirical bases (Beckers 1995).

This lack of communication and transnational exchange of theory and empirical findings in recreational fisheries research is an indicator of the unfinished process of building an interdisciplinary and scholarly field with normal standards of scientific communication. In North America, there exists a significant body of research both in the fields of economics and in social aspects: social research is clearly influenced by a psychological approach, focusing on human demand, attitudes, motives and preferences (Ditton 1996). In contrast, European research is far more fragmented, but is dominated by research in the economic, institutional and legal aspects of recreational fisheries management (Beckers 1995, Aas and Ditton 1998). It appears difficult to gain access to much European research on recreational fisheries, characterised by producing much 'grey literature' inconvenient for colleagues abroad (see for instance the use of references in Hickley and Tompkins 1998). Much has been said about grey literature in science, and not much is positive (Collette 1990). Moreover, lack of adequate review can reduce the quality and reputation of this scientific area as a whole.

Both on the European and North American side we find widespread, applied journals that specialise in presenting research with relevance for scientific management of recreational fisheries. Good examples are *North American Journal of Fisheries Management* and *Fisheries Management and Ecology* (formerly *Aquaculture and Fisheries Management*). They both have editorial policies inviting scientists to publish papers on evaluating recreational fisheries from different perspectives, including research based in economic or social sciences. Despite previous critical comments on fishery professionals for being biased towards biological sciences (Voiland and Duttweiler 1984), both journals, but especially *North American Journal of Fisheries Management,* already have a long history and experience in publishing articles in research areas other than fishery biology. Consequently, a major challenge to those researchers aiming at broadening the scientific perspectives on fisheries management is to publish more of their research in a way that makes it more accessible to colleagues, and which has gone through a process of peer-review. This will increase the attention paid to the human dimensions of fisheries, underlining the importance of human needs, values, and perceptions in resource management.

Few relevant international organisations exist that aim at improving scholarly, scientific communication in recreational fisheries issues. Starting with a European conference in Dublin in 1996, supported by the European Inland Fisheries Advisory Commission (EIFAC: Hickley and Tompkins 1998), we have lately seen the establishment of a world conference on recreational fisheries. The first world conference was held in Vancouver in 1999 (Pitcher 1999), while the next conference is planned for Darwin, Australia in 2002 (http://www.wrfc3.com/). It remains to be seen whether this series will become established as 'The Meeting Place' for recreational fisheries, but the effort is needed to address the type of problems discussed here.

Another example of a recent effort to help researchers in different regions and countries to communicate effectively using modern technology is the Committee on the Human Dimensions of Recreational Fisheries, established in 1993 to promote networking under the Fisheries Management Section of the American Fisheries Society (http://lutra.tamu.edu/hdcom/). The Committee emphasises discussions of applied and methodological aspects of recreational fishing studies via the Internet. The Committee also wants to focus on issues such as urban fishing programmes, demographic implications, recreational fishing/tourism interactions, catch-and-release trends, and resource allocation. They also develop human dimensions materials and programmes that meet the needs of fisheries managers and provide opportunities for managers and social science researchers to discuss recreational fishing topics. Several hundred fisheries scientists and managers from many countries are currently members of the Committee. Both editors of applied, multidisciplinary journals, chairs of conferences, committees and professional academies should play an important role in working toward increased international orientation of research in the assessment of values of recreational fisheries.

Broader involvement in the decision-making process increases demand for research

There is a clear tendency for members of NGOs, landowners, industry and other stakeholder groups to be encouraged to participate in fisheries management. More and more countries formalise such processes; they are labelled transactional, interactive, or community-based (Pikitch *et al.* 1997). Councils, advisory boards and meeting places are established all over the world. However, they are not immune to criticism, such as failing to adhere to conservation standards, or favouring some groups over others. In any case, the trend for broader involvement in fisheries management seems to create increased controversy and conflict in fishery management. These are fuelled by politics, lobbyists and litigation. These processes often lead to increased demand for 'unbiased' research.

Essentially, the process of increased involvement in fisheries management during the 1980s and 1990s might be an important reason for more demand and interest in research in the social and economic aspects of recreational fisheries. At the same time, the scientific basis for fisheries management is subjected to greater scrutiny, which undoubtedly improves the scientific basis for management. However, often different scientific views are posed and primarily used to undermine the overall purpose of the process of increased involvement, where the entire sector including the management profession and science profession may all suffer. The essence of all this is that the trend of increased participation itself rapidly increases the demand for, and the potential contribution from, social and economic sciences in fisheries management. At the same time, it stresses the need for this field to become more professional, producing research of sufficient scientific standard.

The tendency towards broader involvement also underlines the legal and ethical responsibility of scientists towards those from whom they obtained their data: anglers, stakeholder groups and managers. Gathering cultural and social data is an exchange; the researcher takes information from people, and they expect something in return. However, those providing data often feel that more has been given than is received. The findings of the research may alter the regulations of the fishery, limit the freedom of persons and enterprises, change the distribution of the catch between users, and most often result in more rules rather than fewer. Cultural and social data represent a trust that researchers in these areas must carefully manage. Often, too little effort is given to feedback to respondents and the general public. The trend toward broader involvement and delegation of decisions can improve this often-ignored part of research.

More sophisticated research methods

A major factor in improving scientific standards is the application of state-of-the-art methods. Several of the chapters in this book illustrate the usefulness of applying modern and sophisticated methods and techniques in recreational fisheries research. National angling surveys have been a central tool to monitor and assess participation rates, catches and economic value of recreational fishing in many countries, but at the same time have represented a large cost to those responsible for monitoring the fisheries. This book presents several new efforts covering large areas and populations (Lyle *et al.* 2002, Gentner and Lowther 2002, Toivonen 2002) that propose new, effective techniques that can be considered elsewhere.

Thunberg and Milon (2002) illustrate how data from general population censuses can assist in forecasting recreational fishing participation. These models also visualise some of the challenges faced by the recreational fisheries sector, for instance from the growing number of elderly citizens in western countries in the coming 10–30 years. The study also underlines the significant gains from professional scientists with good connections towards general social scientific institutions. Rudd *et al.* (2002) discuss how the relatively new methods of choice experiments (Louviere and Timmermans 1990), originally developed in market research, can assist in economic valuation of attributes of the fishing experience. This technique can also be used in designing fishing opportunities, associated with differences in motivations or attitudes among anglers (Aas *et al.* 2000).

Research needs

The book has touched upon several other topics of emerging or continuing interest to the recreational fisheries community. Cox and Walters (2002) and Sumaila (2002) discuss the challenges in managing common property, open-access fisheries. This is most likely an important issue in many countries. This research must tie into the long tradition that stems from Hardin's (1968) first paper describing the 'tragedy of the commons', so often seen in fishery management lately. This is also an example of how research in recreational fisheries issues can contribute to the more general task of producing knowledge for fair allocation as part of sustainable resource management.

The rapid growth in tourism, with its increased demand for quality and authentic experiences, places fishery managers and resource owners under pressure to identify sustainable management tools in these circumstances. Policansky (2002) presents the seemingly successful history of catch-and-release in North America while Aas *et al.* (2002) describe how it has engendered fierce debate among Europeans. The future of nonconsumptive recreational fishing is intriguing, enabling us to picture some of the biological, economic, cultural and ethical challenges embedded in recreational fisheries management.

Research into recreational fisheries for 'special groups' of people, such as the disabled, children, the elderly and minorities such as indigenous people and those from foreign cultures, needs to be highlighted in the future. Little is known of the role, value and type of recreational fisheries among such groups. Children form an important user group in many countries (Aas 1996b, Rickhards 1986), but we know very little about their preferences, motivations, resource use and special needs. We know that the proportion of elderly people will grow (Thunberg and Milon 2002), and that this new wave in the western world will be different from those of previous generations in terms of health, economic resources and leisure behaviour. Little of this is yet reflected in management-oriented or tourism research. Indigenous people on many continents often have a close relation to renewable resources such as fish, and might be heavily involved in subsistence fishing, angling or businesses related to angling such as guiding. Still we know little about their views and thinking about fishing, for instance what spiritual values and special types of cultural knowledge may be related to fishing.

Recreational fishing has traditionally been a male-dominated activity. Lately, there are signs of increased interest in angling from women in some countries. To understand and maybe enhance further participation from women, research is needed that specifically addresses their preferences and motivations.

The future of research on evaluating recreational fisheries

For more than 20 years, voices have called for more social, economic and anthropological research to supplement the traditional fish biology and freshwater ecology research (Voiland and Duttweiler 1984). 'We manage people, not fish' is a standard statement to underline the need for a multidisciplinary approach to fisheries research.

This book illustrates that there are many efforts to broaden the knowledge base from which to manage recreational fisheries around the world. Much is already in place to expand and increase the quality of research in the values of recreational fisheries, but much remains to be done before research in this topic is established as a scholarly field of investigation. In some countries, maybe most notably in North America, we have lately seen the evolution of research in the social and economic sciences at a record-breaking pace (Socioeconomics section, AFS 1993). In other areas there are clear and identifiable shortages in research on different aspects of recreational fisheries. An important quality of this book is its world-wide perspective. Increased communication between countries and disciplines, more published research results, and a responsible and open attitude concerning the scope of science-based fisheries management is necessary if we are to make progress. This book, together with other central texts (e.g. Guthrie *et al.* 1991, Hickley and Tompkins 1998) and multidisciplinary, applied fishery journals, can contribute to heighten the state-of-the-art of applied research in recreational fisheries, and ease the progress and exchange of knowledge.

This book is in accord with the rest of the literature in the field when it concludes that recreational fishing has many positive values for society and that the effects on the resource may be limited and manageable. If these important values are to be retained and developed further, policy makers must be confident that the research meets the same standards as those used in the natural sciences. Indeed, social and economic scientists have experienced an increasing demand for quality in research, in an area with a long tradition of evaluating results from biological research. Therefore, future research into all aspects of recreational fisheries must be of high quality and be published in refereed journals.

For those planning to commence such studies in countries with a limited research tradition, this book demonstrates that there is much research to build on. By supplementing the empirical base (in some regions somewhat insufficient), by adopting widely-accepted research principles, by using theory and methods from relevant disciplinary research, and by examining research on recreational fisheries issues from other geographic regions, readers will find themselves well on the way towards producing quality research that promotes wiser management and use of fisheries resources.

References

Aas, Ø. (1996a) Recreational fishing in Norway trends 1970–1993 and geographic variation. *Fisheries Management and Ecology* **3**, 107–118.

Aas, Ø. (1996b) Use of two approaches to measure children's motivation to fish in Norway. *Human Dimensions of Wildlife* **1**(3), 15–28.

Aas, Ø. & Ditton, R.B. (1998) Human dimensions perspective on recreational fisheries management: implications for Europe. In: *Recreational Fisheries: Social, Economic and Management Aspects* (eds Hickley, P. & Tompkins, H.), pp. 153–164. Oxford: Blackwell Science, Fishing News Books/Food and Agriculture Organization of the United Nations.

Aas,Ø., Haider, W. & Hunt, L. (2000) Angler responses to potential harvest regulations in a Norwegian sport fishery: a conjoint based choice modeling approach. *North American Journal of Fisheries Management* **20**, 940–950.

Aas, Ø. & Skurdal, J. (1996) Resident and non-resident fishing in a rural district in Norway: subsistence and sport – conflicts or coexistence? *Nordic Journal of Freshwater Research* **72**, 45–51.

Aas, Ø., Thailing, C. & Ditton, R.B. (2002) Controversy over catch-and-release recreational fishing in Europe. In: *Recreational Fisheries: Ecological, Economic and Social Evaluation* (eds Pitcher, T.J. & Hollingworth, C.E.), Chapter 7, this volume. Blackwell Science, Oxford, UK.

Anderson, L.G. (1986) *The Economics of Fisheries Management.* Baltimore and London: John Hopkins University Press.

Beckers, T. (1995) Back to basics: International communication in leisure research. *Leisure Sciences* **17**, 327–336.

Booth, A.L. & Kessler, W.B. (1996) Understanding linkages of people, natural resources and ecosystem health. In: *Natural Resource Management: The Human Dimension* (ed Ewert, A), pp. 231–248. Westview Press, Boulder, CO.

Bottom, D.L. (1997) To till the water: A history of ideas in fishery conservation. *Pacific Salmon and Their Ecosystems: Status and Future Options* (eds Stouder, D.J., Bisson, P.A. & Naimann, R.J.), pp. 569–598. Chapman & Hall, New York, London.

Brown, T. L. (1987) Typology of human dimensions information needed for Great Lakes sport-fisheries management. *Transactions of the American Fisheries Society* **116**, 320–324.

Calvert, B. (2002) The importance of angler motivations in sport fishery management. In: *Recreational Fisheries: Ecological, Economic and Social Evaluation* (eds Pitcher, T.J. & Hollingworth, C.E.), Chapter 9, this volume. Blackwell Science, Oxford, UK.

Collette, B.B. (1990) Problems with grey literature in fishery science. In: *Writing for Fishery Journals* (ed Hunter, J.), pp. 27–32. American Fisheries Society.

Cox, S. & Walters, C. (2002) Maintaining quality in recreational fisheries: how success breeds failure in management of open-access sport fisheries. In: *Recreational Fisheries: Ecological, Economic and Social Evaluation* (eds Pitcher, T.J. & Hollingworth, C.E.), Chapter 8, this volume. Blackwell Science, Oxford, UK.

Ditton, R.B. (1996) Human dimensions in fisheries. In: *Natural Resource Management: The Human Dimension* (ed Ewert, A.), pp. 73–90. Westview Press, Boulder, CO.

Driver, B.L., Dustin, D., Baltic, T., Elsner, G. & Peterson, G. (eds) (1999) *Nature and the Human Spirit. Towards an Expanded Land Management Ethic.* State College, Pennsylvania: Venture Publishing.

Duffield, J.W., Merritt, M.F. & Neher, C.J. (2002) Valuation and policy in Alaskan sport fisheries. In: *Recreational Fisheries: Ecological, Economic and Social Evaluation* (eds Pitcher, T.J. & Hollingworth, C.E.), Chapter 13, this volume. Blackwell Science, Oxford, UK.

Fedler, A.J. & Ditton, R.B. (1994) Understanding angler motivations in fisheries management. *Fisheries* **19**(4), 6–13.

Gentner, B. & Lowther, A. (2002) Evaluating marine sport fisheries in the USA. In: *Recreational Fisheries: Ecological, Economic and Social Evaluation* (eds Pitcher, T.J. & Hollingworth, C.E.), Chapter 14, this volume. Blackwell Science, Oxford, UK.

Glass, R.J., Muth, R.M., More, T.A. & Kruse, J. (1995) Assessing developmental impacts on subsistence fishing. Complications attributable to a mixed economical system. *Society and Natural Resources* **8**, 431–442.

Griffiths, M.H. & Lamberth, S.J. (2002) Evaluating a marine sport fishery in South Africa. In: *Recreational Fisheries: Ecological, Economic and Social Evaluation* (eds Pitcher, T.J. & Hollingworth, C.E.), Chapter 16, this volume. Blackwell Science, Oxford, UK.

Guthrie, D., Hoenig, J.M., Holliday, M. *et al.* (1991) Creel and angler surveys in fisheries management. *American Fisheries Society Symposium* 12.

Hardin, G. (1968) The tragedy of the commons. *Science* **162**, 1243–1248.

Hickley, P. (1998) Comments regarding a code of good practice for recreational fishing. In: *Recreational Fisheries: Social, Economic and Management Aspects* (eds Hickley, P. & Tompkins, H.), pp. 299–304. Oxford: Blackwell Science, Fishing News Books/Food and Agriculture Organization of the United Nations.

Hickley, P. & Tompkins, H. (1998) *Recreational Fisheries: Social, Economic and Management Aspects.* Oxford: Blackwell Science, Fishing News Books/Food and Agriculture Organization of the United Nations.

Iso-Ahola, S. (1980) *The Social Psychology of Leisure and Recreation.* Dubque, IA. Wm. C. Brown Company.

Louviere, J. & Timmermans, H. (1990) Stated preference and choice models applied to recreation research. A review. *Leisure Sciences* **12**, 9–32.

Lowerson, J. (1989) Angling. In: *Sport in Britain* (ed Mason, T.), pp. 12–44. Cambridge University Press.

Lyle, J.M, Coleman, A.P.M., West, L., Campbell, D. & Henry, G.W. (2002) New large-scale survey methods for evaluating sport fisheries. In: *Recreational Fisheries: Ecological, Economic and Social Evaluation* (eds Pitcher, T.J. & Hollingworth, C.E.), Chapter 15, this volume. Blackwell Science, Oxford, UK.

Lyman, J. (1999) Cultural value and change – catch-and-release in Alaska's sport fishery. Paper presented at The National Conference on Catch-and-release in Marine Recreational Fisheries, Virginia Beach, Virginia, USA, December 1999.

Lyons, J., Hickley, P. & Gledhill, S. (2002) An evaluation of recreational fishing in England and Wales. In: *Recreational Fisheries: Ecological, Economic and Social Evaluation* (eds Pitcher, T.J. & Hollingworth, C.E.), Chapter 12, this volume. Blackwell Science, Oxford, UK.

Neulinger, J. (1974) *The Psychology of Leisure.* Charles C. Thomas, Springfield, IL.

Orbach, M. (1980) The human dimension. In: *Fisheries Management* (ed Lackey, R. T. & Nielsen, L. A.), pp. 149–163. Blackwell Scientific, Oxford.

Pikitch, E.K., Sissenwine, M.P., Huppert, D.D. & Duke, M. (1997) Summing up: An overview of Global Trends in Fisheries, Fisheries Science, and Fisheries Management. *American Fisheries Symposium* 20, 275–280.

Pinter, C. & Wolos, A. (1998) Summary report of the symposium topic session on the current status and trends in recreational fisheries. In: *Recreational Fisheries: Social, Economic and Management Aspects* (eds Hickley, P. & Tompkins, H.), pp. 1–4. Oxford: Blackwell Science, Fishing News Books/Food and Agriculture Organization of the United Nations.

Pitcher, T.J. (1999) *Evaluating the Benefits of Recreational Fishing.* University of British Columbia, Fisheries Centre Research Report 1999, 7 (2).

Policansky, D. (2002) Catch-and-release recreational fishing: a historical perspective. In: *Recreational Fisheries: Ecological, Economic and Social Evaluation* (eds Pitcher, T.J. & Hollingworth, C.E.), Chapter 6, this volume. Blackwell Science, Oxford, UK.

Rudd, M., Folmer, H. & van Kooten, G.C. (2002) Economic evaluation of recreational fishery policies. In: *Recreational Fisheries: Ecological, Economic and Social Evaluation* (eds Pitcher, T.J. & Hollingworth, C.E.), Chapter 3, this volume. Blackwell Science, Oxford, UK.

Siegler, W.F. & Siegler, J.W. (1990) *Recreational Fisheries: Management, Theory and Application.* Reno and Las Vegas: University of Nevada press.

Smith, C.L. (1983) Evaluating human factors. In: *Fisheries Techniques* (eds Nielsen, L.A. & Johnson, D.L.) pp. 431–445. American Fisheries Society, Bethesda, MD.

Socioeconomics Section, American Fisheries Society (1993) *Handbook of sources and applications of fisheries social science information*. Bethesda, MD: AFS.

Steffens, W. & Winkel, M. (2002) Evaluating recreational fishing in Germany. In: *Recreational Fisheries: Ecological, Economic and Social Evaluation* (eds Pitcher, T.J. & Hollingworth, C.E.), Chapter 10, this volume. Blackwell Science, Oxford, UK.

Sumaila, U.R. (2002) Recreational and commercial fishers in the Namibian silver kob fishery. In: *Recreational Fisheries: Ecological, Economic and Social Evaluation* (eds Pitcher, T.J. & Hollingworth, C.E.), Chapter 4, this volume. Blackwell Science, Oxford, UK.

Sylvester, C. (1999) The western idea of work and leisure: Traditions, transformations and the future. In: *Leisure Studies: Prospects for the Twenty-First Century* (eds Jackson, E.L. & Burton, T.L.), pp. 17–34. Venture Publishing, State College, Pennsylvania.

Thunberg, E.M. & Milon, W.J. (2002) Projecting recreational fishing participation. In: *Recreational Fisheries: Ecological, Economic and Social Evaluation* (eds Pitcher, T.J. & Hollingworth, C.E.), Chapter 5, this volume. Blackwell Science, Oxford, UK.

Toivonen, A.-L. (2002) A survey of the economic value of Nordic recreational fisheries. In: *Recreational Fisheries: Ecological, Economic and Social Evaluation* (eds Pitcher, T.J. & Hollingworth, C.E.), Chapter 11, this volume. Blackwell Science, Oxford, UK.

Voiland, M.P. & Duttweiler, M.W. (1984) Where's the humanity? A challenge and opportunity for the fisheries community. *Fisheries* **9**(4), 10–12.

Index

Note: Page numbers *in italics* refer to figures and tables.